C0-BXT-548

Geophysical Monograph Series

Including

IUGG Volumes

Maurice Ewing Volumes

Mineral Physics Volumes

GEOPHYSICAL MONOGRAPH SERIES

Geophysical Monograph Volumes

1 **Antarctica in the International Geophysical Year** *A. P. Crary, L. M. Gould, E. O. Hulburt, Hugh Odishaw, and Waldo E. Smith (Eds.)*
2 **Geophysics and the IGY** *Hugh Odishaw and Stanley Ruttenberg (Eds.)*
3 **Atmospheric Chemistry of Chlorine and Sulfur Compounds** *James P. Lodge, Jr. (Ed.)*
4 **Contemporary Geodesy** *Charles A. Whitten and Kenneth H. Drummond (Eds.)*
5 **Physics of Precipitation** *Helmut Weickmann (Ed.)*
6 **The Crust of the Pacific Basin** *Gordon A. Macdonald and Hisashi Kuno (Eds.)*
7 **Antarctica Research: The Matthew Fontaine Maury Memorial Symposium** *H. Wexler, M. J. Rubin, and J. E. Caskey, Jr. (Eds.)*
8 **Terrestrial Heat Flow** *William H. K. Lee (Ed.)*
9 **Gravity Anomalies: Unsurveyed Areas** *Hyman Orlin (Ed.)*
10 **The Earth Beneath the Continents: A Volume of Geophysical Studies in Honor of Merle A. Tuve** *John S. Steinhart and T. Jefferson Smith (Eds.)*
11 **Isotope Techniques in the Hydrologic Cycle** *Glenn E. Stout (Ed.)*
12 **The Crust and Upper Mantle of the Pacific Area** *Leon Knopoff, Charles L. Drake, and Pembroke J. Hart (Eds.)*
13 **The Earth's Crust and Upper Mantle** *Pembroke J. Hart (Ed.)*
14 **The Structure and Physical Properties of the Earth's Crust** *John G. Heacock (Ed.)*
15 **The Use of Artificial Satellites for Geodesy** *Soren W. Henricksen, Armando Mancini, and Bernard H. Chovitz (Eds.)*
16 **Flow and Fracture of Rocks** *H. C. Heard, I. Y. Borg, N. L. Carter, and C. B. Raleigh (Eds.)*
17 **Man-Made Lakes: Their Problems and Environmental Effects** *William C. Ackermann, Gilbert F. White, and E. B. Worthington (Eds.)*
18 **The Upper Atmosphere in Motion: A Selection of Papers With Annotation** *C. O. Hines and Colleagues*
19 **The Geophysics of the Pacific Ocean Basin and Its Margin: A Volume in Honor of George P. Woollard** *George H. Sutton, Murli H. Manghnani, and Ralph Moberly (Eds.)*
20 **The Earth's Crust: Its Nature and Physical Properties** *John C. Heacock (Ed.)*
21 **Quantitative Modeling of Magnetospheric Processes** *W. P. Olson (Ed.)*
22 **Derivation, Meaning, and Use of Geomagnetic Indices** *P. N. Mayaud*
23 **The Tectonic and Geologic Evolution of Southeast Asian Seas and Islands** *Dennis E. Hayes (Ed.)*
24 **Mechanical Behavior of Crustal Rocks: The Handin Volume** *N. L. Carter, M. Friedman, J. M. Logan, and D. W. Stearns (Eds.)*
25 **Physics of Auroral Arc Formation** *S.-I. Akasofu and J. R. Kan (Eds.)*
26 **Heterogeneous Atmospheric Chemistry** *David R. Schryer (Ed.)*
27 **The Tectonic and Geologic Evolution of Southeast Asian Seas and Islands: Part 2** *Dennis E. Hayes (Ed.)*
28 **Magnetospheric Currents** *Thomas A. Potemra (Ed.)*
29 **Climate Processes and Climate Sensitivity (Maurice Ewing Volume 5)** *James E. Hansen and Taro Takahashi (Eds.)*
30 **Magnetic Reconnection in Space and Laboratory Plasmas** *Edward W. Hones, Jr. (Ed.)*
31 **Point Defects in Minerals (Mineral Physics Volume 1)** *Robert N. Schock (Ed.)*
32 **The Carbon Cycle and Atmospheric CO_2: Natural Variations Archean to Present** *E. T. Sundquist and W. S. Broecker (Eds.)*
33 **Greenland Ice Core: Geophysics, Geochemistry, and the Environment** *C. C. Langway, Jr., H. Oeschger, and W. Dansgaard (Eds.)*
34 **Collisionless Shocks in the Heliosphere: A Tutorial Review** *Robert G. Stone and Bruce T. Tsurutani (Eds.)*
35 **Collisionless Shocks in the Heliosphere: Reviews of Current Research** *Bruce T. Tsurutani and Robert G. Stone (Eds.)*
36 **Mineral and Rock Deformation: Laboratory Studies—The Paterson Volume** *B. E. Hobbs and H. C. Heard (Eds.)*
37 **Earthquake Source Mechanics (Maurice Ewing Volume 6)** *Shamita Das, John Boatwright, and Christopher H. Scholz (Eds.)*
38 **Ion Acceleration in the Magnetosphere and Ionosphere** *Tom Chang (Ed.)*
39 **High Pressure Research in Mineral Physics (Mineral Physics Volume 2)** *Murli H. Manghnani and Yasuhiko Syono (Eds.)*
40 **Gondwana Six: Structure Tectonics, and Geophysics** *Gary D. McKenzie (Ed.)*

Maurice Ewing Volumes

IUGG Volumes

1. **Structure and Dynamics of Earth's Deep Interior** *D. E. Smylie and Raymond Hide (Eds.)*
2. **Hydrological Regimes and Their Subsurface Thermal Effects** *Alan E. Beck, Grant Garven, and Lajos Stegena (Eds.)*
3. **Origin and Evolution of Sedimentary Basins and Their Energy and Mineral Resources** *Raymond A. Price (Ed.)*
4. **Slow Deformation and Transmission of Stress in the Earth** *Steven C. Cohen and Petr Vaníček (Eds.)*
5. **Deep Structure and Past Kinematics of Accreted Terrances** *John W. Hillhouse (Ed.)*
6. **Properties and Processes of Earth's Lower Crust** *Robert F. Mereu, Stephan Mueller, and David M. Fountain (Eds.)*
7. **Understanding Climate Change** *Andre L. Berger, Robert E. Dickinson, and J. Kidson (Eds.)*
8. **Evolution of Mid Ocean Ridges** *John M. Sinton (Ed.)*
9. **Variations in Earth Rotation** *Dennis D. McCarthy and William E. Carter (Eds.)*
10. **Quo Vadimus Geophysics for the Next Generation** *George D. Garland and John R. Apel (Eds.)*
11. **Sea Level Changes: Determinations and Effects** *Philip L. Woodworth, David T. Pugh, John G. DeRonde, Richard G. Warrick, and John Hannah (Eds.)*
12. **Dynamics of Earth's Deep Interior and Earth Rotation** *Jean-Louis Le Mouël, D.E. Smylie, and Thomas Herring (Eds.)*
13. **Environmental Effects on Spacecraft Positioning and Trajectories** *A. Vallance Jones (Ed.)*
14. **Evolution of the Earth and Planets** *E. Takahashi, Raymond Jeanloz, and David Rubie (Eds.)*

Mineral Physics Volumes

1. **Point Defects in Minerals** *Robert N. Schock (Ed.)*
2. **High Pressure Research in Mineral Physics** *Murli H. Manghnani and Yasuhiko Syona (Eds.)*
3. **High Pressure Research: Application to Earth and Planetary Sciences** *Yasuhiko Syono and Murli H. Manghnani (Eds.)*

Geophysical Monograph 75
IUGG Volume 15

T.M

Interactions Between Global Climate Subsystems

The Legacy of Hann

G.A. McBean
M. Hantel
Editors

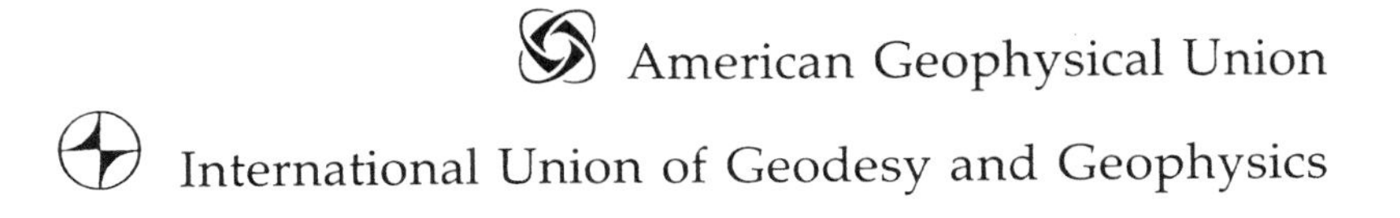
American Geophysical Union
International Union of Geodesy and Geophysics

Published under the aegis of the AGU Books Board

Library of Congress Cataloging-in-Publication Data

Interactions between global climate subsystems : the legacy of Hann / G.A. McBean, M. Hantel, editors.
p. cm. — (Geophysical monograph : 75) (IUGG : v. 15)
ISBN 0-87590-466-1
1. Climatic changes—Congresses. 2. Hydrologic cycle—Congresses. 3. Hann, Julius von 1839-1921. I. McBean, G. A. II. Hantel, Michael. III. Series.
IV. Series : IUGG (Series) : v. 15.
QC981.8.C5I55 1993
551.5—dc20 93-13345
CIP

ISSN: 0065-8448
ISBN 087590-466-1

Printed in the United States of America.

CONTENTS

PREFACE

The global climate system is characterized by exchanges of matter and energy between its various components on a wide range of time and space scales. The essence of understanding, and eventually predicting, climate and global change will depend on our capability to measure and model these exchanges. Energy, water and biogeochemical cycles of all kinds are the focus of this interdisciplinary volume. Some papers deal with theoretical, others with observational and others with the modelling aspects of fluxes of matter and energy between parts of the global climate system. This volume arose out of the Hann Symposium and has been augmented by contributions from two other symposia. The result is a comprehensive set of papers dealing with the interacting components of the climate system.

In planning for the IUGG General Assembly in Vienna, in 1991, it was recommended simultaneously by several groups that major symposia be convened on the global climate and the interactions between its components. At the same time, it was recognized that Vienna had been the home of the distinguished Austrian climatologist Julius von Hann. Therefore, it was decided to link these ideas in an IUGG Union Symposium on the "Fluxes of Matter Between Global Climate Subsystems" and name it in honor of Julius von Hann. Von Hann was a leading scientist of the 19th century who lived in Vienna and helped lay the foundations of modern quantitative climatology.

Appropriately, the Hann Symposium and this volume open with a presentation on the life and contributions of von Hann by P. Kahlig of the University of Vienna. Hann was the first to apply systematically what is now called "Hanning" to climatological data. He also recognized early the importance of measurements at different heights in the atmosphere and for that reason encouraged the establishment of mountain observatories as the only practical way of obtaining such observations at that time. H. Flohn reviews the development of physical three-dimensional climatology from the time of J. von Hann, when data sets were limited primarily to observations of the surface, to the modern era, with satellites giving us a global overview from outer space. It is important that these two viewpoints be assimilated to provide a full coverage of the climate system.

The next group of papers deals with interactions of land surfaces and the atmosphere. M. Hantel reviews the theoretical basis for considering the fluxes of energy across the interface, following in the tradition of Austrian mathematical rigor, applied to geophysical problems. M. H. Zemankovics describes a model study to demonstrate the effects of increasing CO_2 on heat and vapor fluxes over vegetated surfaces. G. Esser and H. Lieth expand the scope to the global biospheric interactions of the carbon cycle. Results of a comprehensive model of both marine and terrestrial biospheres that has been constructed are used to examine variations in the carbon cycle. Water is a critical element of most global cycles, and the paper by A. Zangvil, D. Portis and P. Lamb describes the diurnal variations of the atmosphere's water cycle (surface up to 300 hPa) across the midwestern United States in the summer of 1979.

Another group of papers deals with the oceans. H. L. Bryden reviews the ocean heat transport across 24°N latitude. Due to recent estimates in the North Pacific of 0.76 PW (PW = 10^{15} watts) northward heat transport, and the earlier estimate of 1.22 PW in the Atlantic Ocean at the same latitude, it is now possible to provide an estimate of the total oceanic meridional heat transport through subtropical latitudes, 2 PW. However, since the estimated atmospheric transport is 1.7 PW, there is still an imbalance, as satellite radiation estimates at the top of the atmosphere imply that the oceans and atmosphere together must transport about 5 PW of heat. Bryden shows that the Pacific Ocean heat transport is primarily in the upper 700 m of the ocean, compared with the much deeper transport in the Atlantic. Then R. W. Schmitt and S. E. Wijffels examine the transport of freshwater by the oceans. Flux of freshwater throughout the Bering Strait, which has recently been more directly measured, was used as a reference point. Estimates of Baumgartner and Reichel, who used an assumption of zero flux across the equator in the Atlantic, could then be used to prepare a more consistent global picture of freshwater transport by the oceans. The last oceanic paper deals with chemical aspects. S. W. A. Naqvi, R. S. Gupta and M. D. Kumar present results on carbon dioxide and nitrous oxide and estimated fluxes in the Indian Ocean.

Considerations of climate must expand to global questions, and the global energy balance is of prime concern. A. Ohmura and H. Gilgen have re-evaluated the surface global energy balance through systematic examination of observations of energy fluxes. Linked to the surface energy balance are variations throughout the depth of the atmosphere. H. van Loon and K. Labitzke investigate the interannual variations of the middle atmosphere and explore possible linkages with solar and other forcing. In the following paper, C. Fröhlich examines variations of solar irradiance and considers their possible relevance to climate variations. Although the connection between the mean solar irradiance and climate is clear, the connections between their variations continue to be controversial. In addition to solar irradiance, the earth is recipient to a host of energetic particles from space. This is discussed by C. H. Jackman. The last paper of the volume, by J. O. Dickey, S. L. Marcus, T. M. Eubanks and R. Hide uses space geodesy data to investigate variations in the length of day and its linkages with ENSO.

The Hann Symposium provided an opportunity for interdisciplinary discussions amongst oceanographers, meteorologists and hydrologists. Some of this is reflected in the papers in this volume, but it is difficult to get the sense of interaction through the pages of a book. The need for symposia to bring people together continues to be great.

The co-editors of this volume are pleased to acknowledge the assistance from other co-convenors of the Hann Symposium: S. Dyck (representing IAHS), H. Sigurdsson (IAVCEI), J. R. Toggweiler (IAPSO) and R. G. Prinn (IAMAP). Four of the papers in this volume were presented at Vienna as parts of the symposia "Global Climate Variability: Processes and Prediction," convened by M. Kuhn, and "Anthropogenic and Natural External Forcing of the Middle Atmosphere," convened by M. Geller. The contributions of all authors to these three symposia are gratefully appreciated. All papers were reviewed and revised before publication and the co-editors acknowledge this important assistance.

G. A. McBean
M. Hantel
Editors

FOREWORD

The scientific work of the International Union of Geodesy and Geophysics (IUGG) is primarily carried out through its seven associations: IAG (briefly, Geodesy), IASPEI (Seismology), IAVCEI (Volcanology), IAGA (Geomagnetism), IAMAP (Meteorology), IAPSO (Oceanography), and IAHS (Hydrology). The work of these associations is documented in various ways.

Interactions Between Global Climate Subsystems, The Legacy of Hann is one of a group of volumes published jointly by IUGG and AGU that are based on work presented at the Inter-Association Symposia as part of the IUGG General Assembly held in Vienna, Austria, in August 1991. Each symposium was organized by several of IUGG's member associations and comprised topics of interdisciplinary relevance. The subject areas of the symposia were chosen such that they would be of wide interest. Also, the speakers were selected accordingly, and in many cases, invited papers of review character were solicited. The series of symposia were designed to give a picture of contemporary geophysical activity, results, and problems to scientists having a general interest in geodesy and geophysics.

In view of the importance of these interdisciplinary symposia, IUGG is grateful to AGU for having put its unique resources in geophysical publishing expertise and experience at the disposal of IUGG. This ensures accurate editorial work, including the use of peer reviewing. So the reader can expect to find expertly published scientific material of general interest and general relevance.

Helmut Moritz
President, IUGG

Some Aspects of Julius von Hann's Contribution to Modern Climatology

PETER KAHLIG

Institute of Meteorology and Geophysics, University of Vienna
A-1190 Vienna, Austria

INTRODUCTION

Julius von Hann (1839-1921) was a leading meteorologist and climatologist at the turn of the century who lived and worked in Vienna (Figure 1). In meteorology, a pioneering achievement was his thermodynamic foehn theory (independently of Helmholtz). In climatology, he helped to initiate quantitative methods. Both for meteorology and for climatology, Hann recognized very early the importance of three-dimensional observation systems; consequently he proposed and founded several mountain observatories, and supported their maintenance (cf. Barry, 1981). Today, there are other efficient ways of obtaining data from the third dimension (in particular by satellites, cf. WMO, 1988), yet mountain observatories will continue to play their role as background stations for reference and comparison purposes. Hann was an editor of the internationally respected journal "Meteorologische Zeitschrift" (and its predecessor) for more than fifty years (1866-1920).

In view of Hann's acknowledged contributions to science, his biography can be found in such standard reference works as Encyclopaedia Britannica (1962), World Who's Who in Science (Debus, 1968), or SES (1980). A rather complete review of Hann's life and impact has been given by F. Steinhauser (1951). An appreciation of Hann's life and work can also be found in several obituaries written in English (Mitchell, 1921; Shaw, 1921; Simpson, 1921; Ward, 1922).

Interactions Between Global Climate Subsystems, The Legacy of Hann
Geophysical Monograph 75, IUGG Volume 15

Hann contributed more than 1000 publications to the field of meteorology and climatology. In the present paper, only some special aspects of Hann's monumental work can be treated:

- comments on Hann's epoch-making Handbook of Climatology;
- remarks on the origin of "hanning" (a smoothing operation);
- reflexions on the idea of teleconnections.

THE HANDBOOK OF CLIMATOLOGY

Hann's "Handbuch der Klimatologie" appeared in 1883 in its first edition (2nd edition 1897). It received due attention by the international scientific community and was recognized as a most valuable compendium of climatology (Hann, 1883; Khrgian, 1970). Although German used to be a widely understood language of science at that time, the potential usefulness of an English translation soon became obvious.

The American climatologist Robert De Courcy Ward (1867-1931) translated and edited the Handbook of Climatology after revising and updating the text (Hann, 1903). Ward was the very first professor of climatology in the United States (cf. Debus, 1968). Ward's English translation of Hann's Handbook was widely used and has been cited repeatedly (e.g. Crowe, 1971).

Ward's letter (1903) that accompanied the presentation copy of the Handbook of Climatology is shown in Figure 2. Five years later (1908), Ward asked Hann in a letter for permission to use portions of Volume II and III of the "Handbuch der Klimatologie" in a forthcoming publication (Ward, 1908), namely "Climate Considered

Fig. 1. Julius von Hann.

Especially in Relation to Man" (Figure 3). In the same letter, Ward confirmed also the usefulness of the existing Handbook of Climatology.

Modern climatology comprises several aspects which go beyond Hann's original concept (though they are, in some respect, based on it): for example, contemporary qualitative and quantitative definitions of climate have been reviewed by Hantel et al. (1987), and facts on climate modelling were compiled by Hantel (1989).

The late H.E. Landsberg, well-known climatologist, demonstrated that Hann's sound reasoning and high-quality data may be used successfully even a century later. For example, Landsberg and Albert (1975) presented a comparison of Hann's world temperature chart (Hann, 1887) with WMO Climatological Normals, appreciating Hann's careful work. To quote Landsberg:

> J. Hann was an individual of rare insight, meticulous work habits, and an assiduous collector of data.

Data Smoothing

Treating data from all over the world, Julius von Hann early recognized the great potential of statistical methods. By oral tradition at the Institute of Meteorology in Vienna, Hann's statement on the "purifying power of mean values" has been kept alive as a slogan (H. Reuter, personal communication, 1991). Of course, climatologists and statisticians are well aware of the deficiencies of a mere "mean value climatology", yet the data (and their mean values) still represent the backbone of any climatological/geographical description (Durst, 1951; Essenwanger, 1986).

Hann appears to be the inventor of a certain data smoothing procedure, now called "hanning" (Blackman and Tukey, 1959; Huschke, 1959; Hamming, 1989) or "Hann smoothing" (Essenwanger, 1986). Essentially,

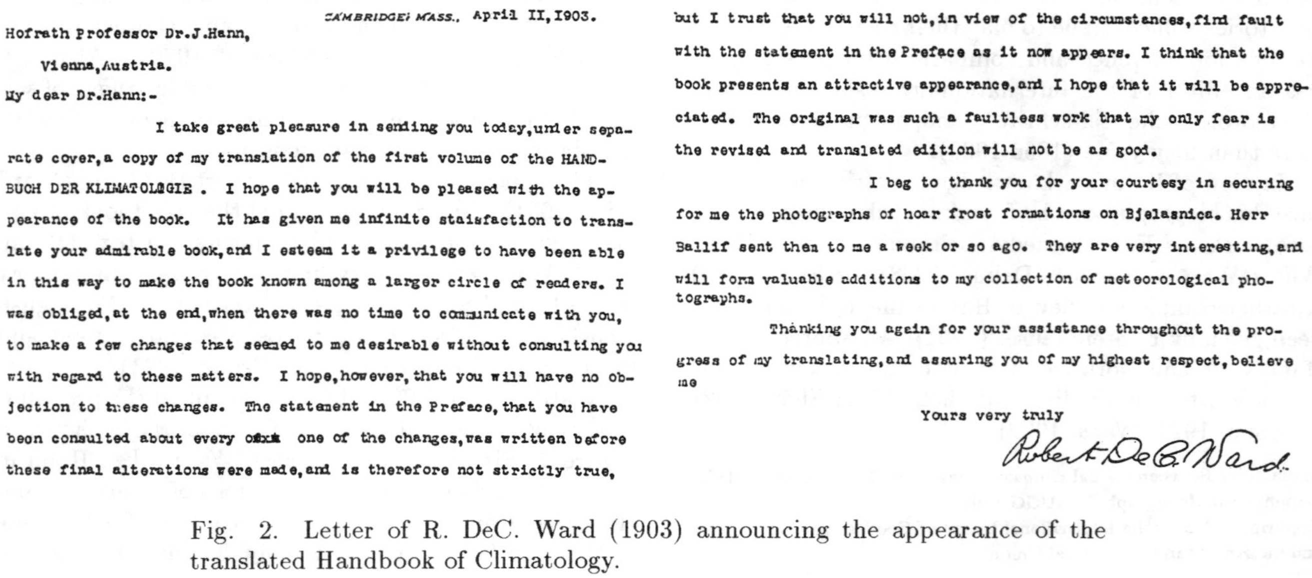

CAMBRIDGE, MASS., April II, I903.

Hofrath Professor Dr.J.Hann,

Vienna,Austria.

My dear Dr.Hann:-

I take great pleasure in sending you today,under separate cover,a copy of my translation of the first volume of the HANDBUCH DER KLIMATOLOGIE . I hope that you will be pleased with the appearance of the book. It has given me infinite staisfaction to translate your admirable book,and I esteem it a privilege to have been able in this way to make the book known among a larger circle of readers. I was obliged,at the end,when there was no time to communicate with you, to make a few changes that seemed to me desirable without consulting you with regard to these matters. I hope,however,that you will have no objection to these changes. The statement in the Preface, that you have been consulted about every one of the changes,was written before these final alterations were made,and is therefore not strictly true, but I trust that you will not,in view of the circumstances,find fault with the statement in the Preface as it now appears. I think that the book presents an attractive appearance,and I hope that it will be appreciated. The original was such a faultless work that my only fear is the revised and translated edition will not be as good.

I beg to thank you for your courtesy in securing for me the photographs of hoar frost formations on Bjelasnica. Herr Ballif sent them to me a week or so ago. They are very interesting,and will form valuable additions to my collection of meteorological photographs.

Thanking you again for your assistance throughout the progress of my translating,and assuring you of my highest respect,believe me

Yours very truly

Robert DeC. Ward

Fig. 2. Letter of R. DeC. Ward (1903) announcing the appearance of the translated Handbook of Climatology.

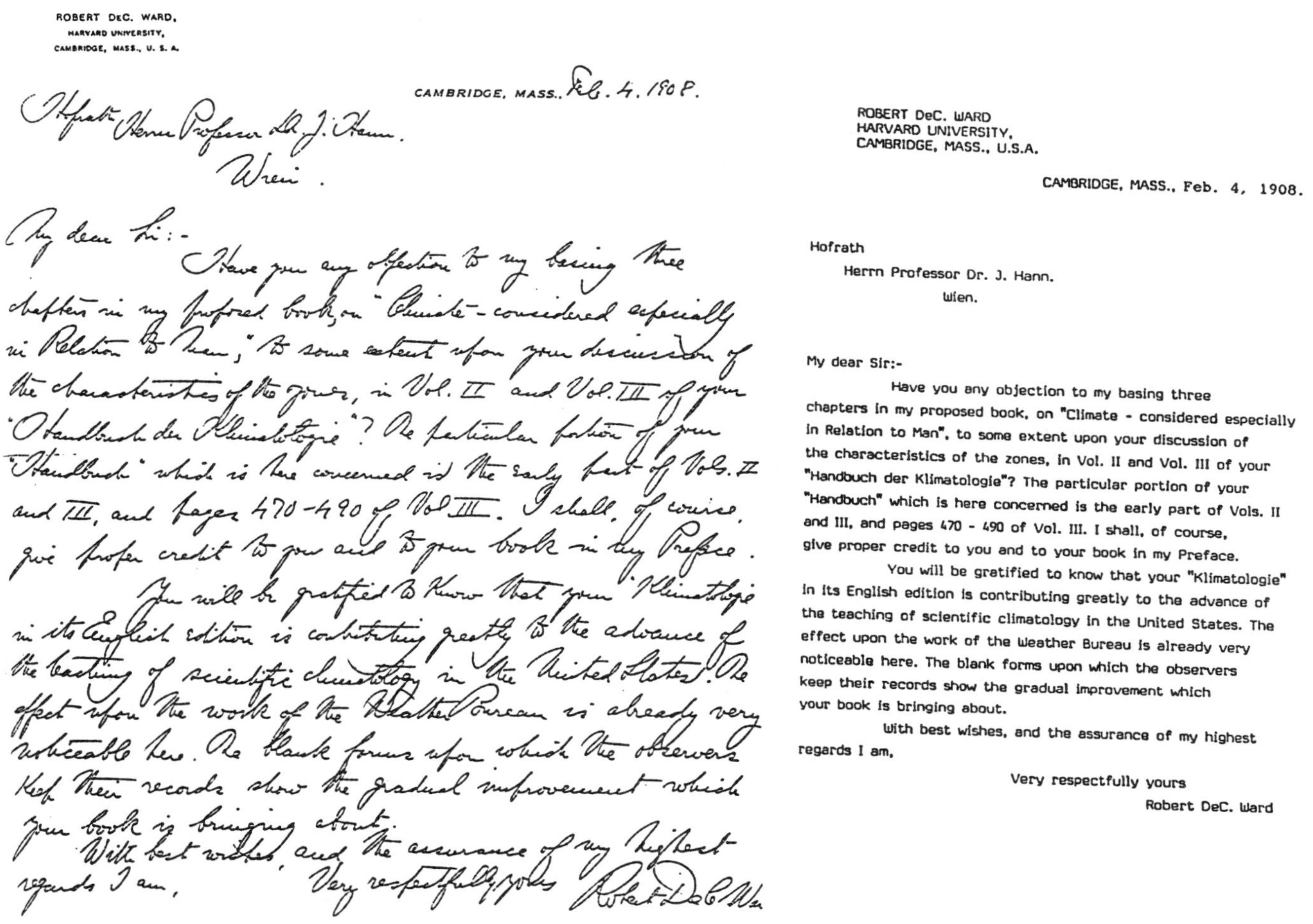

ROBERT DeC. WARD
HARVARD UNIVERSITY,
CAMBRIDGE, MASS., U.S.A.

CAMBRIDGE, MASS., Feb. 4, 1908.

Hofrath
Herrn Professor Dr. J. Hann.
Wien.

My dear Sir:-

Have you any objection to my basing three chapters in my proposed book, on "Climate - considered especially in Relation to Man", to some extent upon your discussion of the characteristics of the zones, in Vol. II and Vol. III of your "Handbuch der Klimatologie"? The particular portion of your "Handbuch" which is here concerned is the early part of Vols. II and III, and pages 470 - 490 of Vol. III. I shall, of course, give proper credit to you and to your book in my Preface.

You will be gratified to know that your "Klimatologie" in its English edition is contributing greatly to the advance of the teaching of scientific climatology in the United States. The effect upon the work of the Weather Bureau is already very noticeable here. The blank forms upon which the observers keep their records show the gradual improvement which your book is bringing about.

With best wishes, and the assurance of my highest regards I am,

Very respectfully yours
Robert DeC. Ward

Fig. 3. Letter of R. DeC. Ward (1908) reporting on the usefulness of the Handbook of Climatology (original and transcription).

it is a three-term moving average (running mean) with unequal weights (1/4, 1/2, 1/4). The method is introduced on p. 199 of Hann's Handbook of Climatology, using data from a table on p. 200 (see Figure 4). The spectral properties of this procedure led to the designations "Hann filter", "von Hann window", "hanning window", or simply "raised cosine window". (A modification, due to Hamming, leads to the "Hamming window" or "raised cosine window with a platform".) Of course, Hann smoothing is re-invented once and again without knowing of Hann's work explicitly (e.g. Le Roy Ladurie and Baulant, 1981).

Early Thoughts on Teleconnections

Julius von Hann is also known for his pioneering use of the idea of teleconnections, expressing some remote linkage of centers of action. For example, low surface pressure at Iceland appeared to correlate well with high surface pressure at the Azores (Hann, 1905; a statistical analysis of Hann's data was given later by Köppen, 1913). A pioneer and a master in this discipline was Sir Gilbert T. Walker (1868-1958), British meteorologist, mathematician by training (cf. Walker, 1906 and 1936; Khrgian, 1970; Barry and Perry, 1973; Henderson-Sellers and Robinson, 1986; Philander, 1990; Rasmusson, 1990). The role of Hann and Walker in recognizing and clarifying mechanisms of monsoon-like circulations was reviewed by Kutzbach (1987).

In 1903, Walker had become director of the India Meteorological Department. Hann (1907) reported on a paper of Walker (1906) concerning the meteorological conditions in the Indian monsoon region. Interestingly, the correspondence of Hann and Walker does not touch the idea of teleconnections but rather the maintenance of observing stations. In his letter, Walker congratulates Hann on his 70th birthday, and suggests to discontinue

TEMPERATURES OF PARALLELS AND HEMISPHERES. 199

of latitude in their relation to the percentage of land and water along each parallel. These temperatures must therefore be considered first.

In his epoch-making work, *The Distribution of Temperature over the Earth's Surface*, published in Berlin in 1852, Dove computed the mean temperatures of the parallels of latitude at intervals of ten latitude degrees for each of the twelve months, and for the year. This was done by taking from the monthly isothermal charts the temperatures at thirty-six equidistant points along any given latitude circle (*i.e.*, at every ten degrees of longitude) and deriving the mean from these thirty-six different temperatures. This mean temperature is also regarded as the *normal temperature* of the given parallel of latitude, although the term is not exact. This mean, however, depends not upon the latitude alone, but also upon the relation of land and water along the given parallel, and this relation changes from one parallel to another. After Dove, Spitaler,[1] and more recently Batchelder,[2] determined the mean temperatures of the parallels of latitude for January, July, and for the year. Spitaler used, as the basis of his work, the author's isothermal charts in Berghaus's new physical atlas (Gotha, 1887), and Batchelder used the isothermal charts prepared by Buchan, and published in the *Report on Physics and Chemistry*, Vol. II. of the *Challenger* expedition.[3]

In view of the importance of these data for climatology in general, and especially for the discussion which follows, they are included here. The second and third columns of the table (*a* and *b*) give the relative distribution of land and water along the different parallels of latitude. These values are taken from Penck's *Morphologie der Erdoberfläche.* The figures under *b* are determined by taking into account the parallels 5° away on either side. Thus, for example, for latitude 60° we have

$$\tfrac{1}{2}[60 + (65 + 55) \div 2].$$

The values given under *b* are the best for use in discussing the relation between the temperature and the amount of land surface along any parallel.

[1] R. Spitaler: "Die Wärmevertheilung auf der Erdoberfläche, *D. W. A.*, LI., 1886, Pt. II., 1-20.

[2] S. F. Batchelder: "A New Series of Isanomalous Temperature Charts," *Amer. Met. Jour.*, X., March, 1894, 451-474. These charts are also published in Bartholomew's *Atlas of Meteorology*, Plate 2.

[3] See also *M. Z.*, XVII., 1900, 36-39.

200 HANDBOOK OF CLIMATOLOGY.

MEAN TEMPERATURES OF THE PARALLELS OF LATITUDE.

Latitude.	Amount of Land Surface, in per cents.		Mean Annual Temperature.		January.	July.	Difference.
	a.	b.	Spitaler.	Batchelder.	Mean, Spitaler and Batchelder.		
			°	°	°	°	°
N. Pole,	—	—	−20·0	−20·0	(−38·0)	(0·0)	38·0
80°,	22	24	−16·5	−16·9	−33·5	1·8	35·3
70°,	55	54	−9·9	−10·2	−26·0	7·0	33·0
60°,	61	64	−0·8	−1·2	−15·8	14·0	29·8
50°,	56	55	5·6	5·8	−7·0	18·1	25·1
40°,	46	47	14·0	13·9	4·9	24·0	19·1
30°,	43	42	20·3	20·2	14·6	27·3	12·7
20°,	33	32	25·6	24·9	21·9	28·3	6·4
10°,	24	24	26·4	27·1	25·8	26·9	1·1
Equator,	22	23	25·9	26·6	26·4	25·6	0·8
10°,	20	23	25·0	25·7	26·3	23·9	2·4
20°,	24	23	22·7	23·3	25·4	20·0	5·4
30°,	20	18	18·5	18·3	21·8	14·6	7·2
40°,	4	5	11·8	12·2	15·6	9·0	6.6
50°,	2	2	5·9	5·3	8·3	2·9	5·4
60°,	0	1	(0·2)	−1·1	1·6	(−3·8)	—

Position of heat equator.—It appears from the above table that the highest mean annual temperature is found at latitude 10° N. This parallel is therefore the *thermal*, or *heat equator.* The equator is the warmest parallel only during the winter of the northern hemisphere, in January; while in July the highest temperature is seen to be somewhat north of latitude 20° N. This displacement of the heat equator into the northern hemisphere shows very strikingly the influence of the greater extent of the land surface in that hemisphere, for land areas in low latitudes are warmer than the ocean. The movement, from the southern into the northern hemisphere, of a considerable body of warm water, under the influence of the trade winds, to which reference has already been made (see page 188), is another factor in causing this displacement. The northern portion of the Indian Ocean, which is well surrounded by land, and the archipelago southeast of Asia and northeast of Australia, in many parts of which the water is comparatively shallow, are likewise sources of warmth which are not present in the southern hemisphere. The oceans of the northern hemisphere, which

Fig. 4. The "hanning" procedure of data smoothing, as applied in Hann's Handbook of Climatology. (Column a: raw data. Column b: smoothed data.)

observations at a small station after nine years, for reasons of economy (Figure 5).

Hann's 1905 paper deserves some detailed comments. Translated into English, the title is "Weather anomalies on Iceland in the period 1851-1900 and their relations to the simultaneous weather anomalies in the northwest of Europe". The introduction reads:

> Teisserenc de Bort coined the felicitous expression "atmospheric centers of action" for the permanent large air pressure maxima and minima of the extratropical zones. For the west of Europe, this role is played by the subtropical high-pressure area near the Azores and by the low barometric minimum near Iceland. The weather anomalies in the west of Europe are closely related to the temporary displacements of these centers of action.

The paper comprises two parts:

> A. The simultaneous anomalies in winter at Stykkisholm, Greenwich, Brussels, and Vienna.
>
> B. Relations between the air pressure fluctuations at Stykkisholm and those at Ponta Delgada, or relations between the two

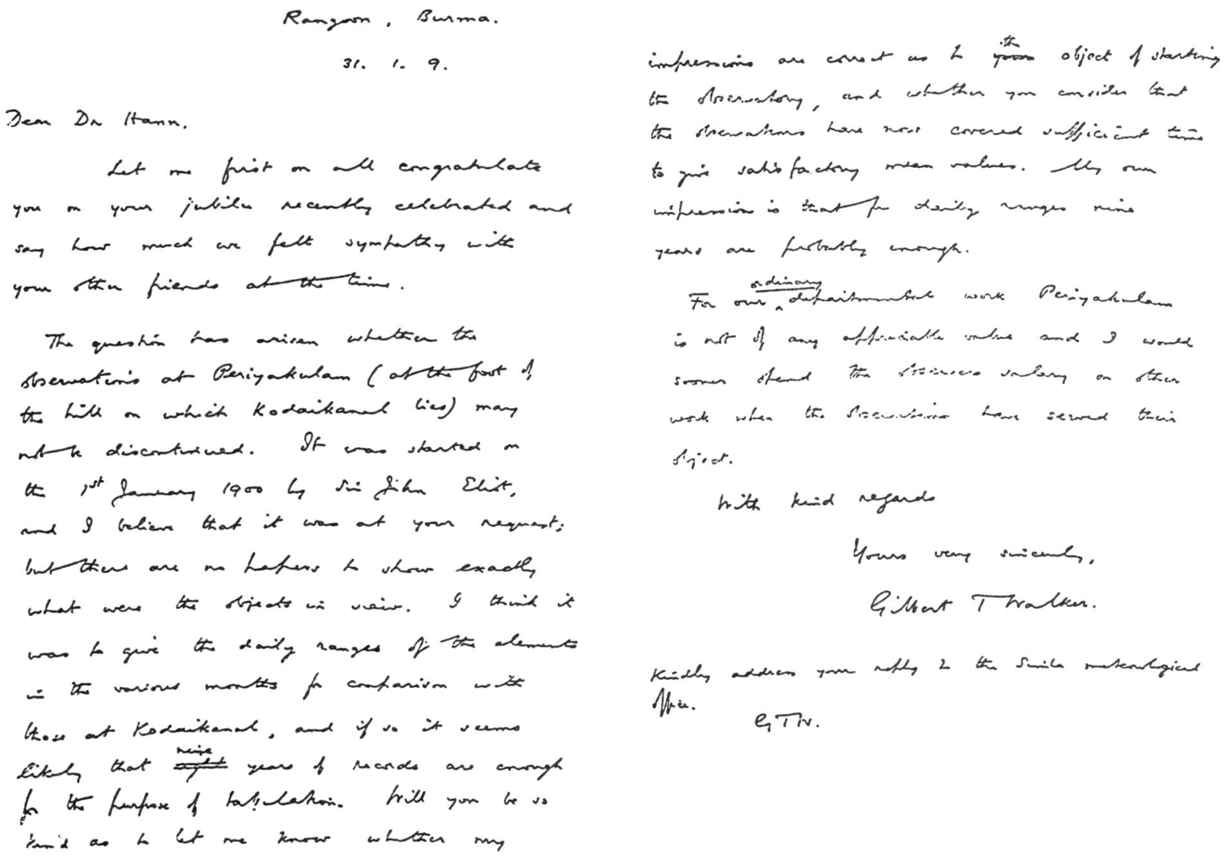

Rangoon, Burma.

31. 1. 9. (Jan. 31, 1909)

Dear Dr Hann,

Let me first on (of) all congratulate you on your jubilee recently celebrated and say how much we felt sympathy with your other friends at the time.

The question has arisen whether the observations at Periyakulam (at the foot of the hill on which Kodaikanal lies) may not be discontinued. It was started on the 1st January 1900 by Sir John Eliot, and I believe that it was at your request; but there are no papers to show exactly what were the objects in view. I think it was to give the daily ranges of the elements in the various months for comparison with those at Kodaikanal, and if so it seems likely that nine years of records are enough for the purpose of tabulation. Will you be so kind as to let me know whether my impressions are correct as to the object of starting the observatory, and whether you consider that the observations have now covered sufficient time to give satisfactory mean values. My own impression is that for daily ranges nine years are probably enough.

For our <u>ordinary</u> departmental work Periyakulam is not of any appreciable value and I would sooner spend the observer's salary on other work when the observations have served their object.

With kind regards

Yours very sincerely,
Gilbert T Walker.

Kindly address your reply to the Simla meteorological office.
GTW.

Fig. 5. Letter of G.T. Walker (1909) suggesting to discontinue observations at a small station (original and transcription).

"atmospheric centers of action" of the North Atlantic Ocean.

A main conclusion of Hann (1905) is:

> If the air pressure at the Azores is higher than its mean value, and if simultaneously the air pressure at Iceland is below its mean value (which is usual in more than 70 % of all cases), the normal air pressure gradient over the Atlantic Ocean is inforced, the atmospheric engine then operates more intensively, and at the same time the climatic favouring of Europe is enhanced. - Conversely, if the barometric maximum at the Azores is reduced, then the air pressure at Iceland is usually higher than its mean value, the normal pressure gradient (in north-south direction) over the Atlantic Ocean is reduced or may vanish entirely, the climatic favouring of northwestern and central Europe is more or less lost.

Conclusions

Julius von Hann's impact on climatology can still be felt. His careful planning of mountain observatories, his editorship of journals, and his authorship of classical standard texts in climatology identify Julius von Hann as an outstanding contributor to the foundation of modern climatology.

Dedicated to the memory of Professor Ferdinand Steinhauser.

Acknowledgments. The kindness of Dr. Gertrud von Hann, making available letters to her grandfather, is appreciated. Lucia Prohaska typed the manuscript.

References

Barry, R.G., 1981: Mountain Weather and Climate. Methuen, London and New York.

Barry, R.G., and A.H. Perry, 1973: Synoptic Climatology. Methuen, London.

Blackman, R.B., and J.W. Tukey, 1959: The Measurement of Power Spectra. Dover, New York.

Crowe, P.R., 1971: Concepts in Climatology. Longman, London.

Debus, A.G. (ed.), 1968: World Who's Who in Science. Marquis Company, Chicago.

Durst, C.S., 1951: Climate - the synthesis of weather. In: Compendium of Meteorology (T.F. Malone, ed.), AMS, Boston.

Encyclopaedia Britannica, 1962: Hann, Julius. Enc.Brit., London/Chicago/Toronto.

Essenwanger, O.M., 1986: Elements of Statistical Analysis. World Survey of Climatology (H.E. Landsberg, ed.), Vol. 1B. Elsevier, Amsterdam/London/New York/Tokyo.

Hamming, R.W., 1989: Digital Filters. Prentice-Hall, Englewood Cliffs.

Hann, J., 1883: Handbuch der Klimatologie. Engelhorn, Stuttgart. (This first edition consisted of one volume only.) - Second edition: 1897 (in three volumes). Third edition: 1908/1910/1911 (in three volumes, but volume I issued separately). - As a kind of continuation, the "Handbuch der Klimatologie" was edited later by W. Köppen and R. Geiger (with R. DeC. Ward contributing as one of many authors), and was published by Borntraeger, Berlin, from 1930 onwards.

Hann, J., 1887: Atlas der Meteorologie. (Berghaus' Physikalischer Atlas, Abt. III.) Perthes, Gotha.

Hann, J., 1903: Handbook of Climatology. (Part I: General Climatology.) Translated from the 2nd German edition by R. DeC. Ward. Macmillan, New York/London.

Hann, J., 1905: Die Anomalien der Witterung auf Island in dem Zeitraume 1851 bis 1900 und deren Beziehungen zu den gleichzeitigen Witterungsanomalien in Nordwesteuropa. Meteor.Z., 22, 64-77.

Hann, J., 1907: Über die Beziehungen des SW-Monsun-Regenfalles in Indien zu den Witterungszuständen entfernter Gegenden. (Report on a paper of Gilbert T. Walker.) Meteor.Z., 24, 74-79.

Hantel, M., 1989: Climate modeling. In: Meteorology: Climatology, Part 2 (G. Fischer, ed.), Landolt-Börnstein, New Series, Vol. V 4 c 2. Springer, Berlin.

Hantel, M., H. Kraus, and C.-D. Schönwiese, 1987: Climate definition. In: Meteorology: Climatology, Part 1 (G. Fischer, ed.), Landolt-Börnstein, New Series, Vol. V 4 c 1. Springer, Berlin.

Henderson-Sellers, A., and P.J. Robinson, 1986: Contemporary Climatology. Longman, Harlow, and Wiley, New York.

Huschke, R.E. (ed.), 1959: Glossary of Meteorology. AMS, Boston.

Khrgian, A.Kh., 1970: Meteorology - a Historical Survey. Vol. I. IPST, Jerusalem.

Köppen, W., 1913: Zusammenhang der Luftdruckabweichungen über Island, den Azoren und Europa. Meteor.Z., 30, 121-125.

Kutzbach, G., 1987: Concepts of monsoon physics in historical perspective: the Indian monsoon (seventeenth to early twentieth century). In: Monsoons (Fein, J.S., and P.S. Stephens, eds.). Wiley, New York/Chichester.

Landsberg, H.E., and J. Albert, 1975: Some aspects of global climatic fluctuations. Arch. Met. Geoph. Biokl., Ser. B, 23, 165-176.

Le Roy Ladurie, E., and M. Baulant, 1981: Grape harvests from the fifteenth through the nineteenth centuries. In: Climate and History (R.I. Rotberg and T.K. Rabb, eds.). Princeton Univ. Press, Princeton.

Mitchell, C.L., 1921: Death of Dr. Julius von Hann. Monthly Weather Review, 49, 510.

Philander, S.G., 1990: El Nino, La Nina, and the Southern Oscillation. Acad. Press, San Diego/New York.

Rasmusson, E.M., 1990: Nature of low-frequency tropical variability. In: Meteorology and Environmental Sciences (R. Guzzi, A. Navarra, and J. Shukla, eds.). World Scientific, Singapore.

SES (Sov.Enzikl.Slovar), 1980: Hann, Julius. Sov. Enzikl., Moskva.

Shaw, N., 1921: Dr. Julius Hann. Nature, 108, No. 2712, 249-251.

Simpson, G.C., 1921: Julius von Hann. The Meteorological Magazine, 56, 300-302. (Abstract in: Bulletin of the Am.Met.Soc., 2, 140-141.)

Steinhauser, F., 1951: Julius von Hann. In: Oest. Naturforscher und Techniker (Oest. Akad. Wiss., ed.). Gesellschaft f. Natur u. Technik, Vienna.

Walker, G.T., 1906: Memorandum on the meteorological conditions prevailing in the Indian monsoon region before the advance of the south-west monsoon of 1906, with an estimate of the probable distribution of the monsoon rainfall in 1906. Government Central Printing Office, Simla.

Walker, G.T., 1936: Seasonal weather and its prediction. Smithsonian Institution, Annual Report for 1935, 117-138. Washington, D.C.

Ward, R. DeC., 1908: Climate Considered Especially in Relation to Man. Putnam, New York, and Murray, London. (2nd ed.: 1918.)

Ward, R. DeC., 1922: Julius von Hann (1839-1921). Proc.Am.Acad.Arts and Sci. (now "Daedalus"), 57, 491-493.

WMO (World Meteorological Organization), 1988: Concept of the Global Energy and Water Cycle Experiment. WCRP-5 (WMO/TD-No. 215). WMO, Geneva.

Peter Kahlig, Institute of Meteorology and Geophysics, University of Vienna, A-1190 Vienna, Austria.

Physical 3D-Climatology from Hann to the Satellite Era

HERMANN FLOHN

Meteorological Institute, University of Bonn

I. INTRODUCTION

I am very grateful for the invitation to review the development of physical climatology since the classical contributions of Hann 125 years ago, here in the town where he wrote his most important work. When I was a young student, more than 60 years ago, Hann's ideas continued to dominate climatology prior to the advent of operational aerology in the 1930's. I have been fascinated with this field ever since we were on a two-week excursion across the Swiss Alps in the Spring of 1931, where I had to prepare local climatological surveys using the wealth of available data. We realized that the climate changed with altitude, at the coasts and on islands of the lakes, as well as the role of dominant winds and cloud motions in high mountains - only here was a view into the third dimension possible. However, at this time, the fundaments of the 3D-climatology of our era already existed: a number of regular airplane ascents, and methods to present their results in a physically sound format had been developed [V. Bjerknes, 1910/11; cf. Flohn, 1955]. The crucial role of a 3D-view for a physical understanding of common atmospheric phenomena such as gales and frontal discontinuities had been formulated by Margules [1905, 1906]. But we young students were unable to imagine that atmospheric sciences, especially climatology, could evolve into a physical science as quickly as they did over the course of the following two decades.

I had been asked to speak on the topic "From Hann to 3D-models". Since I am not a model specialist, this put me into somewhat of an embarrassing predicament. Consequently, I chose the title, "... to the ERBE-Satellite", expecting access to full results from this highly important instrument - unfortunately only initial evaluations [Ramanathan et al., 1989; Raval and Ramanathan, 1989] are available. So, instead, I prefer a more general title (as above), which indeed refers to the most powerful, accurate and complete information source monitoring the climatic system of our planet from space. Since the title of this Symposium deals with fluxes of matter, I prefer to speak mainly, but not exclusively, on the development of physical climatology with emphasis on one of the most important fluxes: that of water between atmosphere, ocean and soil/vegetation as representing the terrestrial biosphere.

Interactions Between Global Climate Subsystems, The Legacy of Hann
Geophysical Monograph 75, IUGG Volume 15

In the climate system, this flux is not only vital at the earth-air interface, where the evolution of micrometeorology necessarily led to the requirements of the budgets of radiation, heat and water vapour. It is also essential for the atmosphere, which became regularly accessible to meteorologists during the 1930's: here the balances of momentum, of potential and kinetic energy formed the central background of scientific evolution. Hann himself succeeded in introducing thermodynamics into meteorology; his former assistants contributed, in the years between 1890 and 1940, much to the development of our science. This ran simultaneously with the fascinating theoretical ideas of V. Bjerknes, the "Bergen School" since 1917, his assistants and their students. During the initial decades of this century, the Austrian and the Scandinavian schools worked independently from each other, both inductively and deductively, simultaneously with some early attempts of mathematical forecasting [F. M. Exner, 1906; L. F. Richardson, 1922] - but, at that time, this still remained "a dream".

Jumping now to present day, in spite of the admirable progress in satellite techniques and in the design of interactive atmosphere/ocean models, our knowledge of the climatic water balance of our planet is still unsatisfactory in a quantitative sense, albeit in consideration of the growing population, its role cannot be emphasized enough. Now, at the beginning of a man-triggered climatic change, we should always remember, that the balances of radiation, heat, water and energy within the climate system are closely interrelated, and that e.g. at the ocean's surface, 88% of the available (net radiational) energy is used for evaporation.

I refer to Baumgartner and Reichel [1975], who, from 37 different authors (since 1881), compiled the global averages of precipitation (P) = evaporation (Ev), varying between 69 and 102 cm/a. In the meanwhile, Russian estimates [Korzun et al., 1974] went even up to 113 cm/a. I refer also to A. Kessler [1985], who summarized the results of 32 authors on radiation and heat budget. I can therefore limit myself to a historical review, including some new data of global parameters at a latitude-season plane.

II. HEAT AND WATER BUDGETS AT THE EARTH'S SURFACE

The scientific evolution of climatology around the turn of the century started mainly from the balances of radiation and water at the earth's surface. Solar radiation was recognized as the

TABLE 1. Annual heat budget (Watt/m²)

	Continents		Oceans		Globe	
	Bu	H&A	Bu	H&A	Bu	H&A
Sensible heat	31	37	12	14	17	20
Latent heat	36	34	109	90	88	74
Net radiation	67	72	121	103	105	95

Bu = Budyko; H&A = Henning 1989, Albrecht 1961

external source of energy operating within the climate system. Instruments were developed to measure the direct radiation of the sun (S), the diffuse radiation of sky (D), the surface albedo (a_{sf}) and the two components of infrared (long-wave) radiation: the terrestrial radiation up to the sky and the counter-radiation downward from the atmosphere ($LW\uparrow$ - $LW\downarrow$). Together with the short-wave radiation balance one obtains the net radiation (Q) at the surface (Table 1):

$$Q = (S + D)(1 - a_{sf}) - (LW\uparrow - LW\downarrow) \quad (1)$$

Here, I can omit all historical details since today this equation is well-known in relation to the greenhouse effect, which is included in the term $LW\downarrow$. I shall only recall a fact which has been too frequently neglected: in a cloud-free atmosphere, about 65% of $LW\downarrow$ is caused by atmospheric water vapour. In the last edition of Hann's famous Handbook of Climatology [1st edition 1883, 4th edition expanded by K. Knoch, 1932, only Vol. 1], the Chapter on "Solar Climate" hardly gives more than a page to the longwave radiation together with the total (net) radiation, based on papers by G. C. Simpson and F. Albrecht (Potsdam). The latter designed many instruments, among them one for the measurement of the whole "Strahlungsbilanz" (net radiation), shortwave and longwave together. In the 1930's, progress was rather rapid - F. Linke [1942] gave a comprehensive review on infrared radiation in the Handbook of Geophysics, which appeared unfortunately during World War II.

Simultaneously, the turbulent fluxes of sensible and latent heat from the surface into the air were investigated - these are the most important processes with which the earth's surface gets rid of the surplus of energy received by the radiational processes (Q). Relatively small are the conductive heat fluxes into the soil (and to the vegetation: B), first studied by von Bezold [1892] and Homén [1897] in Finland (in cold climates, B also contains the heat necessary for melting the snow-cover, which dominates during the melting period over all other quantities). Of much higher importance are the two turbulent fluxes in the atmosphere [W. Schmidt - Vienna, 1925; "Austausch", H. Lettau, 1939] consisting of sensible heat (H) and the latent heat of water vapour (L*Ev), with L = heat of condensation, resulting in the heat balance:

$$Q = H + L * Ev + B \quad (2)$$

While at land, B can be calculated using a set of temperature measurements at different depths in the soil, a direct measurement of the fluxes H and L*Ev needs - because of the turbulent nature of the vertical wind component w - the eddy correlations $w'\Theta'$ and $w'q'$. Here, the prime denotes the instantaneous deviations from a time average (Θ = potential temperature, q = specific humidity). Such measurements imply a rather high instrumental expenditure and remain therefore limited to short series at observatories, during expeditions and field campaigns. A more general derivation of this equation based on the integral principle of energy conservation has been given by Hantel [1992], with some important climatological conclusions.

Approximate results at the oceans can be reached by the assumption that the drag coefficient C is nearly the same for the exchange of momentum, of heat, and of water vapour. Then - details omitted - we can use (c_p = specific heat of air, z = height, Θ = potential temperature, q = specific humidity)

$$H / L^*Ev = c_p\, d\Theta/dz \,/\, L\, dq/dz = Bo \quad (3)$$

The right side of this equation is usually noted as Bowen's Ratio [1926] and is often used at land and sea. Since on land, Q and B could be either measured or approximately calculated, equation (2) leads to the sum H + L*Ev; together with measurements of the Bowen Ratio (3) H / L*Ev, the four terms of the heat balance Q can be estimated, as proposed by Sverdrup [1937]. This simple method gave a quantitative estimate of the fundamental equations (1) - (3) for physical climatology accessible during the 1930's. In 1940, Albrecht's comparative study (written in German) of quite different climates appeared (Fig. 1). But this happened during World War II, it reached only a few German scientists and received no printed echo. After the war, Albrecht (still at Potsdam) wrote some review papers to be translated into Russian and then emigrated to Australia. Budyko - with knowledge of these papers - applied a similar method in the U.S.S.R. during the 1950's[1]. He developed another important dimensionless quantity: the dryness ratio Bu = Q / L * P - which gives a useful physical quantity to classify earth's natural vegetation types.

To calculate the oceanic evaporation, one needs the scalar wind speed u, the gradient of q between the sea surface temperature T_s (assuming saturation) and the air temperature T_a (more exactly Θ_a) and the exchange coefficient $C_E = 1{,}3*10^{-3}$. It should be mentioned, however, that C_E is not a linear function of u and T_s-T_a, with rather important deviations from linearity at very low and high values of u [Isemer and Hasse, 1987]. With $u \geq 12$ m/s, the spray above breaking wind waves increases Ev remarkably. Having this in mind, it becomes obvious that the frequently used "bulk" formula (ρ = density of air) for ocean evaporation:

---

[1] After a serious motion was made to improve climatic conditions in Siberia, through the construction of gigantic reservoirs ("Siberian Sea") and other utopian projects, the 1959 Congress of the Communist Party decided to implement Marx's dogma: "Improve Nature for the Benefit of Mankind", and urged the Academy of Science to direct a great effort towards man-made large-scale climatic changes. Following this, Budyko's research was strongly supported, with the objective of amelioration of Siberia's climate together with the prerequisites of development and traffic.

Early in 1960, I met - upon intervention of H. Ertel - S.P. Chromov and an official from Moscow in the Academy of East Berlin, who intentionally gave me this important information with the obvious presumption that I would pass it on to other western climatologists (which I did).

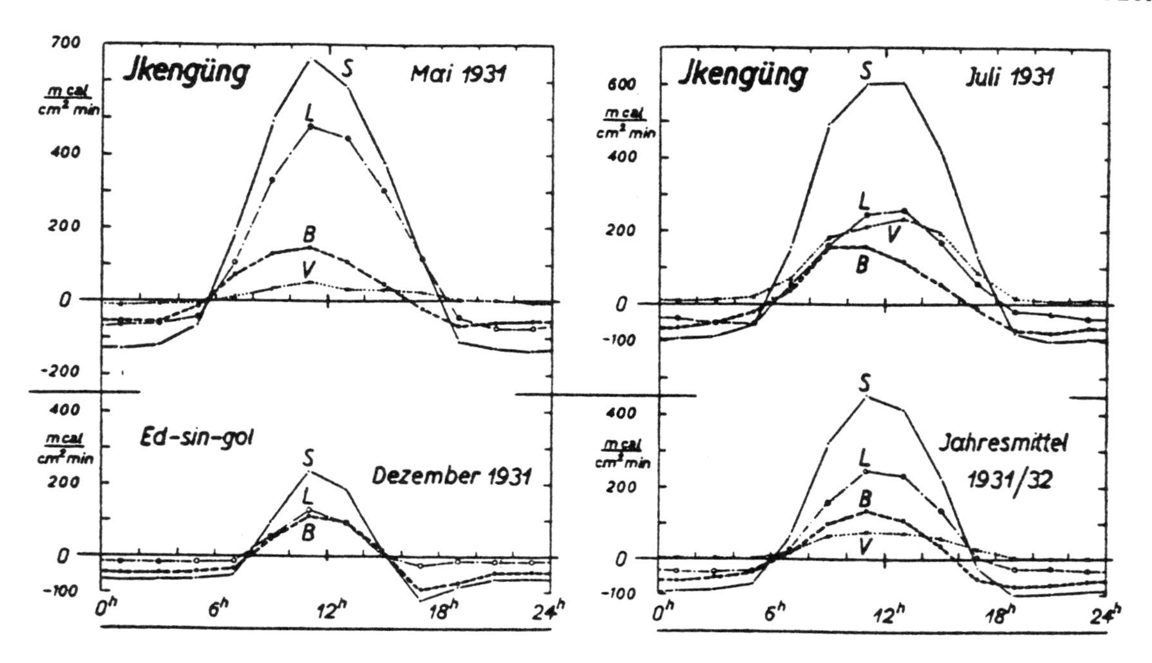

Fig. 1. Diurnal variation of heat budget parameters in the Mongolian Steppe [Albrecht, 1940]. S=net radiation, B=heat flux into soil, L(V) = turbulent fluxes of sensible (latent) heat.

$$Ev = C_E \rho (q_s - q_a) u \qquad (4)$$

cannot give more than a first approximation. One has to use time averages for the highly variable parameters q_s-q_a and (especially) u, and to neglect correlations between them. At the continents, there are two other parameters with episodical or seasonal storage - soil moisture and snow-cover - which are spatially very variable and therefore difficult to handle. One of the most complete early systems of algorithms to solve these problems numerically has been given by Lettau [1969, 1979, 1991]. Lettau [1969] found also a simple relation between three non-dimensional parameters, the Bowen ratio Bo, Budyko's dryness ratio Bu and the runoff factor C = Rf / P used in Hydrology (Rf = runoff, P = precipitation):

$$Bu = (1 + Bo)(1 - C) \qquad (5)$$

In an attempt to improve Thornthwaite's - somewhat oversimplified - methods, Willmott et al. [1985] have given a global review of these parameters and their seasonal variations. Errors in the application of Bowen's method have been discussed by Ohmura [1982].

After Albrecht's return from Australia to a remote village in the Black Forest, still active during his long illness until his death in 1965, A. Baumgartner (Munich) and I then decided to foster a concentrated effort to revise and finalize his work on the main terms of the heat budget for both land and sea. This was to entail a complete evaluation of all available data and as well as a comparison of the results of different algorithms used by Albrecht, Budyko [cf. 1982] and Penman. We received support from the German Research Community, and together with a highly experienced retired climatologist (E. Reichel), Baumgart-

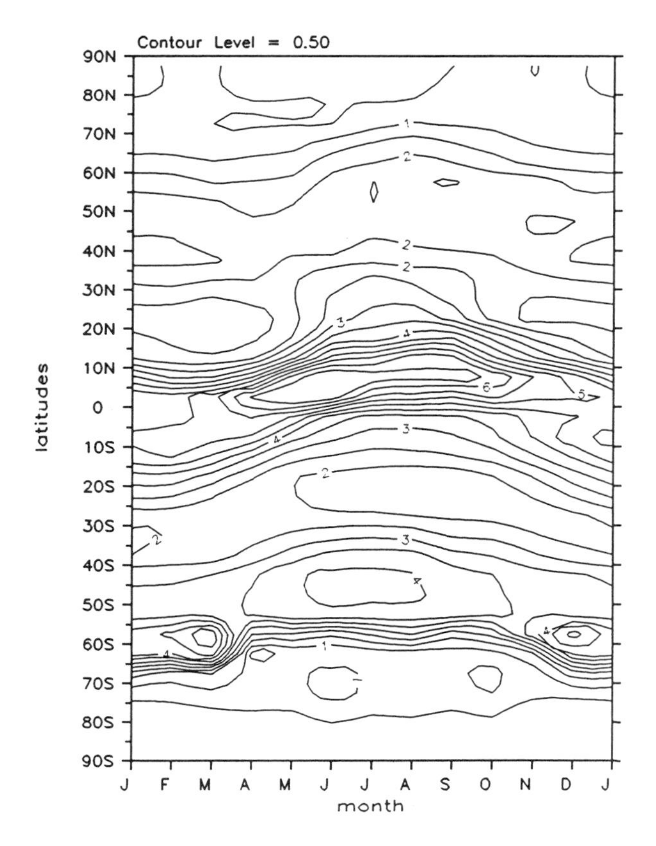

Fig. 2. Seasonal variation of precipitation (land and sea, zonal averages) [Jaeger, 1976].

TABLE 2. Annual water budget (cm/year)

	Continents			Oceans		Globe
	P	E	Rf	P	E	P - E
Budyko, Korzun	80	45	35	127	140	113
German Authors	78	44	34	104	110	97
U.S. Nat. Res. Council	72	48	24	110	120	99

P = precipitation, E = evaporation, Rf = runoff

ner was able to publish the oceanic part with extended tables and maps in 1975. Somewhat later, Jaeger - on the advice of Kessler - drew the first monthly global maps of precipitation, certainly with much generalization (Fig. 2). Due to the increase of precipitation with height in hardly settled mountains and to the reduction of its measurement caused by wind, the absolute figures are most probably somewhat too low (Table 2). In recent years, marine data have become accessible (COADS, British Meteor. Office, Bunker data) and gratefully welcomed by many scientists. New atlases - on sea surface temperatures and ocean heat gain [Bottomley et al., 1990], North Atlantic Heat Budget [Isemer and Hasse, 1987] - have been published [cf. also Hsiung 1986]. These are the most important sources for comparative investigations at a hemispheric or global scale.

At this time, Henning (then in Bonn) had collected climate records from more than 4000 land stations - compared with about 1600 used by Budyko - and had started, with utmost care, the laborious work of comparative calculations. The drawing of many hundreds of maps for each continent and season was not completed until 1978, but the formulation of the introductory text and the calculation of monthly and annual zonal averages (Figs. 3-4) was increasingly delayed by other commitments and responsibilities of the author. Nevertheless, D. Henning - now working alone - finished the manuscript in 1988, finally to be published in 1989, more than 20 years after its beginning. A future comparison with even more complete results of Willmott and Thornthwaite [Willmott et al., 1985] will be interesting.

Fortunately, Henning's atlas appears to have been published at the right time, when broad-scaled international efforts - the Global Energy and Water Experiment (GEWEX) - aim to improve the basic data on which these fundamental budgets have to be based. Indeed, the evaluation results of the earth's surface energy and water budget are still quite different (Tables 1-2). Bearing in mind the fundamental nature of this data at the time of an apparent beginning of a man-triggered climate fluctuation

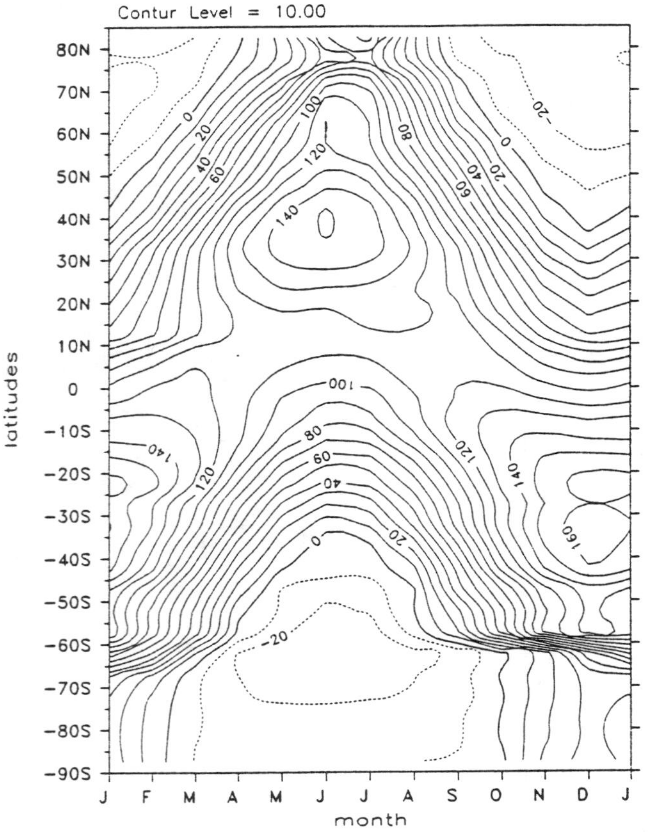

Fig. 3. Seasonal variation of net radiation at the continents (Watt/m²) [Henning, 1989].

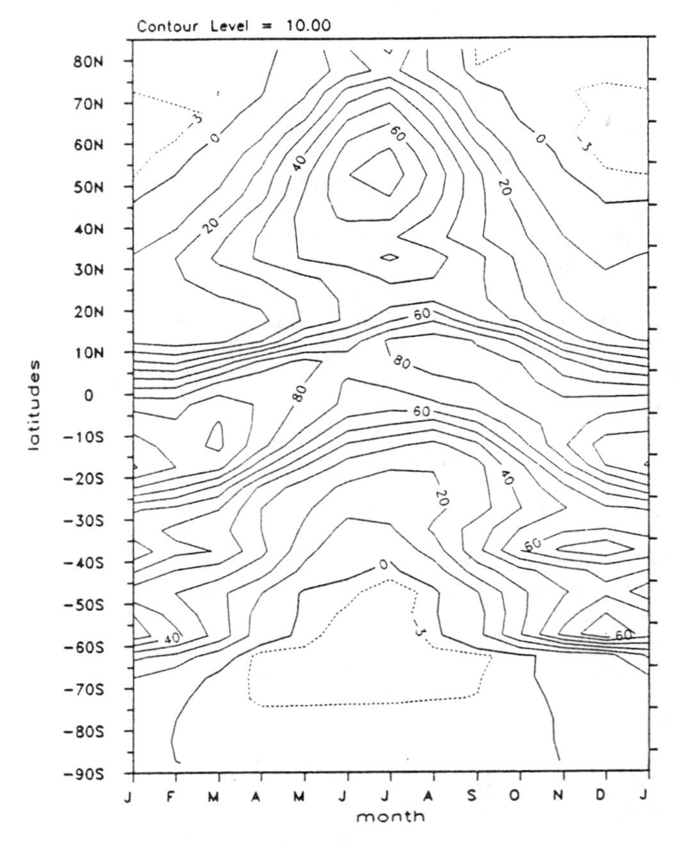

Fig. 4. Seasonal variation of evapotranspiration at the continents (Watt/m²) [Henning, 1989].

of possibly far-reaching consequences, one may even speak of a scandalous lack of elemental knowledge in a virtually vital field for mankind as a whole - much more vital than e.g. the research of outer planets.

III. High mountains initiating 3D-climatology

During the second half of the 19th century, only the fundamentals of climatology as a physical science had been laid. Hann's greatest merit was the application of thermodynamics within the atmospheric processes - the pure existence of the alpine mountains directed his view (comparable to Alexander von Humboldt) into the third dimension of altitude. His thermodynamic interpretation of Föhn - almost simultaneously with Helmholtz - made him well-known; this principle was essential for all kinds of vertical motions in the atmosphere. Compared with contemporaneous publications, this was a real breakthrough, arriving exactly at the right time. It was his initiative which led to an entire chain of around 10 mountain observatories at levels between 800 and 650 hPa to be set up in Central and Southern Europe since the 1880's. These stations were completely isolated during the winter and reached from the Tatra mountains and the Dinarides to the Appennines and the Pyrenees. He founded the "Oesterreichische Zeitschrift für Meteorologie" [Hann, 1866], which later evolved - after the foundation of the German Meteorological Society (1884) - into the "Meteorologische Zeitschrift", a great stimulus of research in Central Europe.

In the 1880's, an important international controversy arose: while in Central Europe, surface anticyclones were warm - which, at first appeared to be contrary to the fundamental hydrostatic equation - American and Russian scientists found their wintertime anticyclones predominantly cold. Only 3-dimensional dynamics could solve the enigma of warm anticyclones, to be interpreted as a 3D-circulation with a large subsiding core.

IV. Atmospheric circulation and 3D-processes

In Chapter II, the physical processes at the earth's surface have been reviewed, leading to flux balances of radiation, energy and water. But one of the main issues of physical climatology needs 3D-investigations. These are the balances of angular momentum and kinetic energy and their relation to the maintenance of the general circulation of the atmosphere. Today this problem appears to be solved in principle, but the quantitative data are still rather uncertain, even after the advent of the essential satellite wind data not yet evaluated for a climatology.

The first physically correct view of the atmospheric circulation was given as early as 1735 from Hadley (Fig. 5A), while 100 years later Maury developed strange ideas on a crossing of upper-air circulations; his book showed him - seen from today - as a sort of creationist, looking into the bible when his limited physical knowledge found no answer. Then we should remember, that before Ferrel [1856], H. W. Dove [1837, Berlin] outlined the role of eddies (Fig. 5B) in the middle and higher latitudes (but without mathematics). At an advanced age, he lost much of his reputation when he blocked the evolution of synoptics and weather forecasting, and his earlier papers sank into obliviation. A. Defant [1921] described the atmospheric circulation as a macroturbulence (Grossaustausch), stimulated by the micro-scale approach of W. Schmidt [1917, 1925] here in Vienna. Today our hemispheric maps demonstrate daily the role of waves and vortices in the great wind systems in the sense of the Norwegian school of V. Bjerknes, further developed by C. G. Rossby and his associates. For all details of this historical evolution reference is made to the excellent reviews by E. N. Lorenz [1967, 1991].

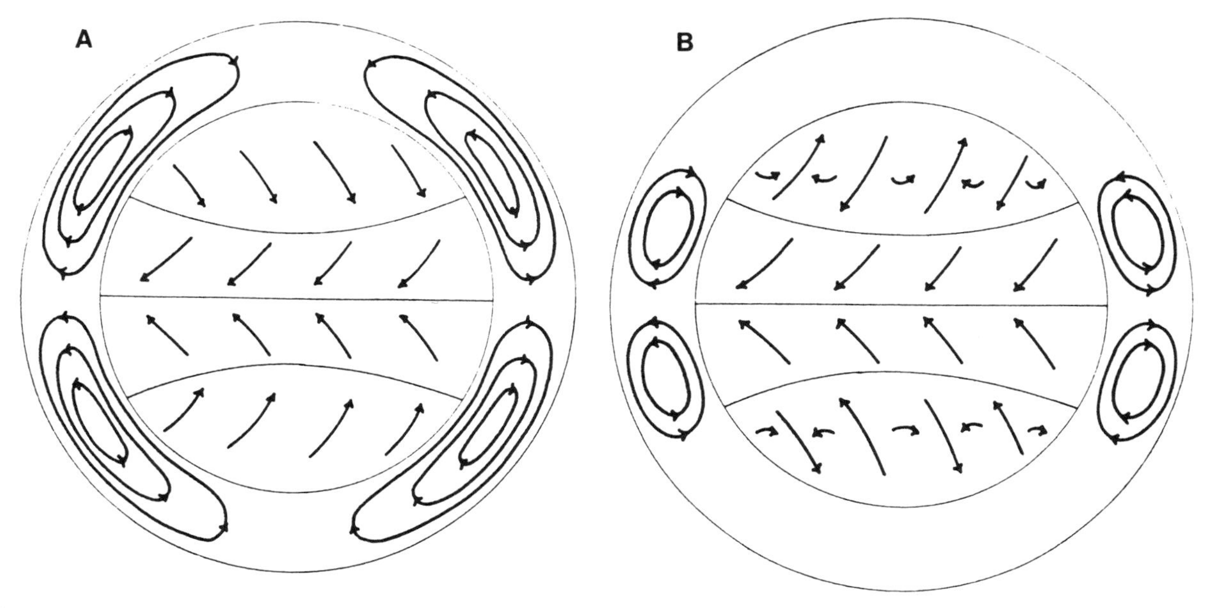

Fig. 5. A) atmospheric circulations conceived by Hadley [1735; cf. Lorenz, 1967]. Today, the mean meridional circulation in this form can only prevail above the Tropics; B) atmospheric circulation as conceived by Dove [1837; cf. Lorenz, 1967].

The recent evolution of physical 3D-climatology begins with papers by V. Starr [1948] and E. N. Lorenz [1955], the letter essentially based on Margules [1905]; for details, the reader should refer to Oort [1964]. Quantitative investigations on the maintenance of the general atmospheric circulation could only be based on data from a network of aerological stations, which became available after World War II.

Using such a fairly dense network, one can derive for each quantity to be transported time averages (noted by bars), zonal averages (noted by brackets [ ]), together with deviations from a time average (prime) and deviations from a zonal average (asterisk). At an isobaric surface, the northward transport of a conservative quantity K (e.g. relative angular momentum, specific humidity, potential temperature) with the meridional wind component v (positive northward) is then:

$$[\overline{Kv}] = [\bar{K}]\ [\bar{v}] + [\overline{K'v'}] + [\overline{K^*v^*}] . \qquad (6)$$

Then the sum of all transports consists of three terms on the right side:

a) transport with the mean meridional circulation,

b) transport with the travelling ("transient") eddies,

c) transport with the stationary eddies.

From historical point of view, a) may represent Hadley's cell, b) and c) together Dove-Ferrel's cell.

While J. Bjerknes and his group used geostrophic winds - which could only give first-order approximations, since their meridional component disappears when averaged along a latitude circle - V. Starr and his group used observed winds, which were much less frequent in the first years. With much more data, Starr's approach was repeated by Oort and Peixóto [1983]; here only results for water vapour (Fig. 6) are shown.

It was now possible to calculate (e.g. as a residual) the latitude-averaged small vertical components and the mean meridional circulations with their unexpectedly large seasonal variations

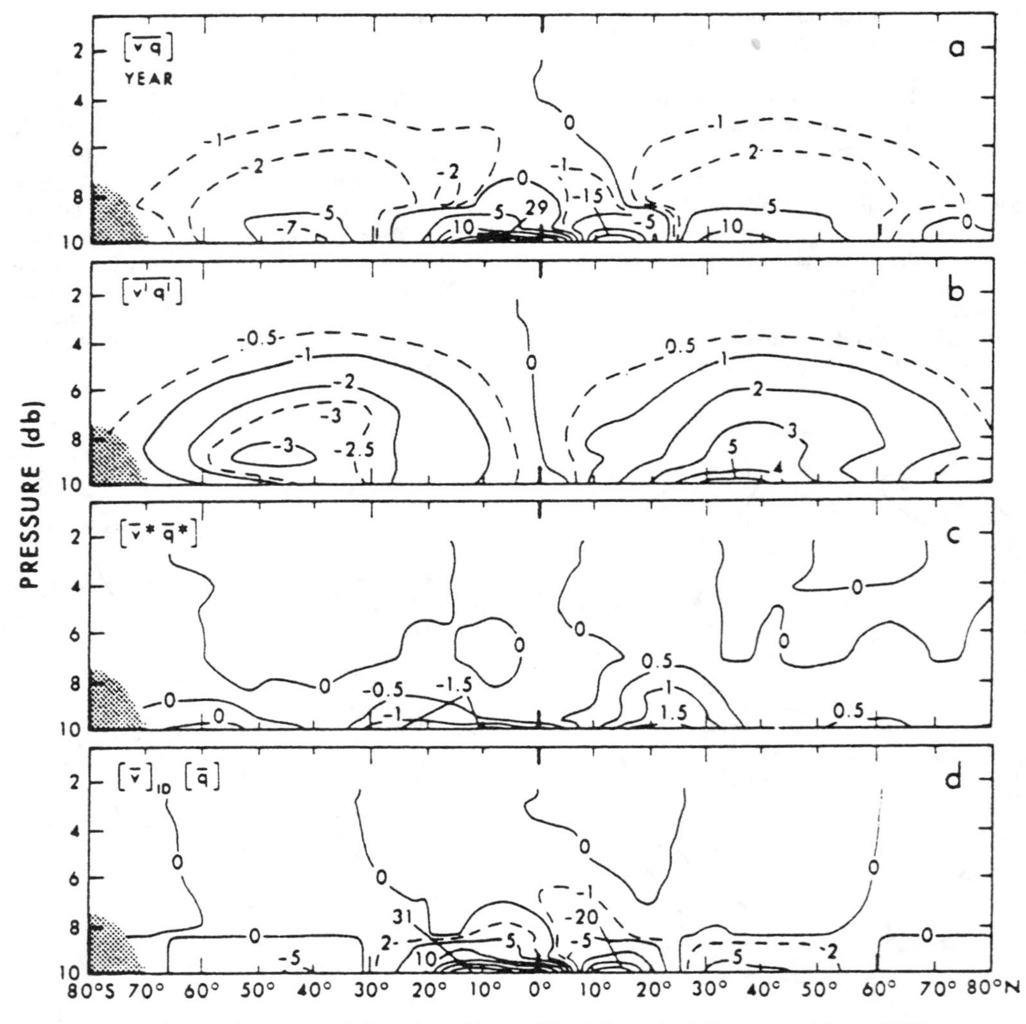

Fig. 6. Zonally averaged cross-sections of the northward flux of specific humidity (g/kg*m/sec) [Peixoto and Oort, 1983]; a = sum, b (c) = transport by travelling (stationary) eddies, d = transport by mean meridional circulation; southern latitude values uncertain. Pressure scale in 10 kPa.

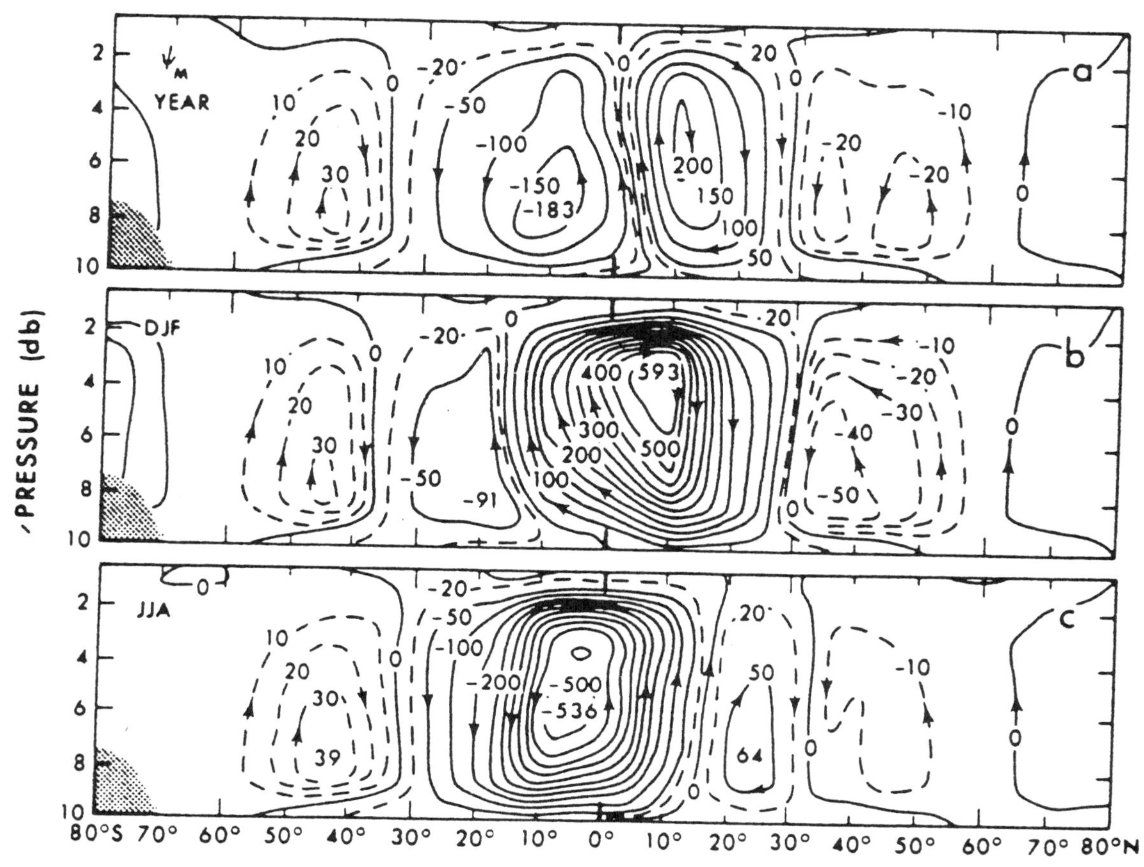

Fig. 7. Streamlines of the zonally averaged mean transport of absolute angular momentum (1963-1973) for year (a), December-February (b) and June-August (c). Transitional seasons similar to year, south of Lat. 40°S uncertain, units 10^{18} kg m^2/sec^2 [Peixoto and Oort, 1983].

(Fig. 7). During the year and in the transitional seasons both Hadley cells are nearly symmetric and much stronger than the mid-latitude Ferrel cell or the quite small polar cells. Whereas, during the two extreme seasons, the (meridional) Hadley cell of the winter hemisphere merges with the summer monsoon cell from the other hemisphere and reaches extraordinary intensity in the λ-p plane (cf. equation (6), first term at right). Here the tropical u-component is disregarded. It is much stronger than the v-component, but consists, in the lower troposphere, of two opposing branches: from E at the winter hemisphere, from W at the summer hemisphere. A powerful meridional flow crosses, especially in some longitudes - East coast of Africa, around 110°E - the equator, leading to this strong equatorial cell with (in a latitudinal average) lifting (subsidence) above the summer (winter) hemisphere. It should be noted that during austral summer a weak monsoon system really exists, running from Indonesia across the Indian Ocean to Africa, in spite of quite different topographical conditions. However, such spatial differences can only be demonstrated with sufficient accuracy if one uses the very large number of satellite-borne winds (i.e. cloud motion vectors) which have already greatly improved our knowledge of tropical wind systems and the role of surface topography for their seasonal variations.

At the time of Hann, it was an enigma, what energy may maintain the surface anticyclones with their continuous outflow near the surface. At the Viennese Zentral-Anstalt, M. Margules (1856-1920) found - based on a local network of stations - that the gradients of surface pressure were insufficient to induce the energy of storms. Looking at the low-level temperature gradients, he discovered a more powerful source in what is now described as the mutual conversion between available potential energy APE (he used the term "available kinetic energy") and kinetic energy KE. Evidence for these fundamental processes could only be found after the advent of operational aerology and the discovery of jet-streams ("Freistrahl"), [Seilkopf, 1939]. Between 1935 and 1945, we learned about the role of ageostrophic components [Philipps, 1939] in the area of the strong baroclinic zones ("frontal zones" of the Norwegian school) and the cross-isobaric mass transports, including the formation of cyclones and anticyclones. In areas with accelerating high-tropospheric winds (i.e. in the entrance region of a jet-stream) this ageostrophic component transports mass to the cold region (with low pressure aloft) north of the jet. In its delta, with decelerating winds, their surplus KE is used for a mass-transport towards the warm side, i.e. against the slope of the isobars (Fig. 8) [Flohn, 1952].

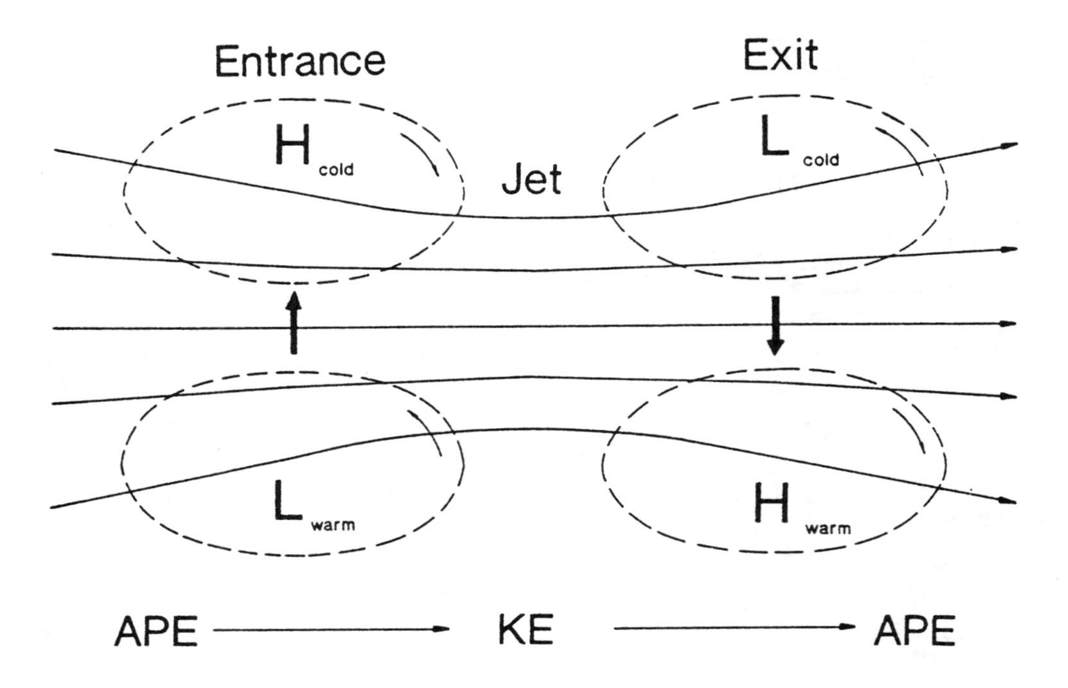

Fig. 8. Schematic view of the jet-stream (Northern Hemisphere, streamlines), the accompanying surface pressure field - note the low level flow (thin arrows, dashed lines) increasing (left) or decreasing (right) the baroclinic flow - and the conversion between APE and KE; thick arrows = cross-isobaric (ageostrophic) mass flow.

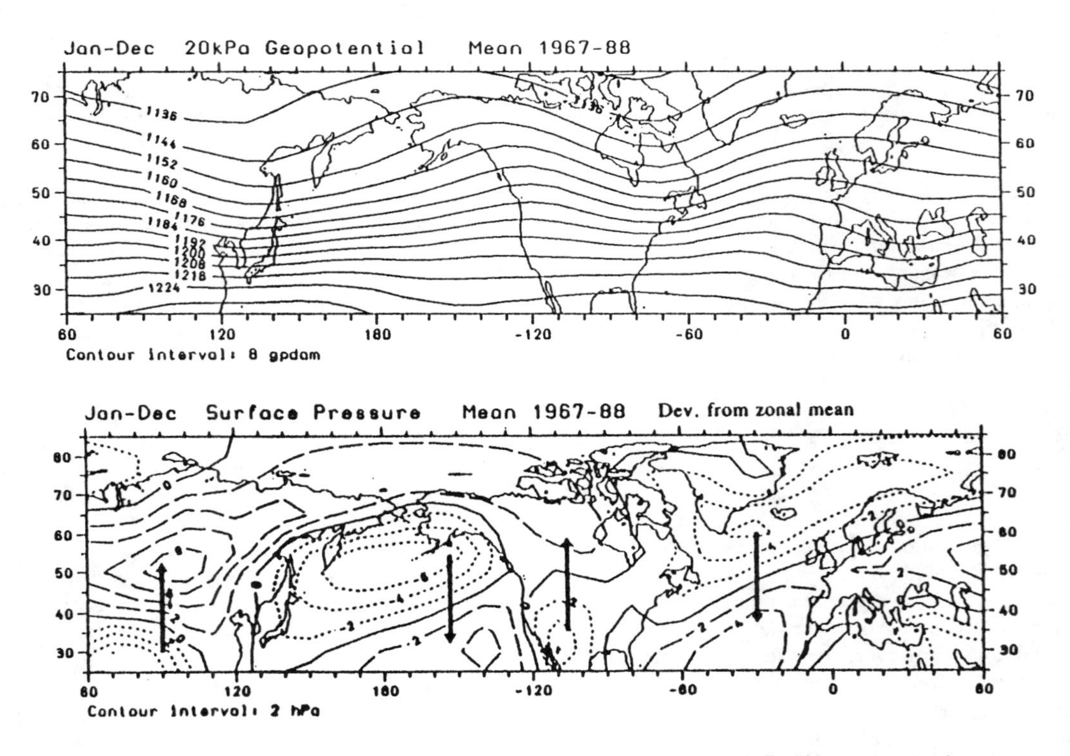

Fig. 9. Above: geopotential 20 kPa with geostrophic flow along the isolines, annual average 1967-1988. Below: Surface pressure, annual average, deviations from zonal mean (dashed lines positive, dotted lines negative values, unit 2 hPa); thick arrows = ageostrophic mass flow as in Fig. 8.

A comparison between the average 20 kPa flow and the deviations of surface pressure from the latitudinal mean (Fig. 9) shows, as a composite, the climatological role of these processes. In the region of convergent upper winds APE is converted into KE, and in the region of divergence KE is reconverted into APE. In the lower part of Fig. 9, these anomalies also indicate, in mid-latitudes, areas of most frequent cyclogenesis and anticyclogenesis. Disregarding such a simplified perceptual view, Lorenz [1955, 1967] and other authors derived a system of equations for an evaluation of this energy cycle. Using the most complete set of radiosonde data, Oort and Peixóto [1983] calculated nearly all terms of the balance equations, except generation of potential eddy energy, which had to be estimated.

V. Discussion: The role of water vapour in climate and climatic change

The vital role of water vapour - more exactly: of its latent heat - in 3D-climatic change is as yet not always sufficiently recognized. A parallel evaluation of maritime surface data (COADS) in the Tropics and of near-homogeneous 3D-analyses of the Northern Hemisphere, together with tropical radiosonde and satellite data, led to the result that the effect of "global warming" - most probably triggered by CO_2 and other "dry" greenhouse gases - is substantially magnified by the role of water vapour in 3D-circulations [Flohn et al., 1990a, b, 1992]. While the total greenhouse effect adds about 2 Watt/m² (IPCC-Report) into the atmosphere, ocean warming needed only about 5 percent of it. But its rising energy input into the atmosphere via water vapour is in the order of 12-15 Watt/m², i.e. an amplification factor of more than 5. An observed increase of tropical water vapour content in the middle troposphere by nearly 30 percent (Fig. 10) [cf. Hense et al., 1988] is accompanied by:

1) an increase of the zonal average evaporation of equatorial oceans (Lat. 10°S-14°N) of nearly 16 percent, equivalent to 15-20 Watt/m² (Fig. 11); it diminishes poleward of Lat. 30°N, but does not disappear,

2) an increase of precipitation in most land areas, except the subtropical belt Lat. 5°-35°N [Diaz et al., 1989],

3) an intensification of oceanic mid-latitude cyclones [Flohn et al., 1992] along with a rather drastic increase of winter gales above the Pacific and the Atlantic (Fig. 12),

4) a hemispheric (Lat. 20-90°N) increase of surface and tropospheric monthly resultant geostrophic winds in the order of 0.4-0.7 m/s, and

5) an increase of hemispherically averaged meridional and vertical tropospheric temperature gradients by about 5% in 22 years.

These evolutions led, since the 1960's, to a growing intensity of three of the four large-scale, thermally forced circulations: the Hadley cell, the Dove-Ferrel cell and the zonal Walker circulation, while the evidence for the seasonal monsoon cells is not sufficiently conclusive.

How are the observed circulation changes 3) to 5) related to the observed changes of water vapour content? It is well known that ascending motion prevailing in cyclones after condensation leads to rainfall, dependent on available water vapour. Release of more latent heat intensifies vertical ascent, and thus the kinetic and potential energy of the cyclone. Due to the asymmetry of mid-latitude frontal cyclones, ascending motions are concentrated in its forward section, subsidence in the rear. This favours convergent inflow near the surface and divergent outflow in the upper troposphere, and thus promotes anticyclogenesis on the eastward (outer) side of the cyclone (see the anticyclonic eddies of the trend around the coasts of Spain/Morocco and Oregon) [Flohn et al., 1990a, 1992].

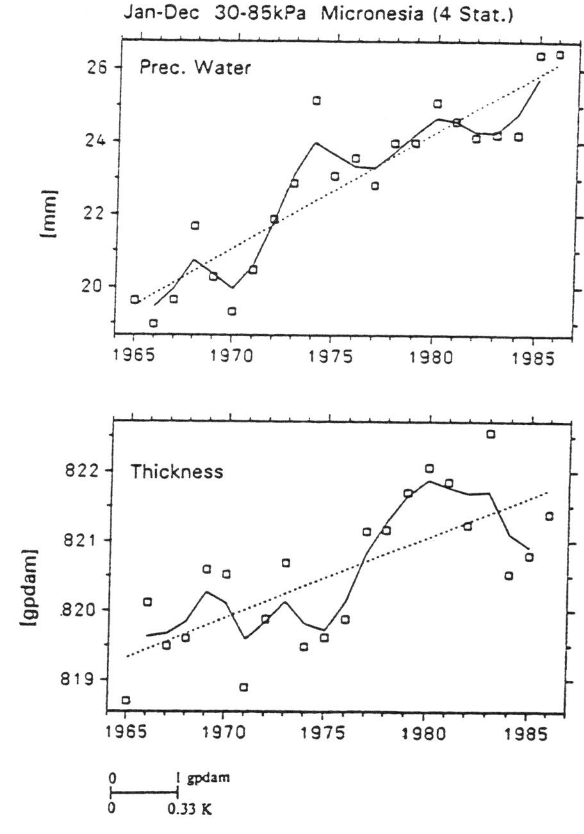

Fig. 10. Changes of precipitable water (mm) of the layer 30/85 kPa, annual values 1965-1986 from 4 Micronesian stations (Lat. 7°-10°N, Long. 135°-180°E). Increase of virtual temperature of 0.80 K is partly due to humidity increase (0.24 K)

A peculiar evolution is observed in the northern parts of both the Atlantic and the Pacific, within the deepening cyclones north of Lat. 40°N. Here, a strong increase of wind speed causes rising evaporation, and the net radiation during winter is insufficient to heat the oceanic mixing layer - which reaches here very deep as compared with the Tropics. Here Ev rises, with only weak changes of q_s-q_a, in spite of decreasing T_s [Flohn et al., 1992].

But such evaporative cooling - which is also observed in tropical hurricanes - contributes through latent heat to the small warming of the atmosphere above the polar cap and to the maintenance of its circulation.

The role of the oceans is indeed essential. The warmest tropical oceans - lacking any upwelling of cooler water between Long. 66 and 160°E - warm fastest, heating and moistening the

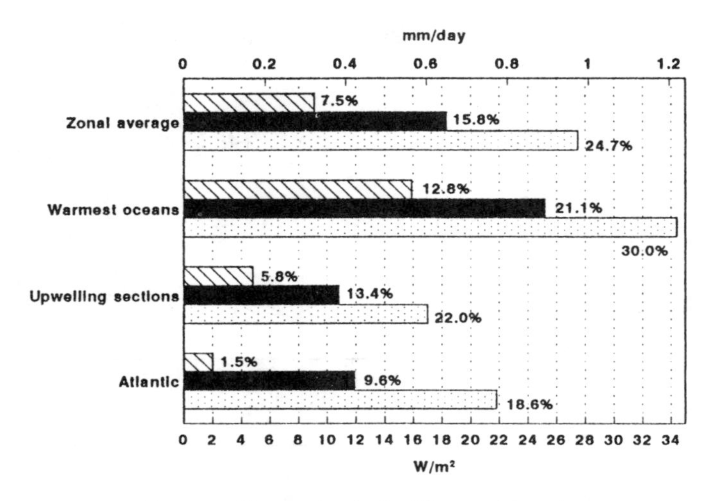

Fig. 11. Changes of evaporation (linear trend 1949-1989, annual values) at tropical oceans, observed wind speed corrected, different trends (0, 50, 100%). Zonal average Lat. 10°S-14°N, warmest oceans Long. 66° -160°E, upwelling sections E-Pacific south of equator, Long. 86°-120°W.

atmosphere. Upwelling "cool" water along several tropical coasts and equatorial belts is also warming, perhaps due to higher absorption of solar radiation as a result of planctonic algal bloom [Sathyendranath et al., 1991]. The rising frequency of gales (Fig. 12) is also evidenced by a distinct increase - near 30% - of significant wave-heights measured around the British Isles [Bacon and Carter, 1991].

These investigations [Flohn et al., 1992] lead to two important conclusions:

a) the climatic system is now in a state of disequilibrium, with substantially increasing fluxes of energy and water between ocean and atmosphere,

b) the longitudinal and regional variations are so large that latitudinal and global averages cannot give more than a first order approximation.

At present, a large portion of the discussion of the "global warming" is limited to surface temperature, which played a primary role in climatology at the time of Hann and before World War II. But now, after 60 years evolution of physical climatology, this limitation appears to be obsolete, to say the least. We have to deal with climatic processes in three or better yet, four dimensions, with all parameters which have been measured over many decades. The satellite data - I mention especially those observing cloud motions resp. winds and estimating rainfall, or measuring T_s and the main components of the radiation budget (ERBE) - are intended to radically improve our knowledge of the global patterns of the most essential parameters. However, the wealth of data is so great that the evaluation for climatological purposes lags far behind. Looking at the most recent climate changes, it is vital that the facilities for this evaluation be vastly improved, as well as made more accessible to scientists.

Our results, briefly presented here in this Chapter, are inter-

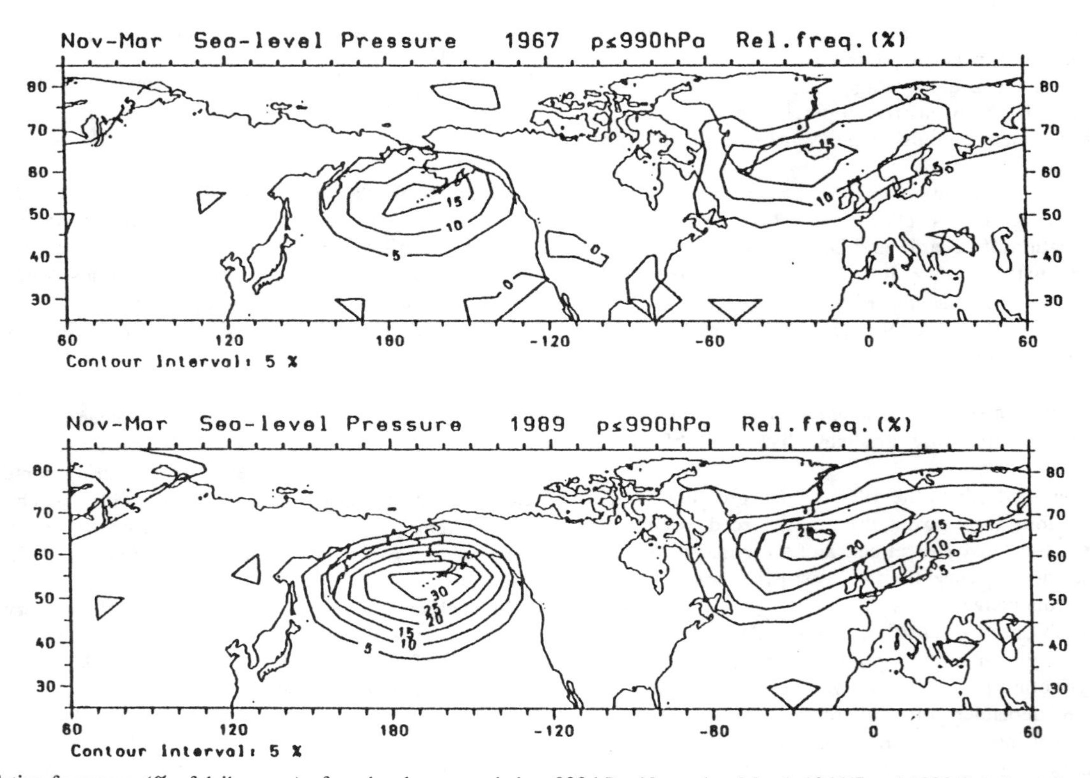

Fig. 12. Relative frequency (% of daily maps) of sea-level pressure below 990 hPa, November-March 1966/7 and 1988/9 derived from linear trend.

nally coherent. A feedback via water vapour and hydrological cycle, with increasing intensity of atmospheric circulations, of potential and kinetic energy - all this can be observed under our very eyes. These facts can (and must) be cross-checked and verified - only then are they more convincing to skeptics than even the most fascinating (but partly contradicting) model results. We need both: observed facts and model background, to solve our problems.

3D-physical-climatology has greatly helped to understand the spatial and temporal coherency of the ongoing changes of climate. However, the author's task was a look backwards into the recent history of our science; he has to withstand the temptation to speculate about the future - this may be postponed to another opportunity.

Acknowledgment. The untiring assistance of Dr. A. Kapala, Dr. H. R. Knoche (now Garmisch-Partenkirchen) and Dipl. Met. H. Mächel, especially with regard to the figures and the literature, and of Mrs. W. Rubin, with regard to the computer text and language, is greatly appreciated. I am particularly grateful to Prof. M. Hantel for his constructive comments.

Selected References

Albrecht F., Untersuchungen über den Wärmehaushalt der Atmosphäre in verschiedenen Klimagebieten, *Reichsamt für Wetterdienst, Wiss. Abhandl. VIII, No. 2*, pp. 82, Berlin, 1940.

Albrecht F., Der jährliche Gang der Komponenten des Wärme- und Wasserhaushaltes der Ozeane, *Berichte des Deutschen Wetterdienstes, 11, 79*, pp. 24, Offenbach a. M., 1961.

Bacon S. and Carter D. J. T., Wave climate change in the North Atlantic and North Sea, *Internat. J. Climatol., 11*, 545-558, 1991.

Baumgartner A. and Reichel E., The World Water Balance - Mean Annual Global, Continental and Maritime Precipitation, Evaporation and Runoff, pp. 179, Munich, Vienna, 1975.

Bezold W., Der Wärmeaustausch an der Erdoberfläche und in der Atmosphäre, *Sitz. Ber. Ak. Wiss., 54*, 543-582, Berlin, 1892.

Bjerknes V. et al., Dynamic Meteorology and Hydrology I, II, Carnegie Institution of Washington, 1910/11.

Bottomley M. et al., Global Ocean Surface Temperature Atlas "Gosta", UK Departments of the Environment and Energy for Intergovernmental Panel on Climate Change, London, 1990.

Bowen I. S., The ratio of heat losses by conduction and by evaporation from any water surface, *Phys. Rev., 27*, 779-787, 1926.

Budyko M. I., The heat balance of the Earth, *in Climatic change*, edited by Gribbin J. et al., 85-113, Cambridge, 1978.

Budyko M. I., The Earth's Climate: Past and Future, *Internat. Geophys. Series, 29*, pp. 317, 1982.

Defant A., Die Zirkulation der Atmosphäre in der gemäßigten Zone, *Geograf. Ann., 3*, 209-266, 1921.

Diaz H. F. et al., Precipitation Fluctuations Over Global Land Areas Since the Late 1800's. *Journ. Geophys. Res., 94*, 1195-1210, 1989.

Dove H. W., Meteorologische Untersuchungen, pp. 344, Berlin, 1837.

Exner F. M., Grundzüge einer Theorie der synoptischen Luftdruckveränderungen, *Sitz. Ber. Ak. Wiss. Wien, Math. Nat. Kl., 115*, 1171-1246, 1906; *116*, 819-854, 1906.

Ferrel W., An essay on the winds and the currents of the ocean, *Medicine and Surgery, 11*, 287-301, 1856; Reprint in *Popular essays on the movements of the atmosphere, Prof. Pap. Signal Serv., 12*, 7-19, Washington, 1882.

Flohn H., Divergenztheorie - Zyklogenese - Antizyklogenese, *Meteor. Rundschau 5*, 8-12, 1952.

Flohn H., Vilhelm Bjerknes, in *Forscher und Wissenschaftler im heutigen Europa - Weltall und Erde*, edited by G. Ställing pp. 303-310, Oldenburg, 1955.

Flohn H. et al., Recent changes of tropical water and energy budget and of midlatitude circulations, *Climate Dynamics, 4*, 237-252, 1990a.

Flohn H. et al., Changes of hydrological cycle and Northern Hemisphere circulation and their initiation at the tropical air-sea surface, in *Internat. TOGA Scient. Conf. Proceedings*, Honolulu 1990, edited by *World Meteor. Org., WCRP 43, WMO/TD 379*, pp. 47-57; also in *Observed Climate Variations and Change: Contributions in Support of Section 7 of the 1990 IPCC Scientific Assessment*, edited by J. T. Houghton et al., Chapter VI, pp. 1-11, 1990b.

Flohn H. et al., Water vapour as an amplifier of the greenhouse effect: new aspects, *Zeitschrift für Meteor., 1*, 122-138, 1992.

Hann J., Zur Frage über den Ursprung des Föhns, *Zeitschr. Österr. Ges. Meteor., 1*, 237-266, 1866.

Hann J., Handbuch der Klimatologie, pp. 764, Stuttgart, 1883.

Hann J., Allgemeine Klimalehre, Bd. I, 4. Auflage, bearbeitet von K. Knoch, pp. 394, 1932.

Hantel M., A note on the energy flux across the earth's surface, (this Volume), 1992.

Henning D., Atlas of the surface heat balance of the Continents, Components and Parameters estimated from climatological data, pp. 402, Berlin, Stuttgart, 1989.

Hense A. et al., Recent Fluctuations of Tropospheric Temperature and Water Vapour Content in the Tropics, *Journ. Meteor. Atmos. Physics, 38*, 215-227, 1988.

Homén Th., Der tägliche Wärmeumsatz im Boden und die Wärmestrahlung zwischen Himmel und Erde, Leipzig, 1897.

Hsiung J., Mean surface energy flux over the global ocean, *J. Geophys. Res., 91*, 10,585-10,606, 1986.

Isemer H-J. and Hasse L., The Bunker Climatic Atlas of the North Atlantic Ocean. Air-Sea Interactions, Vol. 2, pp. 252, Berlin, 1987.

Jaeger L., Monatskarten des Niederschlags für die ganze Erde, *Berichte des Deutschen Wetterdienstes 18, 139*, 38 pp., Offenbach a. M., 1976.

Jaeger L., Monthly and areal patterns of mean global precipitation, in *Variations in the Global Water Budget*, edited by R. Ratcliffe et al., Dordrecht, Boston, Lancaster, pp. 129-140, 1983.

Kessler A., Heat Balance Climatology, in *World Survey of Climatology Vol. 1A*, edited by H. E. Landsberg, Amsterdam, Oxford, New York, pp. 261, 1985.

Korzun V. I. et al., World water balance and water resources of the earth, Leningrad, 1974.

Lettau H., Atmosphärische Turbulenz, pp. 283, Leipzig, 1939.

Lettau H. H., Evapotranspiration Climatonomy I: An approach to numerical prediction of monthly evapotranspiration, runoff and soil moisture storage, *Mon. Weath. Rev., 97*, 691-699, 1969.

Lettau H. H. et al., Amazonia's hydrologic cycle and the role of atmospheric recycling in assessing deforestation effects, *Mon. Weath. Rev., 107*, 227-238, 1979.

Lettau H. H., Evapoclimatonomy III: The Reconciliation of Monthly Runoff and Evaporation in the Climatic Balance of Evaporable Water on Land Areas, *J. Appl. Meteor., 30*, 776-792, 1991.

Linke F. et al., Physik der Atmosphäre I. Atmosphärische Strahlungsforschung, in *Handbuch der Geophysik, Band VIII*, Berlin-Nikolassee, edited by F. Linke and F. Möller, pp. 1102, 1942-1961.

Lorenz E. N., Available potential energy and the maintenance of the general circulation, *Tellus, 7*, 157-167, 1955.

Lorenz E. N., The Nature and Theory of the General Circulation of the Atmosphere, *WMO No. 218, TP, 115*, edited by World Meteorol. Organ., pp. 161, Geneva, 1967.

Lorenz E. N., The general circulation of the atmosphere: an evolving problem, *Tellus, 43 AB*, 8-15, 1991.

Margules M., Über die Energie der Stürme *Jahrb. Zentr. Anst. Meteor. 1903*, 1905.

Margules M., Über Temperaturschichtung in stationär bewegter und in ruhender Luft, *Meteor. Zeitschr., (Hann-Band)*, 243-253, 1906.

Ohmura A., Objective criteria for rejecting data for Bowen ratio flux calculations, *J. Appl. Meteor., 21*, 595-598, 1982.

Oort A. H., On estimates of the atmospheric energy cycle, *Mon. Weather Rev., 92*, 483-493, 1964.

Oort A. H. and Peixóto J. P., Global angular momentum and energy balance requirements from observations, in *Adv. Geophys., 25,* edited by B. Saltzman, New York, pp. 355-490, 1983.

Peixóto J. P. and Oort A. H., The atmospheric branch of the hydrological cycle and climate, in *Variations in the Global Water Budget*, edited by R. Ratcliffe et al., Dordrecht, Boston, Lancaster, pp. 5-65, 1983.

Philipps H., Die Abweichung vom geostrophischen Wind, *Meteor. Zeitschr., 56,* 460-483, 1939.

Ramanathan V. et al., Cloud-radiative forcing and climate: Results from the Earth Radiation Budget Experiment, *Science, 243,* 57-63, 1989.

Raval A. and Ramanathan V., Observational determination of the greenhouse effect, *Nature, 342,* 758-761, 1989.

Richardson L. F., Weather Prediction by numerical processes, pp. 236, Cambridge, 1922.

Sathyendranath Sh. et al., Biological control of surface temperature in the Arabian Sea, *Nature, 349,* 54-56 1991.

Schmidt, W., Der Massenaustausch bei der ungeordneten Strömung in freier Luft und seine Folgen, *Sitz. Ber. Akad. Wiss., Math. Nat. Kl. Abt. 2a, 126,* 757-804, 1917.

Schmidt W., Der Massenaustausch in freier Luft und verwandte Erscheinungen, pp. 118, Hamburg, 1925.

Seilkopf H., Maritime Meteorologie, in *Handbuch der Fliegerwetterkunde, Bd.II,* edited by R. Habermehl, pp. 150, Berlin, 1939.

Sverdrup H. U., On the evaporation from the oceans, *J. Mar. Res., 1,* 3-14, 1937.

Starr V. P., An essay on the general circulation of the earth's atmosphere, *J. Meteor. 5,* 39-43. 1948.

Willmott C. J. et al., Climatology of the terrestrial seasonal water cycle, *J. Climatol., 5,* 589-606, 1985.

H. Flohn, Meteorologisches Institut, Auf dem Hügel 20, D-5300 Bonn, Germany.

A Note on the Energy Flux Across the Earth's Surface

MICHAEL HANTEL

Institute of Meteorology and Geophysics, University of Vienna

The net vertical energy flux approaching the earth's surface from above is equal to the flux leaving the surface in downward direction. This is the continuity requirement for the energy flux vector normal to a material discontinuity; it has, since Hann, served in physical climatology as the fundamental condition at the earth's surface. We demonstrate that, opposite to familiar usage, this continuity of the energy flux does not need to be postulated as a separate boundary condition. Rather, it is implied in the energy principle (the *First Law*) if formulated as an integral (not differential) equation. The significance of this implication for theoretical and applied climatology is sketched.

INTRODUCTION

The surface energy balance is a classical topic in quantitative physical climatology [e.g., Hann, 1897; Sverdrup, 1937; Bowen, 1926; Geiger, 1961; Budyko, 1963; Flohn, 1957, 1969, 1985, 1992]. One important task of this field is to relate the energy fluxes across the earth's surface to the properties of the ground below. This is usually done [e.g., Geiger, 1961; Fleagle and Businger, 1980; Gill, 1982] by postulating a boundary condition: the energy fluxes perpendicular to a boundary must be equal on both sides, or, equivalently: the energy flux must be continuous across the boundary. This property is occasionally motivated by saying "Because the surface has no heat capacity, the fluxes into and out of the earth must balance". While indeed a surface has no heat capacity and thus cannot store energy, the argument in isolated form is inconclusive. For example, a surface cannot store radiation, either, and yet the radiation flux across it may be discontinuous.

In the fields of theoretical and applied climatology the continuity of the energy flux across the earth's surface is traditionally formulated as the heat balance equation [Geiger, 1961] which is equivalent to a boundary condition [Gill, 1980]. The arguments around the heat balance equation since the times of Hann are summarized in Appendix A.

We now ask: on what principle should a boundary condition be based? It is the purpose of this note to indicate that *no separate boundary condition is required at all. The continuity of the energy flux, including continuity across a material boundary, is implied in the integral principle of energy conservation* (the "First Law").

This statement is not new. It has long been accepted in theoretical physics that integral laws are more general than the corresponding differential laws. Boundary conditions for energy and other fluxes across discontinuities are treated in axiomatic texts on fluid dynamics as a consequence of the integral principles [e.g., Sommerfeld, 1964; Whitham, 1974]. However, the connection between this theoretical result and its application in geophysics is not particularly familiar, to say the least. Thus it seems appropriate to revisit the statement above and its consequences for theoretical climatology.

The validity of our statement shall be demonstrated in three steps. The first step consists in noting that it is general indeed; it applies to any variable which obeys a conservation principle similar to the First Law. For example, the formal equivalence between the principles of energy conservation and of mass conservation allows to apply the energy flux results in like manner to the mass flux (specifically to the water flux). Thus the relevance of our problem to large-scale enterprises like the GEWEX [e.g., McBean, 1988] should be obvious.

The second step (section 2) will be to demonstrate that the differential version of the conservation principle is generally not valid at a density discontinuity surface although the flux across this surface is continuous. In other words, differential conservation principle and flux continuity across a discontinuity surface are separate and independent properties.

In the third step (section 3) we shall show that *the integral version of the conservation principle does imply the continu-*

Interactions Between Global Climate Subsystems, The Legacy of Hann
Geophysical Monograph 75, IUGG Volume 15

ity of the flux everywhere, specifically *across a discontinuity* surface. This will be done for the one-dimensional case. The proof of the general 3D-case will be sketched in Appendix B.

The main argument of this note is restricted to a *stationary discontinuity.* This case is the simplest. If the discontinuity surface moves the advective component of the flux turns to be discontinuous. We consider this case in Appendix C for one dimension; the additional flux discontinuity can be removed so that *the flux relative to the moving discontinuity becomes continuous* again. A further generalization to three dimensions with moving discontinuity is sketched in Appendix D.

The continuity of the energy flux vector and its relation to the energy principle is far from being an academic problem. Its significance for climatology will be addressed in section 4. Similary, the climatological implications of the mass flux continuity will be touched upon in section 5. Some general conclusions are drawn in section 6.

LIMITATIONS OF THE DIFFERENTIAL CONSERVATION PRINCIPLE

Let us consider the differential energy conservation principle in one space coordinate (Fig. 1):

$$\frac{\partial u}{\partial t} + \frac{\partial F}{\partial z} = 0 \qquad (1)$$

u denotes energy density (with respect to volume), F the vertical component of the energy flux[2] vector; independent arguments are time t and vertical coordinate z.

Eq. (1) is assumed valid on both sides of a horizontally uniform discontinuity surface located at $z = 0$. Mathematically: at the level $z = 0$, the density $u(z,t)$ is discontinuous with respect to the argument z (Fig. 1).

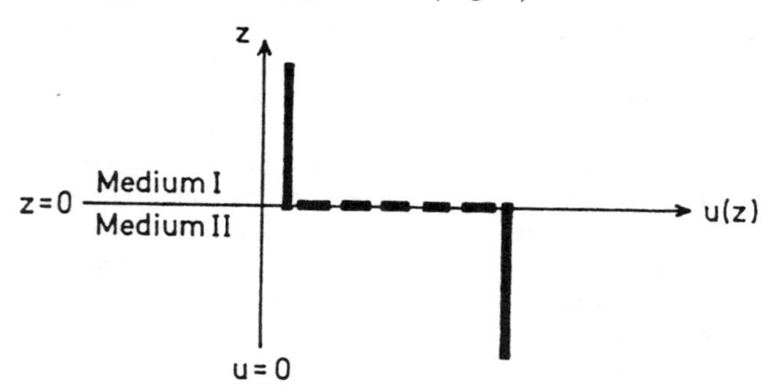

Fig. 1. Model profile of energy density $u(z)$ across boundary surface between two media (e.g., I = air, II = water). u is continuous for all z except for $z = 0$.

[2]For clean terminology the term energy *flux* should be restricted to the integral quantity (a scalar) with physical unit $Js^{-1} = W$ [e.g., Glansdorff and Prigogine, 1971], while the term enery *flux density* should be used for the corresponding differential quantity (a vector) with physical unit $Jm^{-2}s^{-1} = W/m^2$. For example, the sensible heat flux SH should correctly be termed *sensible heat flux density*. However, since no fluxes in the strict sense are involved in the present paper we shall follow usual convention and drop the word *density* througout.

This obviously implies that eq. (1) is generally undefined in this level. Our assumptions do imply that F is differentiable (and thus continuous) for all z *except for* $z = 0$. At $z = 0$, however, neither storage $\frac{\partial u}{\partial t}$ nor divergence $\frac{\partial F}{\partial z}$ are necessarily continuous; they can be continuous, but they do not need to be.

More information concerning F cannot be drawn from the assumptions made. In particular, *the continuity of F at* $z = 0$ *cannot be derived from eq. (1).*

For example (Fig. 2, profile a) a vertically constant $F(z)$ would have zero divergence everywhere. This $F(z)$ is differ-

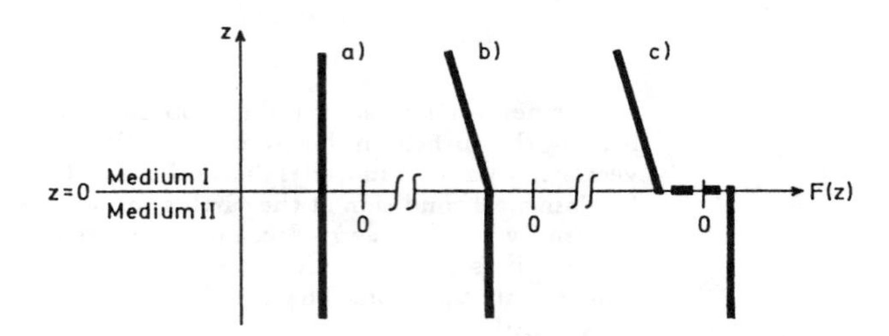

Fig. 2. Model profiles of vertical energy flux $F(z)$ perpendicular to density discontinuity. Profiles a), b) are continuous at $z = 0$, c) is not. Profile a) is differentiable at $z = 0$, profiles b), c) are not.

entiable (and thus continuous), including $z = 0$; further, $\frac{\partial u}{\partial t}$ is continuous, including $z = 0$, while $u(z)$ in Fig. 1 remains constant with time. Profile b) is convergent in medium I but has zero divergence in medium II; this $F(z)$ is not differentiable at $z = 0$ but at least continuous. Profile c) has the same properties as b) with the only difference that $F(z)$ is not continuous anymore. There is no way of distinguishing between profiles b) and c) if one has nothing but eq. (1).

In short, the u-profile of Fig. 1 with its discontinuity at $z = 0$ is consistent with F-profiles (Fig. 2) that are either continuous at $z = 0$ (cases a,b) or discontinuous (case c). Eq. (1) is valid in all three cases and for all z except for $z = 0$.

The reason for the impossibility to demonstrate continuity of F at a discontinuity of u is the differential form of the conservation equation (1). It may be tempting to cure this by simply postulating that eq. (1) holds also at $z = 0$. This however would be equivalent to postulating that $F(z)$ is differentiable at $z = 0$ - a condition typically violated in many cases of practical interest (e.g., Fig. 2b). What is needed is to somehow match the profiles of F across the boundary. This can be done in a rather formal way and in a more physical way.

The formal way can in the present simple example be sketched as follows. The profiles b) and c) in Fig. 2 represent the same solution of eq. (1), because any solution $F(z)$ of (1) is arbitrary to the extent of an additive constant; due to the discontinuity at $z = 0$, this constant may be different in media I, II. What criterion shall be applied to fix these constants? For example, we may combine all solutions $F(z)$, which differ only by the two constants, into one class. This class contains one distinguished member that is continuous at $z = 0$ while all other members are discontinuous. We now could pick this unique member as representative of its class since all other (discontinuous) members of the class give essentially the same solution. This way of removing the arbitrariness in the F-profile is entirely formal and does not use physical reasoning. It simply postulates, but does not explain, the continuity of F.

A more physical way to match the profiles of F in media I, II across the boundary is to employ the integral form of the conservation principle.

THE INTEGRAL CONSERVATION PRINCIPLE

The integral energy conservation principle in one space coordinate, analogous to eq. (1), reads (Fig. 3):

$$\frac{\partial}{\partial t}\int_{z_{II}}^{z_I} u(z,t)dz + F(z_I,t) - F(z_{II},t) = 0 \qquad (2)$$

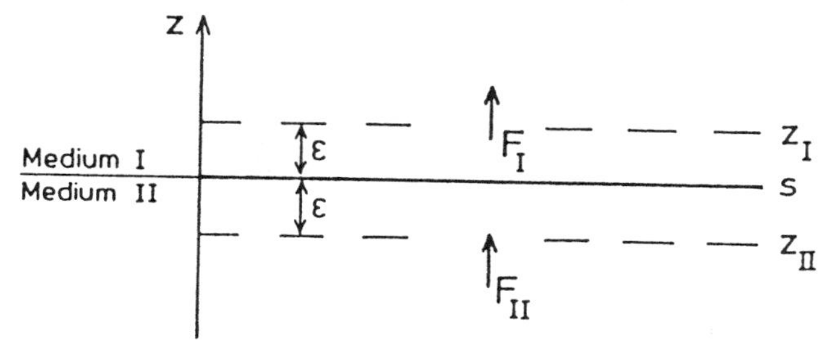

Fig. 3. Principal sketch for deriving one-dimensional energy balance equation across discontinuity surface located at $z = s$. For further explanation see text.

It states that the time-rate of change of total energy in any interval $z_{II} \leq z \leq z_I$ must be balanced by the net inflow across z_{II} (a level in medium II) and z_I (a level in medium I). Eq. (2) is postulated to apply for each finite interval (i.e., $z_{II} < z_I$), however big or small; further, $u(z,t)$ may have discontinuities within the interval.

At first glance it may seem that the integral law (2) follows from the differential law (1) and vice versa. For example, van Mieghem [1973, Chs. 2, 3] discusses the extensive and intensive versions of the energy balance equations without reference to a possible limitation of their equivalence. Eqs. (1), (2) are indeed equivalent *as long as no discontinuities occur in the interior.*

However, if we allow for discontinuities (not considered by van Mieghem) the equivalence breaks down. Suppose there is a discontinuity at $z = s$ somewhere in the interior, i.e., $z_{II} < s < z_I$. Due to the postulated general validity of eq. (2) we may specify the limits of the interval such that:

$$z_I - s = s - z_{II} = \epsilon \qquad (3)$$

where ϵ is a time-independent positive real number but otherwise arbitrary. Suppose further that u and $\frac{\partial u}{\partial t}$ are continuous in the subintervals $z_{II} \leq z < s$, $s < z \leq z_I$, and have finite limits as $z \to s$ from below and above. Then, (2) may be written:

$$\begin{aligned} F(z_{II},t) - F(z_I,t) &= \qquad (4) \\ = \frac{\partial}{\partial t}\int_{z_{II}}^{s} u(z,t)dz &+ \frac{\partial}{\partial t}\int_{s}^{z_I} u(z,t)dz = \\ = \int_{s-\epsilon}^{s} \frac{\partial u(z,t)}{\partial t}dz &+ \int_{s}^{s+\epsilon} \frac{\partial u(z,t)}{\partial t}dz \end{aligned}$$

Since $\frac{\partial u(z,t)}{\partial t}$ is bounded in each of the intervals separately, both integrals in (4) tend to zero as $\epsilon \to 0$. Thus:

$$\lim_{\epsilon \to 0} \{F(z_{II},t) - F(z_I,t)\} = 0 \qquad (5)$$

which represents the continuity of the energy flux across the level $z = s$.

This consideration demonstrates that eq. (2) provides the required matching principle across the energy density jump. The argument is simple but not trivial. It is based upon the well known fact that a function which in an interval is continuous (except for a finite number of discontinuities) is integrable over that interval but not necessarily differentiable everywhere within. Note that eq. (2) does not imply validity of eq. (1) at $z = s$ although it does imply continuity of $F(z,t)$ at $z = s$.

The simple one-dimensional stationary discontinuity problem just considered can be generalized to three dimensions (Appendix B) and to a moving discontinuity (Appendix C). Both generalizations require essentially no new physical arguments except that in case C the flux must be redefined.

Concerning this latter case the following remark seems in order. Moving discontinuities have been considered, for example, in the quite different context of nonlinear kinematic waves with shocks [Whitham, 1974, Ch. 2]. The generalization there (i.e., the shock moves) is that the discontinuity is located at $z = s(t)$. This leads to:

$$F[s_{II}(t),t] - F[s_I(t),t] = \{u[s_{II}(t),t] - u[s_I(t),t]\}\dot{s} \qquad (6)$$

Here, $F[s_I(t),t]$, $u[s_I(t),t]$ are the values of F and u as $z \to s(t)$ from above; equivalently, the limiting values with index II apply as $z \to s(t)$ from below. Eq. (6) replaces (5) in the case of a moving discontinuity; it expresses the discontinuity of F that comes exclusively from the motion of s.

We consider this type of flux discontinuity as almost trivial in the present context and ask for the conditions *relative to a coordinate system in which the discontinuity surface is stationary.* This program is carried out in Appendix C; it essentially reproduces the result (5).

THE SURFACE ENERGY BALANCE

The continuity of the vertical energy flux component across the horizonal air/ground boundary has served for decades as an established working hypothesis, both in physical climatology [e.g., Sverdrup, 1937; Albrecht, 1940; Budyko, 1963; Kessler, 1985] and in boundary layer meteorology [e.g., Geiger, 1961; Kraus, 1987]; for some further historical remarks see Appendix A. The flux components at the levels just above and below the interface are, respectively:

$$F_I = RAD + SH + LH \tag{7}$$

$$F_{II} = CON \tag{8}$$

RAD denotes vertical *radiation* flux, SH *sensible heat* flux, LH *latent heat* flux, and CON *conductive* heat flux; all fluxes are counted positive upward. Sufficiently away from the surface, SH and LH are turbulent while very close to the surface they are due to molecular diffusion and conduction; CON may also include turbulent (if the "ground" is water) and latent heat contributions (if there is water diffusion in the ground). If the "ground" is ice, one has to add RAD on the right of eq. (8) in the topmost layers. For practical purposes, F_{II} is often replaced by the energy storage of the ground; this is equal to the time change of the vertically integrated heat content of the column below the surface (the thickness of which is of the order of 1 m over land, 10 m in glaciers and 100 m in the ocean).

The continuity condition now couples eqs. (7) and (8) through:

$$RAD + SH + LH - CON = 0 \tag{9}$$

This energy balance equation states that *the total flux is continuous across the earth's surface* (examples sketched in Fig. 4) *while its components RAD, SH, LH, CON are generally discontinuous.* Thus eq. (9) can be a useful category for classification of various different boundary layer processes [e.g., Kraus, 1987, Flohn, 1992]. This is also valid for the horizontal air-water surface, even if it moves vertically due,

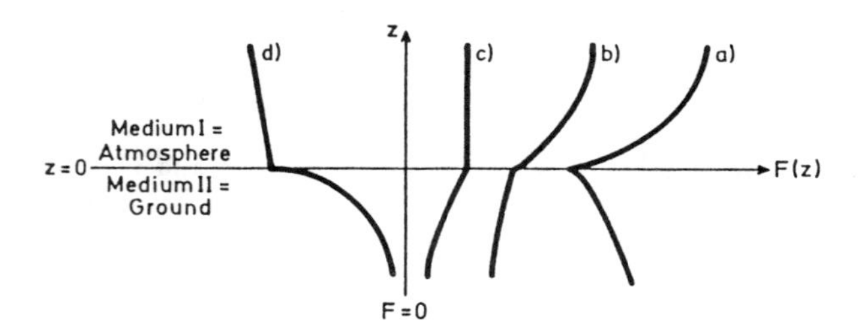

Fig. 4. Model profiles of energy flux component F perpendicular to planetary surface. Flux $F(z)$ is continuous (but not differentiable) at $z = 0$. The profiles illustrate the following cases: a) Hypothetical profile with flux divergence (*cooling*) in medium I and convergence (*heating*) in medium II; b) nightly cooling of air and soil; c) cooling of planetary surface (medium II) without atmosphere (medium I = vacuum) after sunset; d) daily warming of air and ocean surface layer. Note that heating rate (*warming, cooling*) is proportional to flux convergence/ divergence only after modification with heat capacity of medium.

for example, to evaporation; the argument of section 3 is invariant with respect to shifting the discontinuity $z = s(t)$ in time, provided all quantities are defined relative to the moving discontinuity (Appendix C).

Our continuity condition (5) is not only valid for climatological fluxes as given in eqs. (7), (8) but for fluxes on short space and time scales as well. For example, when it comes to the surfaces of droplets and ice particles, $F_I = F_{II}$ applies equally well but may be of limited use since the fluxes involved are hard to measure experimentally. The same is valid for ocean waves, specifically breaking waves which are accompanied by heavy turbulence [Kraus, 1972]. While the energy principle with its implications remains always valid it appears to be of limited practical use on the shortest time scales. After all, the most widely accepted application of (9) has been in climatology on and above the daily time-scale.

WATER BALANCE

The continuity of the water flux across the horizontal air/water or air/ground boundary is likewise an established working hypothesis in physical climatology [e.g., Wüst, 1920; Albrecht, 1940; Lettau, 1969], boundary layer meteorology [e.g., Geiger, 1961] and hydrology [e.g., Baumgartner and Reichel, 1975; Willmott et al., 1985]. The flux components of the levels just above and below the interface are, respectively:

$$F_I = LH + PREC \tag{10}$$

$$F_{II} = SUR \tag{11}$$

LH denotes *latent heat* flux as before, $PREC$ *precipitation*, SUR the *surplus* vertical water flux relative to, and immediately below, the surface; all fluxes are in energy units and are counted positive upward. The flux SUR [Willmott *et al.*, 1985] is usually converted into a equivalent *runoff = discharge* plus *water storage* [Lettau, 1969; Baumgartner and Reichel, 1975]; in the ocean, SUR consists only of runoff (= equivalent horizontal ocean current divergence) with no storage equivalent. The continuity condition requires:

$$LH + PREC - SUR = 0 \tag{12}$$

This water balance equation states that *the total flux is continuous across the earth's (and ocean) surface while its components LH, $PREC$, SUR are generally discontinuous.*

Further, the mass conservation law would strictly apply to the total mass of the continuum (comprising gas, fluid and fixed components, specifically air plus water). However, to the extent that no chemical reactions but only physical phase changes may happen (which we have implicitly assumed above), the water conservation law can be separated from the total mass conservation law. At a still higher level of generalization, flux conditions similar to (12) may be formulated for chemical elements (e.g., carbon), provided proper account is taken of the chemical reactions involved.

The role of the flux SUR may not be immediately obvious although its use by Willmott *et al.* [1985] is completely consistent. The important point is that SUR must be counted relative to the surface. If the discontinuity is land, this is self-evident. For the ocean suppose the water surface moves upward since there is precipitation and suppose further for the sake of simplicity that LH vanishes. Now consider Fig. 3 with upward moving s which implies that the levels z_I, z_{II} are also rising.

If the water density of the rain is ϱ_R and the water density in the ocean is ϱ_W, we have:

$$\varrho_R w_R + \varrho_W \dot{s} = 0 \tag{13}$$

$w_R < 0$ is the fall speed of the rain drops, eq. (13) defines the rising of the ocean surface (we assume horizontal homogeneity). Obviously,

$$F_I = \varrho_R w_R \quad \longrightarrow \quad \dot{s} = -F_I/\varrho_W \tag{14}$$

On the other hand, level z_{II} moves upward across the water (density ϱ_W) with speed $\dot{s}$. This implies a net downward flux:

$$SUR = -\varrho_W \dot{s} \tag{15}$$

and completes our interpretation of SUR for water surfaces (ocean and lakes).

CONCLUSIONS

This note has demonstrated that the well-established continuity of the total energy flux vector everywhere in the continuum, specifically across boundaries like density discontinuities, can be derived from the integral (but not the differential) version of the energy principle; no separate boundary condition is required. The pertinent continuity theorem applies both to the energy flux and the mass flux, because the respective conservation principles are formally equivalent.

What is the benefit of this demonstration? Why bother with mathematical subtleties as long as the energy principle in differential and integral forms are practically equivalent as implied, e.g., by van Mieghem [1973]? The answer may be summarized by the following "equation":

Energy principle in *integral* version for all volumes (with and without discontinuity surfaces)	=	Energy principle in *differential* version everywhere except at discontinuity surfaces	+	Continuity of energy flux across discontinuity surfaces

If the energy principle is postulated in the differential version (which implies that it cannot be postulated at the discontinuity itself) one needs to adopt an independent postulate concerning flux continuity across the discontinuity surface; it cannot be derived from the differential energy principle. On the other hand, if no boundaries are involved, differential and integral versions of the energy (or mass) principle are equivalent and can be converted into each other.

Thus the benefit of this note may have been to shed further light upon the high level of generality of the global energy and mass conservation principle. Batchelor [1967, p. 74] notes that the mass equation "has been called the 'equation of continuity', although not for any evident good reason". The arguments above may furnish good reasons to refer not only to the mass, but also to the energy conservation principle as to generalized equations of continuity.

The higher level of generality of integral laws over their differential counterparts has long been noted in physics. For example, Sommerfeld [1964, section 3] starts his lectures on electrodynamics with the integral version of the Maxwell equations, seemingly because the experimental results (in this case, the laws of Faraday and Ampère) can only be formulated for finite volumes and not in the differential idealization. Interestingly, Sommerfeld refers to the integral conservation law for the electric charge as the "continuity equation of electricity". And Whitham [1974] maintains that physical problems should be "first formulated in the basic integrated forms from which both the partial differential equations and the appropriate jump conditions follow". The reason for Whitham is his result that ambiguities in the solutions of the differential equations appear when shocks (=discontinuities) are involved; this can be avoided by postulating the integral law. The integral conservation principle is more general and more robust than the differential principle which is just a specialization of the integral principle for media free of discontinuities.

The present study has been restricted to quantities (energy, mass) which are strictly conservative. Generalizations for quantities with sources/sinks (e.g., entropy; chemical constituents; momentum or vorticity) have not been intended at this stage.

Today, physical climatology comprises more than just the surface energy and water balance. Climate is the result of complex interactions within the three-dimensional climate system. However, no single surface is more significant for the physics of climate than is the two-dimensional discontinuity surface between the atmosphere above and the litho-/hydro-/cryospheres below. Thus to relate the flux continuity condition across the earth's surface to the most basic principles of physics may have been a satisfying exercise. Hann would have been happy to see his early tedious computations well represented in this elegant and, after all, simple framework.

APPENDIX A: THE ENERGY BALANCE EQUATION - HISTORICAL REMARKS

In his Handbook of Climatology, in the section on continental and maritime climate, Hann [1897, 1908] discusses the reaction of either land or sea with respect to the net radiation flux (*insolation* and *heat radiation*). According to Hann this reaction is represented by the heat storage in either medium and by the latent heat take-up due to evaporation. Hann interpreted these heat balance terms correctly as fluxes. In his textbook on Meteorology [Hann, 1901] he determines the coefficient of temperature conduction and equates it to a vertical heat flux. Hann (p. 741): *Die durchschnittlich im Laufe eines Tages aus dem Königsberger Boden strömende Wärmemenge beträgt daher pro Centimeter Fläche 387.5 g-Kalorien unter der Annahme eines Temperaturgefälles von 1 Grad/Centimeter.* This figure is equivalent to a driving radiation flux of 188 W/m^2 which may be contrasted with the climatological global radiation value for Königsberg (115 W/m^2 [Budyko, 1963], reproduced in Hantel [1989, p. 342]). This example demonstrates that (i) Hann was principally aware of a heat flux balance behind his observations, and that (ii) he could not yet quantitatively formulate this law; after all, Hann represents the dawn of modern climatology [Kahlig, 1992].

The first to express the heat flux balance at the earth's surface as a consequence of the energy principle were Schmidt [1915, 1916] and Sverdrup [1937]. Sverdrup, at the opportunity of the Maud Expedition, developed the concept of a flux budget in a form in which it became the basis of modern climatology.

In his classical textbook *Das Klima der bodennahen Luftschicht* Geiger [1961] expressed the heat balance equation in what he called the *complete equation of the heat budget at a plane soil surface with no vegetation:*

$$S + B + L + V + Q + N \quad = \quad 0 \qquad \text{(A-1)}$$

Here, S = radiation, B = soil heat flux, L = sensible heat flux, V = evaporation, Q = horizontal advective heat flux, N = heat flux due to temperature difference between precipitation and soil. According to Geiger, eq. (A-1) follows from the consideration that the earth's surface is a 2D-plane with no mass; thus it cannot store heat. According to Geiger it then follows that the fluxes approaching and those leaving the earth's surface must add up to zero.

This argumentation is suggestive but not consistent. The consequence is that a quantity slips into (A-1) which actually does not belong there: it is the horizontal advective heat flux Q. This flaw does not diminish the lasting value of Geiger's book. It nevertheless demonstrates the need for internally consistent principles upon which the theoretical sciences are to be based.

In the field of global climatology, Albrecht [1940] based his climatology upon the concept of Sverdrup [1937] and Bowen [1926]; in a sense, Albrecht is the predecessor of the work of Budyko [1963], the master of what has been termed *physical climatology* [Flohn, 1957, 1985]. For reviews on this subject see, e.g., Malkus [1962] and Flohn [1969]. A comprehensive review on the global flux climatology has been given by Kessler [1985], a new global calculation of the fluxes has been published by Henning [1989].

In summary, the heat balance condition as consequence following from the energy principle has been, beginning with a preliminary approach of Hann, the basis of the earth-bound climatology, regional or global, up to today.

APPENDIX B

The general balance equation for a conserved quantity in global form reads [e.g., Glansdorff and Prigogine, 1971]:

$$\frac{\partial U(V,t)}{\partial t} + \int_{\Omega} \mathbf{F}(p,t) \cdot d\Omega(p) = 0 \qquad \text{(B-1)}$$

U is an extensive variable (mathematically, a *set-function*; e.g., total energy = internal plus mechanical energy) and assumed to be a countably additive function of volume, V, and a differentiable function of time, t (Fig. 5); this means

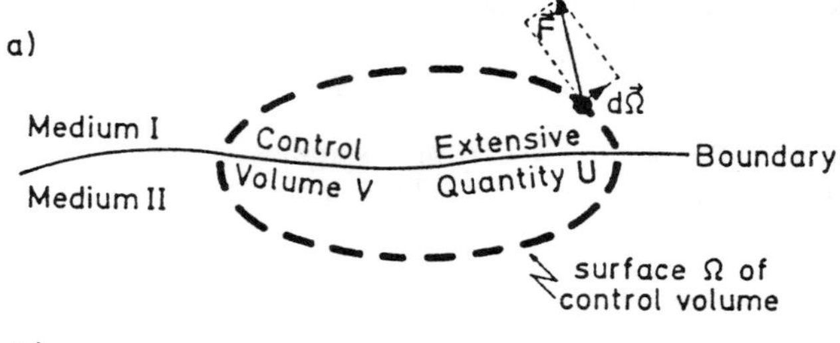

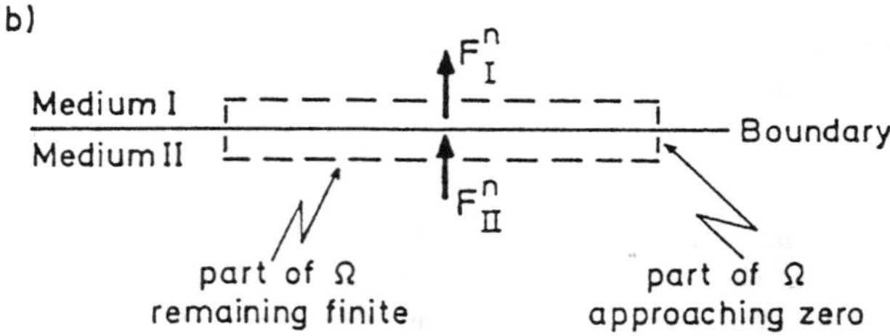

Fig. 5. Sketch of finite control volume V with surface Ω (dashed) containing finite total energy U. $\mathbf{F}$ = energy flux vector across Ω. Boundary between two different media (e.g., air/soil) may be located inside or may even cross V but has no impact upon conservation principle. $d\Omega$ = infinitesimal vector representing direction outward normal to Ω. a) Arbitrary control volume for which global conservation principle is postulated; b) specific control volume used for proof of continuity theorem.

that the value of U belonging to the union of a sequence of pairwise disjoint values equals the value of the infinite series of the $U's$ belonging to the individual volumes of that sequence. Further, U is assumed to be a volume-continuous function of V; this means that every volume-null set is an energy-null set. Both assumptions together imply the physically reasonable property that U goes to zero if V does. Further, they imply that the *extensive* variable U has an *intensive* counterpart, the energy density u, which is related to U through:

$$U \quad = \quad \int_V u dV \qquad \text{(B-2)}$$

This implication follows from the Radon-Nikodym theorem

[Dombrowski, 1985]. u may have discontinuities within V.

The second term in eq. (B-1) represents the surface Ω of the control volume V (Fig. 5). The energy flux density $\mathbf{F}(p,t)$ is a time-dependent vector field where p denotes a point in space; $\mathbf{F} \cdot d\mathbf{\Omega}$ is its (infinitesimal) projection upon the outside normal of Ω.

Eq. (B-1) states first that there exists a vector field $\mathbf{F}(p,t)$ which is integrable over the surface Ω constituting the outer boundary of V and second that the time change of U is uniquely determined by the total flux of $\mathbf{F}(p,t)$ across Ω.

The generality of this conservation principle can be recognized in that eq. (B-1) is valid irrespective of the specific nature of the medium (or media) inside V. The vector $\mathbf{F}$ is defined everywhere, no matter if the medium is vacuum, gas, liquid or solid or if there are (like in Fig. 5) discontinuities embedded in V. The conservation principle applies to any volume, however big or small. The only limiting assumption we make is that Ω itself is independent of time; this, however, is no serious restriction of generality in the present context.

For ease of notation, let us stipulate that the term *surface* is reserved for the surface Ω while the term *boundary* is reserved for the boundary between media I, II notwithstanding the fact that both terms denote two-dimensional surfaces in the mathematical sense; in particular, both *surface* and *boundary* are assumed piecewise differentiable surfaces. The *boundary* in this Appendix represents the *discontinuity surface* in the main text.

The general continuity theorem, on basis of the integral conservation principle, along with the assumptions formulated above can now be expressed as follows:

The limiting values of the normal components of $\mathbf{F}$*, when approaching a piecewise differentiable boundary from either side, are equal.*

This is the desired continuity of the flux across the *discontinuity* of the intensive property of u. It rests upon the volume *continuity of the extensive property* U which represents the physical matching principle postulated in section 2).

The rigorous proof of this theorem [Dombrowski, 1985] is rather involved (requiring some measure theory); the mathematical details are beyond the present note. A key step in the proof is to use the *volume-continuity* of U (and thus of $\frac{\partial U}{\partial t}$) for a specific control volume around the boundary, see Fig. 5b: the control volume is made arbitrarily small by diminishing only the parts of the surface perpendicular to the boundary while the parts parallel to the boundary are kept finite. This can be expressed by saying that the storage term $\frac{\partial U}{\partial t}$ goes to zero which, by virtue of eq. (B-1), forces any limit values F_I^n, F_{II}^n of the flux $\mathbf{F}$ approaching the boundary from either side to be equal. The boundary, like the surface Ω, is assumed independent of time. For generalizations see Appendices C, D.

APPENDIX C

The integral energy conservation principle for moving boundaries $\zeta_I(t)$, $\zeta_{II}(t)$ reads:

$$\frac{d}{dt}\int_{\zeta_{II}(t)}^{\zeta_I(t)} u(z,t)dz + \Phi[\zeta_I(t),t] - \Phi[\zeta_{II}(t),t] = 0 \qquad \text{(C-1)}$$

The $\zeta's$ are differentiable functions of time. This is formally a generalization of the simple model of section 3 because (C-1) can be specialized by choosing $\zeta_I(t) \equiv z_I$, $\zeta_{II}(t) \equiv z_{II}$, which reproduces eq. (2). In order to distinguish eq. (C-1) from (2) we have switched from the symbol F for the flux vector to the symbol Φ.

However, there is not much physical difference because (C-1), like (2), states that the time-rate of change of total energy in the interval $\zeta_{II}(t) \leq z \leq \zeta_I(t)$ must be balanced by the net inflow across the level $\zeta_{II}(t)$ located in medium II and the level $\zeta_I(t)$ located in medium I (Fig. 3; replace z_I by $\zeta_I(t)$ and z_{II} by $\zeta_{II}(t)$). Further, $u(z,t)$ may have discontinuities within the interval.

We consider one such discontinuity located at $z = \sigma(t)$ somewhere in the interior, i.e., $\zeta_{II}(t) < \sigma(t) < \zeta_I(t)$. We specify the limits such that:

$$\zeta_I(t) - \sigma(t) \quad = \quad \sigma(t) - \zeta_{II}(t) = \epsilon \qquad \text{(C-2)}$$

ϵ, as before, is a time-independent positive real number but otherwise arbitrary. Suppose that $u(z,t)$ and $\frac{\partial u(z,t)}{\partial t}$ are continuous (with respect to both z and t) in the subintervals $\sigma(t) < z \leq \zeta_I(t)$, $\zeta_{II}(t) \leq z < \sigma(t)$, and have finite (but generally different) limits as $z \rightarrow \sigma(t)$ from below and above. This means physically that the densities and their derivatives with respect to time can be continuously extended (for each of the media I, II separately) to the discontinuity located at $z = \sigma(t)$.

The latter convention may be formalized as follows. We can always find a function $u_I(z,t)$ which is continuous with respect to both z and t and has a partial derivative with respect to time so that one has for all t: If z is located within the interval $\sigma(t) < z \leq \zeta_I(t)$, then $\lim[u(z,t)] = u_I[\sigma(t),t]$ and $\lim[\partial u(z,t)/\partial t] = \frac{\partial u_I(z,t)}{\partial t}$, where lim refers to $z \rightarrow \sigma(t)$.

With this specification (and the equivalent specification for a proper function $u_{II}(z,t)$ belonging to the subinterval $\zeta_{II}(t) \leq z < \sigma(t)$) eq. (C-1) can be written:

$$\begin{aligned} \Phi[\zeta_{II}(t),t] - \Phi[\zeta_I(t),t] \quad &= \\ \frac{d}{dt}\int_{\zeta_{II}(t)}^{\sigma(t)} u(z,t)dz + \frac{d}{dt}\int_{\sigma(t)}^{\zeta_I(t)} u(z,t)dz \quad &= \qquad \text{(C-3)} \\ \int_{\sigma(t)-\epsilon}^{\sigma(t)} \frac{\partial u(z,t)}{\partial t}dz + \int_{\sigma(t)}^{\sigma(t)+\epsilon} \frac{\partial u(z,t)}{\partial t}dz \quad &+ \end{aligned}$$

$$[u_{II}(\sigma(t),t) - u(\zeta_{II}(t),t)]\dot{\sigma}(t) - [u_I(\sigma(t),t) - u(\zeta_I(t),t)]\dot{\sigma}(t)$$

Since $\frac{\partial u(z,t)}{\partial t}$ is bounded in each of the intervals separately, the integrals, as before, tend to zero as $\epsilon \rightarrow 0$.

The difference to section 3 is represented by the two terms in brackets. However, due to the coupling assumption (C-2) between the moving discontinuity $\sigma(t)$ and the moving boundaries $\zeta_I(t)$ and $\zeta_{II}(t)$ both brackets tend also to zero as $\epsilon \rightarrow 0$. Thus:

$$\lim\{\Phi[\zeta_{II}(t),t] - \Phi[\zeta_I(t),t]\} = 0 \qquad \text{(C-4)}$$

which represents the continuity of the energy flux relative to the level $z = \sigma(t)$.

APPENDIX D

The approach of Fortak [1987] in three dimensions applies to a moving discontinuity and requires slightly modified assumptions with essentially the same physical reasoning. The argument goes as follows. One starts with the general conservation principle in integral form but *for a material outer boundary. A moving discontinuity is assumed inside. The fluid velocity vector across the discontinuity is assumed continuous, and the velocity of the discontinuity itself is assumed to be equal to this fluid velocity.* It then follows that the nonconvective flux is continuous across the discontinuity. Fortak's argument is attractive because it allows to treat the case of non-material discontinuities in a fairly general manner.

ACKNOWLEDGMENTS

Various discussions on the points made in this note were necessary to clarify the arguments. Of particular help were talks with Prof. Dr. P. Dombrowski, Prof. Dr. H. Fortak, Dipl.-Phys. K. Keuler, Dipl.-Met. R. Knoche and Prof. Dr. H. Kraus. Prof. Dr. H. Fortak kindly provided unpublished material for the argument sketched in Appendix D. Mr. D. Wilinski drafted the figures. Mrs. C. Frese and Mrs. B. Berger took care of the manuscript.

REFERENCES

Albrecht, F., 1940: Untersuchungen über den Wärmehaushalt der Erdoberfläche in verschiedenen Klimagebieten. *Reichsamt für Wetterdienst, Wiss. Abh., 8,* **2**, 82 pp.

Batchelor, G.K., 1967: *An Introduction to Fluid Dynamics.* Cambridge University Press, 615 pp.

Baumgartner, A., and E. Reichel, 1975: *Die Weltwasserbilanz. Niederschlag, Verdunstung und Abfluß über Land und Meer sowie auf der Erde im Jahresdurchschnitt.* R. Oldenbourg Verlag München-Wien, 179 pp. + Karten (in German and English language).

Bowen, I. S., 1926: The ratio of heat losses by conduction and by evaporation from any water surface. *Phys. Rev.,* **27**, 779-787.

Budyko, M. I. (Ed.), 1963: *Atlas Teplovogo Balansa Zemnogo Shara (Atlas of the heat balance of the earth).* Mezhved. geofiz. komitet, Moskva.

Dombrowski, P., 1985: *Personal communication.*

Fleagle, R. G., and J. A. Businger, 1980: *An Introduction to Atmospheric Physics.* Academic Press, New York, London, Toronto, Sydney, San Francisco, 432 pp.

Flohn, H., 1957: Zur Frage der Einteilung der Klimazonen. *Erdkunde* **11**, 161-175.

Flohn, H., 1969: General Climatology. *2. Vol. 2 of World Survey of Climatology,* 266 pp. (Articles by: H. Riehl, E. L. Deacon, R. Geiger, H. Flohn, H. H. Lamb).

Flohn, H., 1985: *Das Problem der Klimaänderungen in Vergangenheit und Zukunft.* Wiss. Buchges., Darmstadt, 228 pp.

Flohn, H., 1992: Physical 3D-Climatology from Hann to the Satellite Era. (this volume).

Fortak, H., 1967: *Vorlesungen über Theoretische Meteorologie.* Veröff. Inst. f. Th. Met. d. Freien Universität Berlin.

Fortak, H., 1987: *Personal communication.*

Geiger, R., 1961: *Das Klima der bodennahen Luftschicht.* Vieweg, Braunschweig, 4. Aufl., 646 pp.

Gill, A. E., 1982: *Atmosphere-Ocean Dynamics.* Academic Press, 662 pp.

Glansdorff, P., and I. Prigogine, 1971: *Thermodynamic Theory of Structure, Stability and Fluctuations.* Wiley-Interscience, 306 pp.

Hann, J., 1897: *Handbuch der Klimatologie. Band I: Allgemeine Klimatologie.* Bibliothek Geographischer Handbücher (Hrsg.: Friedrich Ratzel), Verlag von J. Engelhorn, 404 pp.

Hann, J., 1901: *Lehrbuch der Meteorologie.* (11 Abb., 8 Tafeln, 15 Karten), Tauchnitz-Verlag Leipzig, 805 pp.

Hann, J., 1908: *Handbuch der Klimatologie. Band I: Allgemeine Klimalehre.* Bibliothek Geographischer Handbücher, N.F., Verlag von J. Engelhorn, 394 pp.

Hantel, M., 1989: *The Present Global Surface Climate.* In: Landolt-Börnstein - Meteorology (Ed. G. Fischer), New Series, **V/4c2**, Springer-Verlag, Berlin, 117-474 pp.

Henning, D., 1989: *Atlas of the Surface Heat Balance of the Continents.* (327 continent maps and 15 tables), Gebrüder Bornträger Berlin-Stuttgart, 402 pp.

Kahlig, P., 1992: Some Aspects of Julius von Hann's Contribution to Modern Climatology. (this volume)

Kessler, A., 1985: *Heat Balance Climatology.* In: (ed.: O. M. Essenwanger) *General Climatology,* **1A**, Elsevier Amsterdam-London-New York-Tokyo, 224 pp.

Kraus, E. B., 1972: *Atmosphere-Ocean Interaction.* Oxford monographs on meteorology, Oxford, Clarendon Press 1972, 275 pp.

Kraus, H., 1987: *Specific Surfaces Climates.* In: *Meteorology,* New Series, Landolt-Börnstein, **4/c1/Part 1**, Springer-Verlag Berlin, Heidelberg, New York, London, Paris, Tokyo.

Lettau, H., 1969: Evapotranspiration Climatonomy. I. A new approach to numerical prediction of monthly evapotranspiration, runoff, and soil moisture storage. *Mon.Wea.Rev.,* **97**, 691-699.

Malkus, J. S., 1962: *Large-scale interactions.* In: M. N. Hill (Editor), *The Sea,* **Vol. 1**, Interscience Publishers, New York - London - Sydney, pp. 88-294.

McBean, G., 1988: *Concept of the global energy and water cycle experiment.* Report of the JSC Study Group on GEWEX (Montreal, Canada, 8-12 June 1987 and Pasadena,USA, 5-9 January 1988), WCRP-5, WMO/TD - No. 215, 70 pp.

Schmidt, W., 1915: Strahlung und Verdunstung an freien Wasserflächen. *Ann. Hydrogr., Berl.,* **43**, 111-124, 169-178.

Schmidt, W., 1916: Zur Frage der Verdunstung. *Ann. Hydrogr., Berl.,* **43**, 136-145.

Sommerfeld, A., 1964: *Elektrodynamik.* Vorlesungen über Theor. Physik, **Bd. III**, 4. Aufl., Akad. Verlagsges. Geest & Portig KG., Leipzig, 345 pp.

Sverdrup, H. U., 1937: On the evaporation from the oceans. *J. Mar. Res.,* **1**, 3-14.

van Mieghem, J., 1973: *Atmospheric Energetics.* Oxford Monogr. on Meteorology (ed.: P. A. Sheppard), Clarendon Press, Oxford, 306 pp.

Wüst, G., 1920: Die Verdunstung auf dem Meere. *Veröff. Inst. Meereskd. N.F., Reihe A,* **H.6**, 1-96.

Whitham, G. B., 1974: *Linear and Nonlinear Waves.* (eds.: L. Bers, P. Hilton, H. Hochstadt), John Wiley and Sons, New York, London, Sydney, Toronto, 636 pp.

Willmott, C. J., Rowe, C. M., Mintz, Y., 1985: Climatology of the terrestrial seasonal water cycle. *J. Climatology,* **5**, 589-606.

Michael Hantel, Institute of Meteorology, University of Vienna, A-1190 Vienna, Austria.

The Effect of Increasing the Level of Atmospheric CO_2 on Heat and Vapour Fluxes over Vegetated Surfaces

M. H. ZEMANKOVICS

Meteorological Service of the Hungarian Republic, Agrometeorological Research Station, Keszthely, Hungary

Vegetation plays an important role in the processes of matter and energy flow between the surface and the atmosphere. The Crop Micrometeorological Simulation Model (CMSM) completed with the response function to external CO_2 concentration was used to investigate the effects of the doubled amounts of CO_2 on the rate of photosynthesis, sensible and latent heat.
The output of CMSM with respect to normal and doubled CO_2 concentration is different. The rates of photosynthesis and sensible heat are increased and the rate of latent heat is decreased both in C_3 and C_4 plants due to the doubled CO_2 concentration.

INTRODUCTION

Atmospheric systems gain the main part of their energy from the processes of energy exchange taking place at the earth's surface, where the sun's radiation is transformed into chemical forms of energy and into heat. In global circulation models the surface energy exchange is governed by the main physical characteristics of the surface like ocean, continent or ice. However, the surface layer has also biological characteristics. The most part of terrestrial surfaces is covered by vegetation and even the ocean is a biological system. Some micrometeorological models contain energy exchange processes taking into account some biological characteristics of a given surface. Atmospheric CO_2 plays a significant role in the radiation balance of the atmosphere. The global amounts of CO_2 tend to increase as measurements prove it all over the world. The most obvious reason for the rise of atmospheric CO_2 is the burning of fossil fuel, although many aspects of this phenomen have not been completely revealed. Recently a great deal of investigations were carried out not only about the reasons but about the possible consequences: the effects on the climate, on the agricultural production etc. Since vegetation is both a source and a sink of CO_2 and the biomass fixes and redistributes energy, the direct effects of CO_2 on plants, matter and energy fluxes deserve attention.

The Crop Micrometeorological Simulation Model (CMSM) developed by Chen [1984] gives a detailed description of the matter and energy flow inside the plant canopy and between the canopy and the low atmosphere. Therefore it seems to be suitable for incorporating the response function of photosynthesis to the external CO_2 concentration of ambient air. This will make it possible to study the effect of increasing CO_2 concentration on the rate of photosynthesis and heat fluxes.

Interactions Between Global Climate Subsystems, The Legacy of Hann
Geophysical Monograph 75, IUGG Volume 15

THEORETICAL BACKGROUND

CO_2 concentration of ambient air can be taken as environmental factor of photosynthesis. The effects of factors like radiation, temperature, water and nutrient supply are resolved in detail because of their great variability in nature. The response functions are used in primary production and micrometeorological models. In these models the level of CO_2 is assumed to be constant; thus there is no response function to CO_2 concentration involved. Nevertheless the effects of CO_2 on the rate of photosynthesis were investigated independently by plant physiologists. In general the normal partial pressure of ambient CO_2 of about 330 ppm is suboptimal for C_3 plants, which form 95% of the biomass. Seasonal dry weight gain is stimulated by increasing ambient CO_2 up to 1000 ppm. Over a large range of CO_2 (200-1000 ppm) the response of dry weight gain to CO_2 can be described by a logarithmic response function [Goudriaan, 1990]. In the case of C_4 plants the direct effects of increasing CO_2 are negligible above 300 ppm (Fig. 1). Chartier [1970] illustrates the response function of net CO_2 assimilation to CO_2 concentration of ambient air. C_3 and C_4 plants are different considering their response functions to light and temperature, especially they are different accord-

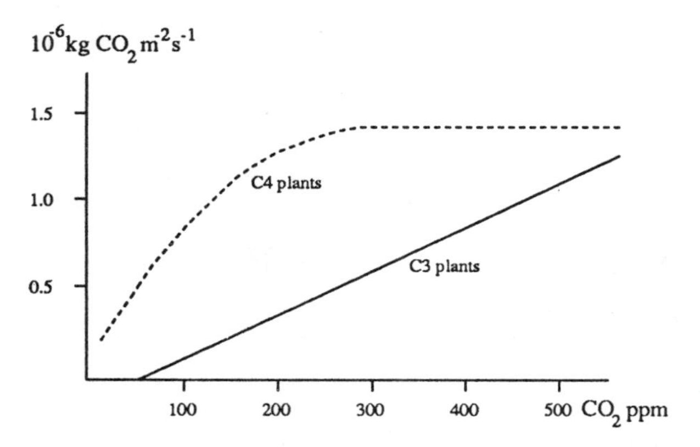

Fig. 1. Response functions of net assimilation to external CO_2 concentration for C_3 and C_4 plants at $T = 25°C$ and $PhAR = 300 Wm^{-2}$.

ing to numerical values of the parameters of the response functions. As van Laar and Penning de Vries [1972] found by an empirical representation of measured curves the net CO_2 assimilation is:

$$F_n = (F_m - F_d)\left(1 - e^{R_v e_s / F_m}\right) + F_d \tag{1}$$

where F_n is the net CO_2 assimilation, F_m is the maximum rate of net CO_2 assimilation (at a given temperature), F_d is the dark respiration (at a given temperature), R_v is the absorbed photosynthetically active radiation, e_s is the slope of F_n with respect to the photosynthetically active radiation (PhAR).

F_m, F_d are plant parameters and they are determined by the temperature response functions, F_m is an optimum function and F_d is also calculated from air temperature with a reference value of $-0.17 \cdot 10^{-6} kgCO_2 m^{-2} s^{-1}$ at the optimum temperature. e_s is different for C_3 and C_4 plants, the approximate values are $11.4 \cdot 10^{-9}$ and $17.2 \cdot 10^{-9}$ kg CO_2 per J of visible radiation for C_3 and C_4 plants, respectively. R_v depends on the radiative characteristics of the plant tissue (reflectance, absorptivity, transmittance) and the canopy structure. Otherwise the rate of photosynthesis can be described as CO_2 diffusion:

$$F_n = a \frac{C_e - C_i}{r_{s,c} + r_{b,c}} \tag{2}$$

where the conversion factor $a = 1.83 \cdot 10^{-6}$ converts the CO_2 concentration from ppm to kg CO_2 m^{-3}, C_e is the external value of CO_2 concentration, C_i is the internal value of CO_2 concentration, $r_{s,c}$ is the stomatal resistance for CO_2, $r_{b,c}$ is the boundary layer resistance for CO_2.

The resistances against CO_2 diffusion are parameterized by the resistances against water vapour and heat according to:

$$r_{s,c} = 1.66 \cdot r_{l,v} \quad \text{and} \quad r_{b,c} = 1.32 \cdot r_{b,h} \tag{3}$$

where $r_{l,v}$ is the leaf resistance against water vapor and $r_{b,h}$ is the boundary layer resistance against heat. By combining eqs. (2), (3) and solving for $r_{l,v}$ one obtains the leaf resistance:

$$r_{l,v} = 1.83 \cdot 10^{-6} \frac{C_e - C_i}{1.66 F_n} - 0.783 r_{b,h} \tag{4}$$

The value of $r_{b,h}$ is given by Goudriaan [1977]. Its value is related to the Nusselt number N_u, the diffusivity for heat in the air D_h and a characteristic leaf dimension w (taken as the width of the leaves) as

$$r_{b,h} = 0.5 \frac{w}{D_h N_u} \quad . \tag{5}$$

Under field conditions $r_{b,h}$ can be calculated as

$$r_{b,h} = 0.5 \cdot 1.8 \cdot 10^2 \sqrt{\frac{w}{u}} \quad , \tag{6}$$

where u is the local wind speed in ms^{-1}. In the calculations to follow eq. (6) was used.

The micrometeorological models MICROWEATHER [Goudriaan, 1977] and CMSM [Chen, 1984)] use the equations listed above for calculating the net photosynthesis and resistances. The value of the external CO_2 concentration is assumed constant in both models. To investigate the effects of an increased external CO_2 concentration the response function of net photosynthesis to CO_2 had to be implemented into the model. The CMSM model was used because of its simpler requirements concerning computer implementation. In the case of a C_4 plant, (in the present study: maize) the maximum rate of photosynthesis does not change with respect to CO_2 concentration if the CO_2 concentration is higher than 300 ppm; in this case we can keep the original response functions with respect to temperature and light for the doubled CO_2 value. In the case of a C_3 plant (in the present study: wheat) the maximum rate of photosynthesis increases according to the logarithmic function

$$F_n^* = F_n \left(1 + B \log \frac{CO_2^*}{CO_2}\right) \quad , \tag{7}$$

where F_n^* and F_n is the net CO_2 assimilation at the level of CO_2^* (simulated level) and CO_2 (actual level) respectively; the parameter B is taken as 1.5. This value comes from the curve given by Chartier [1970]. It is than the value 0.7 suggested by Goudriaan [1990], but it is quite close to the value found by Esser [1987]. The present high value of B was applied because in our case we intended to describe the maximum response under optimal conditions. In this sense the duplication of CO_2 means 1.45 times as much assimilates.

ENERGY PARTITIONING ACCORDING TO THE CROP MICROMETEOROLOGICAL SIMULATION MODEL

According to the meteorological conditions above the plant canopy the CMSM model gives a sophisticated description on the processes of energy exchange taking place inside the canopy. Besides these elements the soil water potential and the soil temperature profile are needed as environmental variables to begin the simulation. The soil physical characteristics, the plant biological characteristics and the morphological structure of the plant stand are taking into account as parameters. The plant canopy is divided into layers. The net radiation absorbed by an individual layer is the final energy source of photosynthesis, sensible and latent heat. Radiation balance for each layer is calculated also for the photosynthetically active near infrared and long wave range. Penetration of direct and diffuse radiation for both directions (up and down) is determined using the vertical distribution of leaf area and the extinction coefficients for the different ranges of radiation according to Beers law. In each layer there are sources and sinks of the different forms of energy. For uncoupling the sensible (C) and latent heat (LE) fluxes Chen [1984] introduced the idea of enthalpy (H) and saturation heat (J) flux densities.

$$H = C + LE \qquad \text{and} \qquad J = C - \frac{y}{s} LE, \qquad (8)$$

where y is the psychrometric constant and s is the slope of the saturation vapour pressure curve. The profile of H is determined with the help of radiation balance of the layers and the profile of J is calculated with recurrent formulae in which the driving force is the vapour pressure deficit. The source of sensible heat flux density and latent heat flux density in the i-th layer is C_i and LE_i, respectively:

$$C_i = qH_i + (1 - q)\, J_i \qquad LE_i = (1 - q)(H_i - J_i) \qquad (9)$$

where $q = \frac{y}{y+s}$.

The transport of sensible and latent heat is determined by the morphological structure of the plant canopy calculating the turbulent resistances between the layers. For example water vapour comes into the layer via stomata (the stomatal resistance has to be overcome) and the boundary layer (the boundary layer resistance has to be overcome). Water vapour from one layer to the next has to overcome the turbulent resistance between the layers. The sensible heat has a similar way except the stomatal phase. The sensible and latent heat flux density above the canopy is an accumulation of sources/sinks of the layers below. According to the model there is a hierarchy among the different energy forms. First of all the energy fixation by photosynthesis takes place, secondly the latent heat flux forms, it depends on the water supply of the plant and the drying power of ambient air, and the rest is the sensible heat flux.

SIMULATION EXPERIMENTS

The response function of photosynthesis to external CO_2 concentration was built into the CMSM model. Since maize is a C_4 plant the CO_2 level is saturated above 300 ppm. In the optimum temperature range the maximum net photosynthesis is $1.67 \cdot 10^{-6}$ kg $CO_2 m^{-2} s^{-1}$ both for 330 ppm and

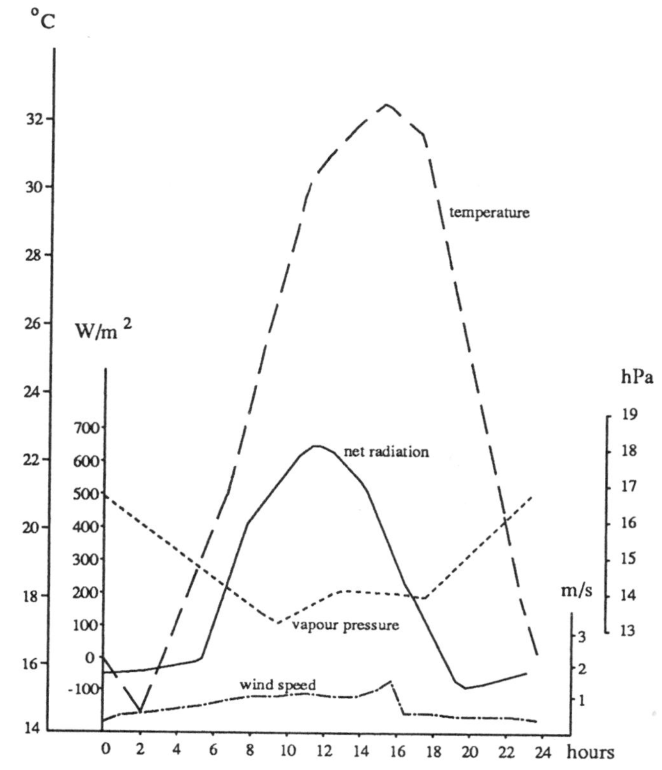

Fig. 2. Daily course of meteorological inputs at the reference height above the maize canopy (Keszthely, date 27.7.1989).

660 ppm CO_2. Internal CO_2 concentration was taken to 120 ppm [Goudriaan, 1977]. Wheat is a C_3 plant, so the increasing external CO_2 stimulates the net CO_2 assimilation. The maximum rate of net photosynthesis for the present level of CO_2 (330 ppm) was chosen $1.03 \cdot 10^{-6} kgCO_2 m^{-2} s^{-1}$ (that is the mean value of several measurements reported by van Heemst, [1988]) and the effect of temperature was taken

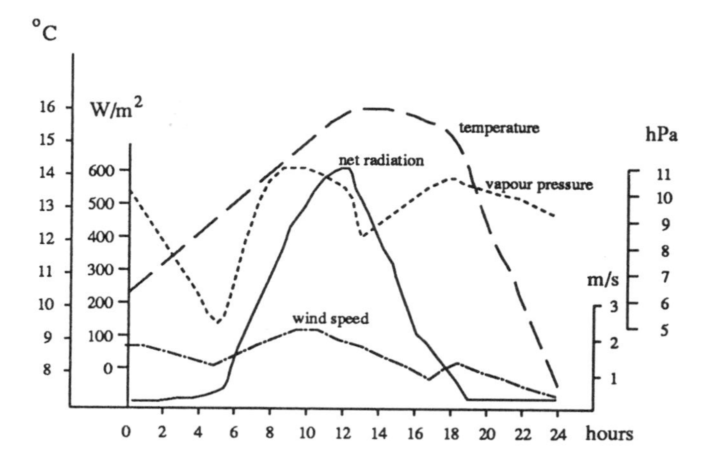

Fig. 3. Daily course of meteorological inputs at the reference height above the wheat canopy (Keszthely, date 11.5.1989).

after Vong and Murata [1977]. At the level of 660 ppm CO_2 the maximum rate of net photosynthesis was taken $1.5 \cdot 10^{-6} kgCO_2m^{-2}s^{-1}$. Internal CO_2 concentration was chosen 210 ppm [Goudriaan, 1977]. The CMSM model adjusted according to the above was used to investigate the effects of an increased level of CO_2 on photosynthesis, sensible and latent heat flux. Scenarios were as follows:

No_1/maize, No_1/wheat, No_2/maize, No_2/wheat.

No_1 means the present level of CO_2 (330 ppm), and No_2 means the doubled (660 ppm) CO_2 concentration. Fully developed maize and wheat canopies were chosen for the simulation experiments. The date of simulated day was 27 July, 1989 and 11 May, 1989 respectively. The input meteorological data were measured at 5 m above the soil surface for maize and 2 m above the soil surface for wheat. The daily courses of meteorological inputs are shown in Figs. 2 and 3. These days were representative clear days with moderate wind speed. Daily courses of the rate of photosynthesis, the latent heat flux and the sensible heat flux densities were compared for the No_1 and No_2 scenarios.

At night time there were no differences in the scenarios, but at daytime significant differences occured. The rate of photosynthesis increased in both plant canopies (see Figs.

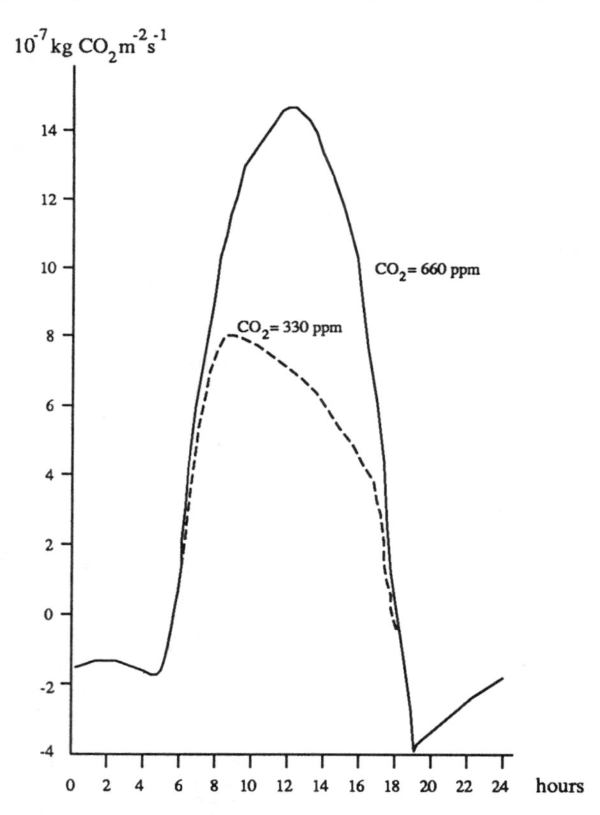

Fig. 4. Rate of photosynthesis of maize (for $CO_2 = 330$ *ppm* and $CO_2 = 660$ *ppm* daily totals are 0.0154 and 0.0359 $kgCO_2m^{-2}$ respectively).

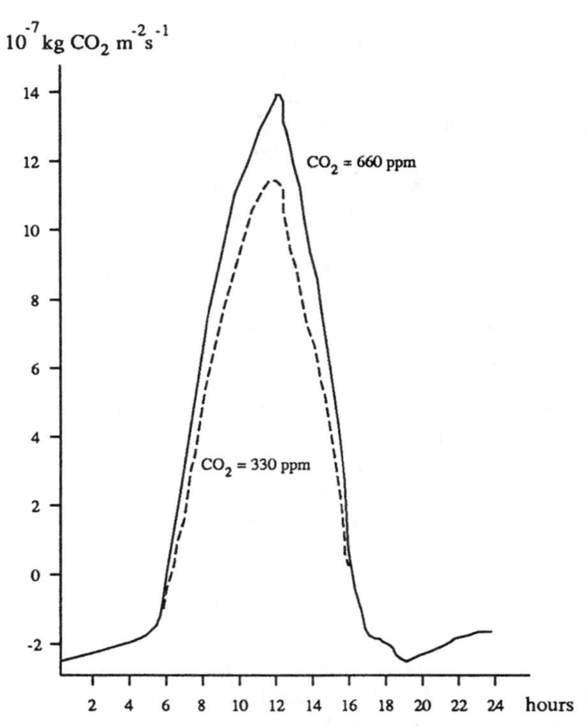

Fig. 5. Rate of photosynthesis of wheat.

4,5). In the case of maize the increment in CO_2 assimilation was strikingly great, in spite of the lack of a direct response to CO_2. As mentioned before the C_4 plant are at saturation level with respect to CO_2 concentration of ambient air. However the soil was extremely dry on the simulated day, the water potential in the soil was -8 bar. This dry condi-

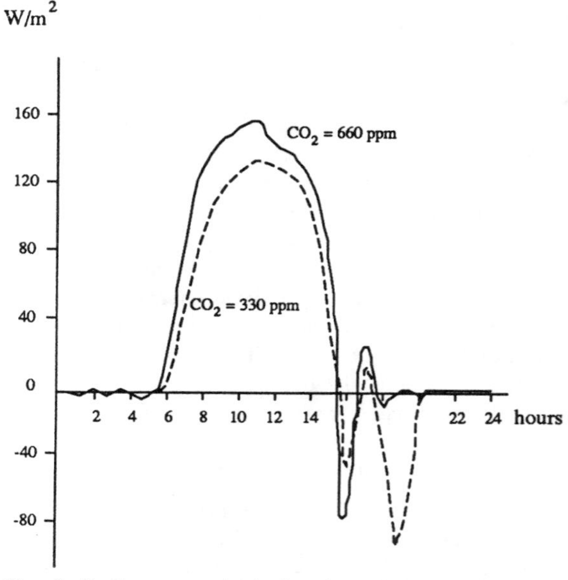

Fig. 6. Daily course of calculated sensible heat flux density above the maize canopy.

tion limited the CO_2 assimilation, the photosynthesis curve of No_1/maize shows identitation in the afternoon, but the same dry soil together with high level of CO_2 does not mean so strong limitation in the rate of net photosynthesis, the curve of No_2 is quite regular (Fig. 4). The increment of net photosynthesis is presumably due to the changed micrometeorological conditions caused by stomatic closure. The less water loss reduces the water stress effects and the higher leaf temperature is favourable for maize.

The sensible heat flux density also increased in No_2 experiments (Figs. 6, 7). That is an outcome of decreased latent heat flux density (Figs. 8, 9) caused by the higher stomatal resistance. The decrease of the latent heat flux is quite resonable, because the external CO_2 concentration is taken account in the calculation of stomatal resistance, as it was presented in eq. (4). The values calculated of stomatal resistances for the individual layers are shown in Table 1. It can be seen that the doubled CO_2 causes significantly higher stomatal resistances both in maize and wheat canopy. The increments are larger in wheat than in maize. As the latent heat flux density decreases, the sensible heat flux density increases both in maize and wheat.

The relative differences in the daily totals of latent heat fluxes are -12.7% and -33.5% for the maize and wheat canopies respectively. In daily totals of sensible heat flux

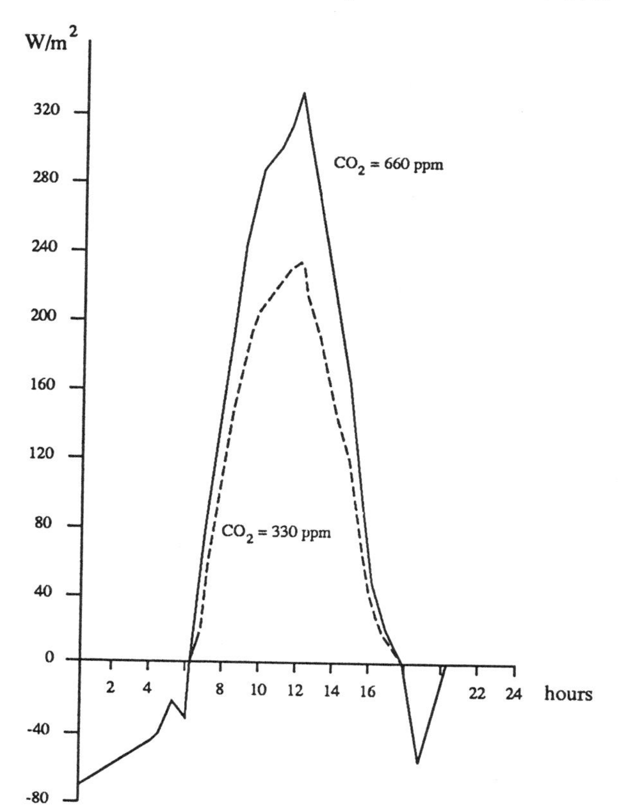

Fig. 7. Daily course of calculated sensible heat flux density above the wheat canopy.

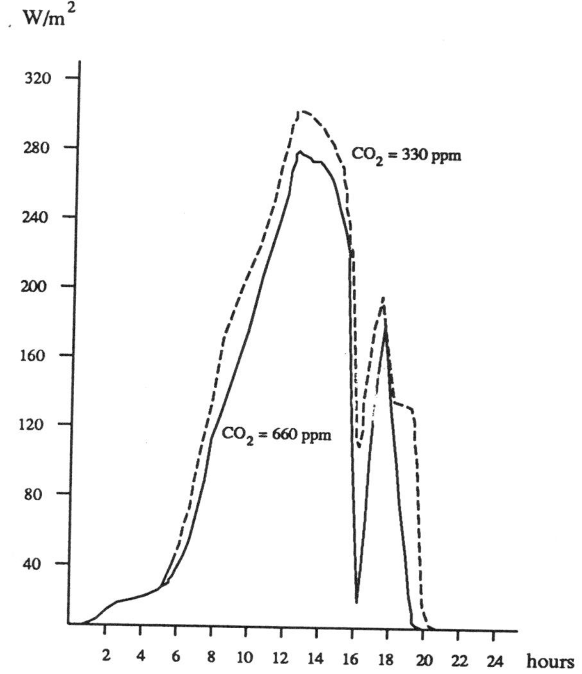

Fig. 8. Daily course of calculated latent heat flux density above the maize canopy.

the relative differences are 27,6% and 49,7% for the maize and wheat canopies, respectively.

CONSEQUENCES

Sensible and latent heat flux across a surface above the canopy provide the energy for the planetary boundary layer

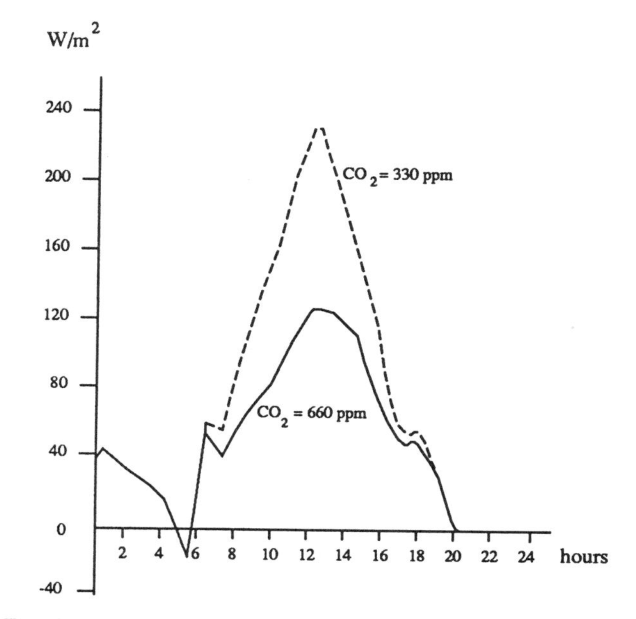

Fig. 9. Daily course of calculated latent heat flux density above the wheat canopy.

TABLE 1. Calculated stomatal resistances (s/m) according to different scenarious No_1 - present level of CO_2 (330 ppm); No_2 - doubled concentration of CO_2 (660 ppm). 6,00, 9,00 etc. hours of the day.

Maize	6,00		9,00		12,00		15,00		18,00	
layers	No_1	No_2	No_1	No_2	No_1	No_2	No_1	No_2	No_1	No_2
upper	591	1304	425	693	522	640	605	754	683	1193
mid	1432	1848	754	1221	837	1023	865	993	1549	1806
lower	1861	1968	1544	2314	983	1071	1189	1293	1907	1956
Wheat	6,00		9,00		12,00		15,00		18,00	
upper	1177	1999	330	1125	260	864	507	1690	1452	2144
mid	dew	dew	1094	3894	635	2155	1011	1709	1979	2000
lower	dew	dew	1312	1821	982	1619	1540	1900	1995	1995

of the atmosphere. The ratio of sensible and latent heat fluxes, the Bowen ratio, is an important parameter in climate formation. If the Bowen ratio increases over a large area due to the response of vegetation to high CO_2 concentration of the air, it means a positive feedback to global warming. This kind of feedback process should be interesting also in climate change models.

This simulation study can be taken as a demonstration of the activity of plant canopies in energy exchange between the surface and the atmosphere. The relationship between the stomatal resistance and the CO_2 concentration has been known to biologists; our simulation experiments have reproduced this relationship. Thus we have demonstrated that the Crop Micrometeorological Simulation Model (CMSM) coupled with CO_2 response function is applicable for quantification of changing heat fluxes and photosynthesis as the effects of increasing atmospheric CO_2.

Numerical values of the increased sensible and the decreased latent heat fluxes due to high CO_2 concentration of ambient air depend on several environmental factors like meteorological conditions, soil and plant conditions. It would be a challenging matter of further investigations.

REFERENCES

Albrecht, F., 1940: Untersuchungen über den Wärmehaushalt der Erdoberfläche in verschiedenen Klimagebieten. *Reichsamt für Wetterdienst, Wiss. Abh.*, *8*, **2**, 82 pp.

Chartier, P., 1970: (in: Hortobágyi, T. and Simon,T. /eds./ Növényföldrajz, társulástan és ökológia (Plant Geography, Theory of Association and Ecology), 1981 Tankönyvkiadó, Budapest).

Chen, J., 1984: *Mathematical Analysis and Simulation of Crop Micrometeorology.* Ph. D. Thesis, Pudoc, Wageningen.

Esser, G., 1987: Sensitivity of global carbon pools and fluxes to human and potential climatic impacts. *Tellus* 39B:245-260.

Goudriaan, J., 1977: *Crop Micrometeorology: A simulation study*, Pudoc, Wageningen.

Goudriaan, J., 1990: Atmospheric CO_2, global carbon fluxes and the biossphere. (in: Theoretical Production Ecology: reflection and prospects, eds: Rabbinge R., Goudriaan J., van Keulen H., Penning de Vries F.W.T., van Laar H.H.), Pudoc, Wageningen.

van Heemst. H. D. H., 1988: Plant data values required for simple crop growth simulation models: review and bibliograph, Simulation Reports CABO-TT, Wageningen.

van Laar H. H., and Penning de Vries F. W. T., 1972: CO_2 assimilation light response curves of leaves, some experimental data, Versl.Inst. biol.scheik.Onderz.LandbGewassen, 62, Wageningen.

Vong, N. Q., and Murata, Y., 1977: Studies on the physiological characteristics of C_3 and C_4 crop species. Part 1: The effects of air temperature on the apparent photosynthesis, dark respiration and nutrient absorption of some crops, *Japanese Journal of Crop Science* 46:45-52.

M. H. Zemankovics, Meteorological Service of the Hungarian Republic, Agrometeorological Research Station, Keszthely, Hungary.

Interactions between Biosphere and Atmosphere analysed with the Osnabrück Biosphere Model

GERD ESSER

Department of Botany, University of Giessen, FRG

HELMUT LIETH

Department of Biology, University of Osnabrück, FRG

The steadily increasing injection of trace gases into the atmosphere and the large scale land use changes by human activities is the basis for the discussion of present and future climate changes with possibly significant feedbacks to the biosphere.

The major component of the greenhouse gases is CO_2. The atmospheric content of which is largely controlled by the biosphere. In order to calculate with sufficient enough accuracy sources, fluxes and sinks of the global carbon circulation a biosphere model was constructed which contains regional net CO_2 assimilation levels, allocation of biomass in woody and herbaceous portions, litter production and decomposition, carbon transfer into soil and rivers, and land use changes from natural vegetation to different agricultural crops.

These biosphere properties were cast into numerical equations of which 33 are discussed in this paper. Together with the known assessments of annual carbon dioxide release by the technosphere, the exchange balance atmosphere ocean, and changes of soil carbon content this model was used successfully to simulate the annual averages of the CO_2 content of the atmosphere since 1860. The model known under the name Osnabrück Biosphere Model is being improved continuously. Its results are in several cases supported by field observations. Better validation is expected, however, when seasonal CO_2 transfers of the biosphere can be simulated.

INTRODUCTION

The discussion about the possibility of global warming by the continuous injection of CO_2 into the atmosphere through excessive burning of fossil fuel stimulated the scientific community to analyse the global carbon budget as accurately as possible. The possibility itself was already known to the geochemists and geophysicists of the past generation. The junior author received the first stimulation to study the carbon pools and budgets of vegetation from his chemistry teacher Walter Noddack in 1948, who outlined the basic components of the world carbon budjet in the way as shown in Table 2. In his time they had a reasonable knowledge about the principles of carbon circulation between atmosphere and biosphere but the quantity of pools and transfer rates were not elaborated adequately. Noddack suggested, however, already that the danger may lay before us when the human population would utilize relentlessly the fossil fuel source by burning and suggested to his students a variety of scenarios for a world with a shortage of carbon sources. The historical aspects of the carbon circulation knowledge has been described by Lieth 1975 and 1978. When the junior author presented at the Hann symposium in 1991 the quantitative aspects of the biospheric contribution to the global carbon flux it concluded for him about 30 years of research and modelling, much of which was done in conjunction with international projects like the IBP, SCOPE and more recently with the IGBP. After the lecture at the Hann symposium had been given a "Festschrift" edited by Esser and Overdieck [1991] was presented to him as gift for the 65th birthday. Prof. Esser contributed to this volume chapter 31, p. 679: "Osnabrück biosphere model, structure, construction, results". Professor Esser had worked with the junior author on the global carbon modelling for more than 10 years. After reading that paper the speaker found it nec-

Interactions Between Global Climate Subsystems, The Legacy of Hann
Geophysical Monograph 75, IUGG Volume 15

TABLE 1. Most important anthropogenic trace gases ("greenhouse gases", column 1), their effectiveness for the greenhouse effect relative to CO_2 (2), half-life time in the environment (3), estimated sources (4), present atmospheric concentration increase (5), and their contribution to the presently observed additional greenhouse effect (6). CO_2 carbon dioxide, CH_4 methane, N_2O dinitrogen-monoxide, FCKW fluoro-chloro-hydrocarbons, O_3 (troposph.) tropospheric ozone, UV ultraviolet radiation. Data sources: Guderian [1985], Ramanathan [1987], Wuebbles et al. [1988].

gas (1)	rel. eff. (2)	$t_{1/2}$ [a] (3)	sources (4)	trend [$\%\cdot yr^{-1}$] (5)	contribution [%] (6)
CO_2	1	$\simeq$ 140	80% fossil 20% deforest.	0.5	50
CH_4	32	5–7	ruminants rice fields garbage, mining natural gas	1.1	19
N_2O	150	150	deforestation agriculture burning	0.25	4
FCKW	15,000	75–110	aerosols plastic foams refrigerants	5	15
O_3 (troposph.)	3,000	0.1	waste gas +UV	0.5	8

essary to ask him to coauthor this paper. He kindly agreed. This paper is, therefore, largely based on his treatment, since it contained in several instances newer aspects which Esser developed after he had left Osnabrück for Vienna to work in IIASA and finally to the University of Giessen, known to all carbon modellers as the place where Liebig worked during the middle of the last century on plant fertilizers and global primary productivity assessments.

The fact that Lieth 1964/65 had assessed the primary productivity of the biosphere by first extrapolating yield data and then by predicting it from climatic variables in the late sixties and published it as a computer model based map called Miami model in 1972, brought him into close contact with the climate modellers, among them Prof. Flohn, with whom we had the honour to contribute several times to the same symposium, like the one we are having here in Vienna. The frequent discussions with him influenced the focus of our own research, especially the regionalization of our modelling.

Various trace gases are known to reduce the permeability of the atmosphere for long-wave infrared radiation, while the permeability for sunlight is essentially unchanged. Therefore, an elevated global mean temperature is necessary to restore the balanced global fluxes of short and long wave radiation. Although processes may exist, and some are already known, which enhance or reduce the primary effects of the climate relevant trace gases, those gases are suspected to cause changes of the clobal climate, if their concentration in the atmosphere continues to increase. Table 1 gives an overview over the most important anthropogenic "greenhouse gases" and their environmental behavior.

In addition to the gases in Table 1 water vapor is an important greenhouse gas, although its atmospheric concentration is not likely to be changed directly by human activities.

Carbon dioxide is presently the most important greenhouse gas, and in contrast to the other greenhouse gases, the human potential to rise its future atmospheric concentration seems to be very high since the fossil carbon resources (coal, oil, natural gas) probably exceed $6.5\cdot 10^{12}$ tons (6.500 Gt) of carbon.

The atmospheric level has risen considerably since preindustrial times as shown by Friedli et al. [1991] in siple ice core and Keeling [1986] at the Mauna Loa observatory. In the same period the mean global temperature seemed to have risen somewhat. Several time series presented in Figure 1 show that no clear picture of an increase in temperature is visible through the background variability, but an upward trend seems to exist in temperature and temperature related aspects of the climate for the second half of this century. The atmospheric CO_2 level and the temperature rise seem to be correlated. But correlations never establish cause/effect relationships. While we are unsure whether the observed CO_2 rise has caused the observed increase of the global mean temperature, it can be excluded with a high degree of reliability for the last few hundred years that the temperature rise has caused the CO_2 increase. We know the releases from fossil sources within a limit of uncertainty of about 10%. Those emissions are greater than, but parallel in time to the buildup of atmospheric CO_2. Our knowledge further rests on the sophisticated models of the global carbon cycle and on data acquired by deconvolution of the atmospheric CO_2-trend and the ratio of carbon isotopes in the CO_2. All known processes which influence the global energy budget may be put together quantitatively to create a comprehensive global climate model. Those models have been constructed by the scientific community in recent years. They are among the most complex products of human intellect ever developed. The temperature rise pre-

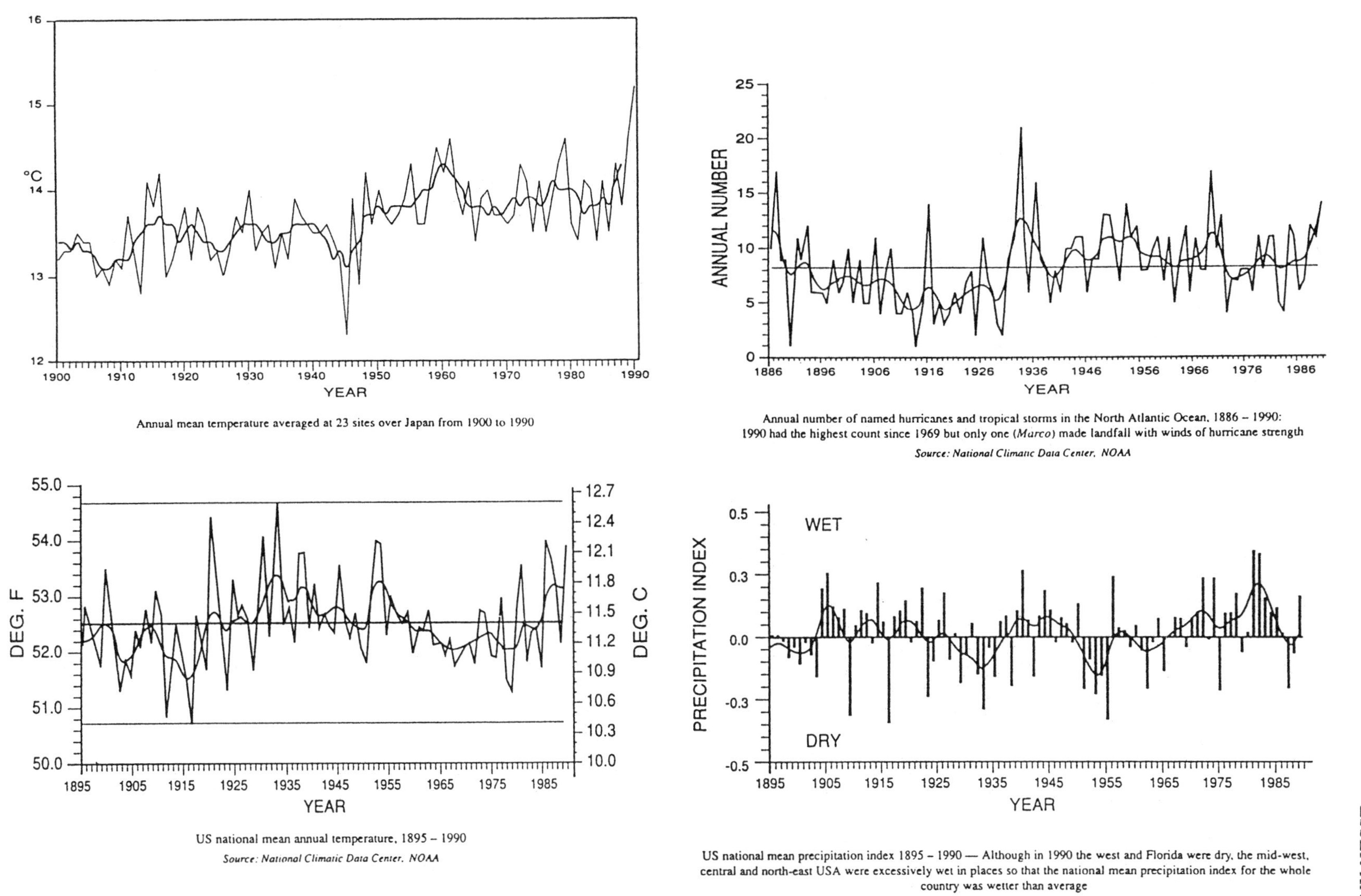

Annual mean temperature averaged at 23 sites over Japan from 1900 to 1990

Annual number of named hurricanes and tropical storms in the North Atlantic Ocean, 1886 – 1990: 1990 had the highest count since 1969 but only one (*Marco*) made landfall with winds of hurricane strength

Source: National Climatic Data Center, NOAA

US national mean annual temperature, 1895 – 1990

Source: National Climatic Data Center, NOAA

US national mean precipitation index 1895 – 1990 — Although in 1990 the west and Florida were dry, the mid-west, central and north-east USA were excessively wet in places so that the national mean precipitation index for the whole country was wetter than average

Source: National Climatic Data Center, NOAA

Fig. 1. Changes of temperature, precipitation and tropical storms during the last hundred years (taken from Limbert [1991]).

dicted by those models is in some regions of the world the same order of magnitude as the observed, see Figure 1 in others not yet noticable, however. For the next decades the predicted climate change is 10 to 100 times faster than the change at the end of the last glaciation. Even though the geographical pattern of temperature changes cannot be predicted accurately, it must be the task of the scientists to make the models more reliable. The interest of scientists and policy makers is concentrating on CO_2 since it has the strongest greenhouse potential. The total amount of contributing carbon is little less than $5 \cdot 10^{13}$ tons, while the atmospheric pool presently amounts to only about $7.5 \cdot 10^{11}$ tons carbon. Therefore, small changes in the global carbon budget influence the atmosphere considerably and feedback effects are considered very important. Estimates for the carbon content of major carbon pools and the annual turnover of the fluxes between them are shown in Table 2. Small size variations of very large fluxes create sources or sinks for carbon in the related pools. This has been demonstrated by many authors.

It has been the aim of the scientific community concerned with the climate problem since the beginning of the relevant discussion, to develop quantitative hypotheses for the dynamic functioning of the global carbon cycle. Those hypotheses could then be tested against independent data (i.e. data not used for the calibration of the model functions) to validate or disprove the hypotheses. One of those models, the Osnabrück Biosphere Model[2], is comprehensively reported here. It has been developed to investigate the carbon budget of the terrestrial biosphere. It was developed at the University of Osnabarück, FR Germany, in the years 1980–1990 [Esser, 1984, 1986, 1987, 1990; Lieth, 1978, 1984, 1985; Lieth & Esser, 1982, 1985; Esser, et al. 1982]. The latest model version is No. 3. Version No. 4, which is presently programmed, will introduce considerable improvements, namely higher time resolution (1 month instead of 1 year), and higher spatial resolution (0.5 degrees latitude and longitude). This paper will, therefore, be the last one which uses mean annual temperature and precipitation sum values as predictors for NPP.

TABLE 2. Major global pools of carbon and the respective annual fluxes which contribute to the global carbon budget. Values are given in Gt (10^9 t) respectively $Gt \cdot yr^{-1}$ of carbon. Sources: Maier-Reimer & Hasselmann [1987], Esser [1990], based on model studies. The pools deep ocean and soil organic carbon mainly contribute to the carbon budget on long time scales (hundreds of years).

pools	ocean	39,000	Gt
	upper mixed ocean layer	800	
	atmosphere	750	
	fossil fuel reserves	6,500	
	soil organic carbon	1,500	
	litter	100	
	live terrestrial phytomass	650	
fluxes	atmosphere → ocean	70–87	$Gt \cdot yr^{-1}$
	ocean → atmosphere	70–87	
	net primary productivity (atmosphere → green plants)	~45	
	litter production (green plants → litter)	~45	
	soil organic carbon production (litter → SOC)	~10	
	litter depletion (litter → atmosphere)	~35	
	soil organic carbon depletion (SOC → atmosphere)	~10	
	fossil fuel burning (fuel → atmosphere)	~5.5	
	deforestation (green plants → atmosphere)	~1	

Before we go into the details of the OBM we describe the basic pools of the global carbon cycle and their properties used in the Osnabrück Biosphere Model.

The Basic Pools and Fluxes of the Global Carbon Cycle

If one wants to model the global carbon circulation one needs to differentiate the major categories listed in Table 3.

In these generalized categories several subunits are distinguished. The atmosphere is taken as one single state variable with annual changes. The anthropogenic burning is also taken as one pool and uses the long existing tabulation of annual fossil fuel consumption by Rotty [1983] for the last 140 years. The transfer of CO_2 between ocean, atmosphere and rivers uses the calculations provided by Maier-Reimer and Hasselmann [1987] and Oeschger et al. [1975]. Land use changes in time were taken from Richards et al. [1983], validated partially by Esser and Lieth [1986]. For river carbon flow the assessment by Degens [1982], Degens et al. [1983, 1985] was accepted.

TABLE 3. Generalized pools and fluxes distiguished in the Osnabrück Biosphere Model

Carbon Pools	Carbon Flows
Fossil Carbon	Anthropogenic burning
Atmospheric Carbon	Photosynthesis
Live Biomass	Biomass accumulation
Dead (litter)Biomass	Respiration
Soil organic Carbon	Litter accumulation/depletion
	Vegetation changes
	Soil changes
	River carbon flow
	Ocean uptake
	Sedimentation

[2]The model was named after the city of Osnabrück, FR Germany, since it was presented the first time during a Symposium held in Osnabrück sponsored by the Commission of the European Communities in March 1983. Lieth is the initiator of the work. He contributed his MIAMI model, an empirical statement to calculate the flux net primary productivity of the potential natural vegetation from climate variables [Lieth, 1972].

TABLE 4. List of variables used in the model description in alphabetical order. Most of the variables are arrays which is indicated in the equations (taken from Esser [1991]).

symbol	meaning	units
c	field crop	
i	plant material (woody, herbaceous), $i = h, w$	
j	model year A.D., $j = 1860, \ldots, 2100$	
m	grid element, $m = 1, \ldots, 2433$	$2.5^\circ \cdot 2.5^\circ$
o	soil unit, $o = 1, \ldots, O(m)$	
AA	agricultural area of m in a model year j	m^2
AG	total area of m	m^2
AGE	mean stand age of a vegetation unit	years
AS	area of soil type $o(m)$	m^2
AV	area covered by natural vegetation in m	m^2
c_1, c_2	parameters to determine slope and turning point of the clearing function	–
CO_2	carbon dioxide concentration in the atmosphere	$\mu l \cdot l^{-1}$ (= ppmv)
DIS	annual discharge	$mm \cdot yr^{-1}$
DOC	dissolved organic C export	$g \cdot m^{-2} \cdot yr^{-1}$
f	soil factor of soil type o	–
F	weighted mean of all f of (m)	–
fa	conversion factor yield to NPP	–
FA	weighted mean of the $fa(c)$ of m	–
FCO_2	fertilizing factor of CO_2 on NPP	–
H	factor for sharing NPP into (i) compartments	–
j_{TR}	year of the turning point of the logistic clearing function	–
kd	decay coefficient for litter	yr^{-1}
klp	litter production coefficient	yr^{-1}
$KORR$	factor to correlate $PART(j)$ $j = 1860, \ldots, 1980$ with $PART(1970)$; $0 \leq KORR \leq 1.5$	–
$ksocp$	soil organic carbon production coefficient	yr^{-1}
$ksoc$	soil organic carbon depletion coeficient	yr^{-1}
k_T	kd derived from T	yr^{-1}
k_{Pp}	kd derived from Pp	yr^{-1}
L	litter pool	$g \cdot m^{-2}$
LD	litter depletion	$g \cdot m^{-2} \cdot yr^{-1}$
LP	litter production	$g \cdot m^{-2} \cdot yr^{-1}$
NPP	net primary productivity	$g \cdot m^{-2} \cdot yr^{-1}$
$NPPT$	NPP derived from T	$g \cdot m^{-2} \cdot yr^{-1}$
$NPPPp$	NPP derived from Pp	$g \cdot m^{-2} \cdot yr^{-1}$
$PART$	part of m under agricultural use	%
P	total above and below ground phytomass	$g \cdot m^{-2}$
$SOCP$	soil organic carbon production	$g \cdot m^{-2} \cdot yr^{-1}$
POC	particulate organic C export	$g \cdot m^{-2} \cdot yr^{-1}$
Pp	average annual precipitation	$mm \cdot yr^{-1}$
T	mean annual temperature	°C
WL	clearing probability derived from land–use in the period 1950–1980	–
WP	clearing probability derived from NPP	–
WS	clearing probability derived from soil fertility F	–
WU	clearing probability derived from land–use in surrounding grid elements	–
$yield$	yield of crops in m (weighted mean)	$g \cdot m^{-2} \cdot yr^{-1}$

The major emphasis on subdivisions was placed on all possible pools and transfers of the biosphere. The variables required to run the biospheric subroutine of the OBM are listed in Table 4 together with the symbols used in the model and the relevant quantitative unit if applicable.

The following symbols in Table 4 require more explanation:

1. Grid element, m: The OBM is geographically referenced. The area of the globe is divided in $2.5^\circ \cdot 2.5^\circ$ elements, the size of which varies with latitude. Each grid element size is kept in a file and subdivided into carbon flux relevant portions for vegetation, soils, agriculture, and land use parameters. Only grid elements containing land surfaces are included yielding 2433 elements.

2. Climate variables, Pp and T: Mean annual temperature

TABLE 5. Soil factors $f(o)$ of 45 soil types of the FAO–UNESCO Map which characterize the fertility of the main soil units of the world.

soil unit o	$f(o)$	soil unit o	$f(o)$
Gleyic Acrisol	0.87	Chromic Luvisol	1.04
Humic Acrisol	0.22	Ferric Luvisol	1.65
Orthic Acrisol	0.70	Gleyic Luvisol	2.78
other Acrisol	0.60	Orthic Luvisol	0.85
Dystric Cambisol	0.94	Dystric Histosol	1.39
Eutric Cambisol	1.69	Humic Podzol	0.56
Humic Cambisol	1.58	Orthic Podzol	0.61
Gelic Cambisol	0.76	other Podzols	0.55
Luvic Chernozem	0.99	Calcaric Regosol	1.61
Dystric Podzoluvisol	0.83	Eutric Regosol	1.14
Xanthic Ferralsol	0.55	Gelic Regosol	0.91
Humic Gleysol	0.47	other Regosol	1.20
Gelic Gleysol	0.57	Orthic Solonetz	0.59
other Gleysol	0.50	Vitric Andosol	1.65
Lithosol	0.52	Haplic Xerosol	0.42
Lithosol–Yermosol	1.14	Yermosol	0.30
Fluvisol	0.49	Haplic Yermosol	0.66
Eutric Fluvisol	0.61	Luvic Yermosol	0.23
other Fluvisol	0.55	Takyric Yermosol	0.09
Haplic Kastanozem	1.96	Orthic Solonchak	0.44
Luvic Kastanozem	1.61	Takyric Solonchak	0.03
other Kastanozem	1.80	other Solonchak	0.20
Albic Luvisol	0.34		

TABLE 6. Mean stand ages $AGE(w)$ of woody material and factors $H(h)$ for separating herbaceous NPP for the 31 formations of the potential natural vegetation. $H(w)$ is simply calculated from $1 - H(h)$. Values were derived using our data base DATAVW, gaps were filled using the method of ranking. Ranked units carry an *. From Esser [1984].

formation	mean stand age $AGE(w)$, years	herbaceous factor $H(h)$
tropical moist lowland forest	200	0.37
tropical dry lowland forest	*80	0.4*
tropical mountain forest	*80	0.37*
tropical savanna	5	0.98
tropical paramo woodland	*10	0.95*
tropical paramo grassland	*1	1.0*
Puna formation	*2	1.0*
subtropical evergreen forest	200	0.37
subtropical deciduous forest	150	0.44
subtropical savanna	5	0.90
subtropical halophytic formation	5	0.9
subtropical steppe and grassland	1	1.0
temperate steppe and meadow	1	1.0
subtropical semidesert	15	0.85
xeromorphic formation	20	0.4
desert (tropical, subtropical, cold)	*5	0.85*
Mediterranean sclerophyllous forest	*100	0.4*
Mediterranean shrub and woodland	15	0.47
temperate evergreen (coniferous) forest	130	0.29
temperate deciduous forest	150	0.38
temperate woodland	25	0.53
temperate shrub formation	*10	0.85
temperate bog and tundra	5	0.48
boreal evergreen coniferous forest	100	0.34
boreal deciduous forest	100	0.38
boreal woodland	*15	0.6*
boreal shrub formation	*10	0.85*
woody tundra	10	0.7
herbaceous tundra	2	1.0
azonal formation	*5	0.6*
mangrove	50	0.29

and average annual precipitation were derived from the WMO standard net of climate stations (data set from NCAR, Boulder), the World Atlas of Climate Diagrams [Walter & Lieth, 1960 ff.], and a data collection with climatic zone maps by Müller [1982]. The data were interpolated and corrected for the mean elevation of the grid element. Mean annual values for temperature and sum total of precipitation were used to calculate Net Primary Productivity.

3. Soil, O: The Soil Map of the World [FAO–Unesco, 1974 ff.] was digitized on the grid. Areas covered by the soil units were expressed as percentage of grid element area. For 45 soil units the "soil factors" changing the NPP values calculated from climatic variables are shown in Table 5.
4. Potential natural vegetation, AV, AVE: The Vegetation Atlas published by Schmithüsen [1976] was digitized in percentage of each grid element. The 172 vegetation units of the atlas were summarized to the 31 formations in Table 6. The Table contains also the values for mean stand age and the percentage of herbaceous and woody litter derived from the produced biomass.
5. Land use, field crops, and crop yield, c, AA, ***yield***: World Atlas of Agriculture [Instituto Geographico de Agostini, 1969, 1971, 1973], FAO Production Yearbooks [FAO–Unesco, 1980 ff.], FAO Agro- Ecological Zones Project Results [FAO–Unesco, 1978 ff.]. The country-related statistical data were distributed on the grid by use of the information supplied by the Agro–Ecological Zones Project. Factors to convert crop yields to NPP are shown in Table 7.
6. Land–use changes: Factor array for 121 countries and each year of the period 1860–1980, based on data published by Richards et al. [1983], and the evaluation of 934 Landsat images for the years 1972–1980 carried out by Esser & Lieth [1986], see Table 8.

TABLE 7. Factors $fa(c)$ for the calculation of NPP (dry weight) of some agricultural crops from the yields (fresh weight). From Aselmann & Lieth [1983].

crop	$fa(c)$
wheat	2.15[1]
barley	2.12[5]
oats	3.44[1]
rye	2.65[5]
maize	2.46[1]
sorghum	3.44[1]
pearl millet	3.44[1]
paddy rice	2.86[1]
sugar beets	0.32[6]
cotton	5.00
sugar cane	0.44
mandioka (cassava)	0.64[4]
potato	0.54[2]
batate	0.55[3]
soybeans	2.46[1]
Phaseolus beans	2.86[1]
rape	2.81[5]

dry matter content in yield:
[1] 86%, [2] 32.5 %, [3] 30%, [4] 35%, [5] 87%, [6] 23%

The Model Concept and Numerical Equations

The Osnabrück Biosphere Model (OBM) is geographically referenced. It rests on a global grid of 2.5 degrees longitude and latitude. For the land area except Antarctica 2433 grid elements were obtained. All model functions were specified to be valid for all for all driving forces over the entire span of the global array.

The OBM does not follow the common practice of other models to assign certain flux values to vegetation units etc. for two reasons:

1. The number of vegetation units is much higher than the number of available data for the fluxes. Even for the key flux, net primary productivity, only about 100-1500 values were available globally, very few of them in remote regions.
2. the correlation of ecological fluxes with vegetation units is, therefore, much weaker than the correlation of such fluxes with the array of environmental influences (climate, soil, human impacts, and others). Therefore the OBM uses those influences as driving functions for fluxes. Then the extensive global climate data sets (bases on more than 7,000 WMO standard stations) can be used to drive the model.

The model is coupled to ocean models through a single–pool atmosphere or each grid element exchanges CO_2 with the respective atmospheric pool of a compartmented atmosphere, depending on the problem to be investigated. For complex experiments the OBM is coupled to the ocean and atmosphere models developed at the MPI for Meteorology in Hamburg. A generalized flow chart fot the OBM including the coupling to technosphere, atmosphere, ocean and lithosphere is shown in Figure 2.

TABLE 8. Areal changes of the natural vegetation of South American countries as derived from the evaluation of 934 Landsat scenes. Column (3) gives the percentage of the country area covered by the overlapped parts of scene pairs. Column (5) gives the respective estimate by Richards et al. [1983] as mean of the period 1958–1978. From Esser & Lieth [1986].

country	net area changed [$km^2 \cdot a^{-1}$]	evaluated area [%]	related phytomass changes (C) [$10^6 t \cdot a^{-1}$]	mean 1958–78 [Richards et al. 1983] [$km^2 \cdot a^{-1}$]
Brasil	6,689	76	80.	7,708
Argentina	1,516	85	12.	1,880
Paraguay	790	80	9.5	227
Venezuela	357	65	3.	960
Bolivia	76	40	1.	713
Chile	66	90	0.3	104
Columbia	3	30	0.02	280
Peru	0	34	–	649
Ecuador	0	6	–	187
Uruguay	0	95	–	−48
total	9,497	72	106.	12,660

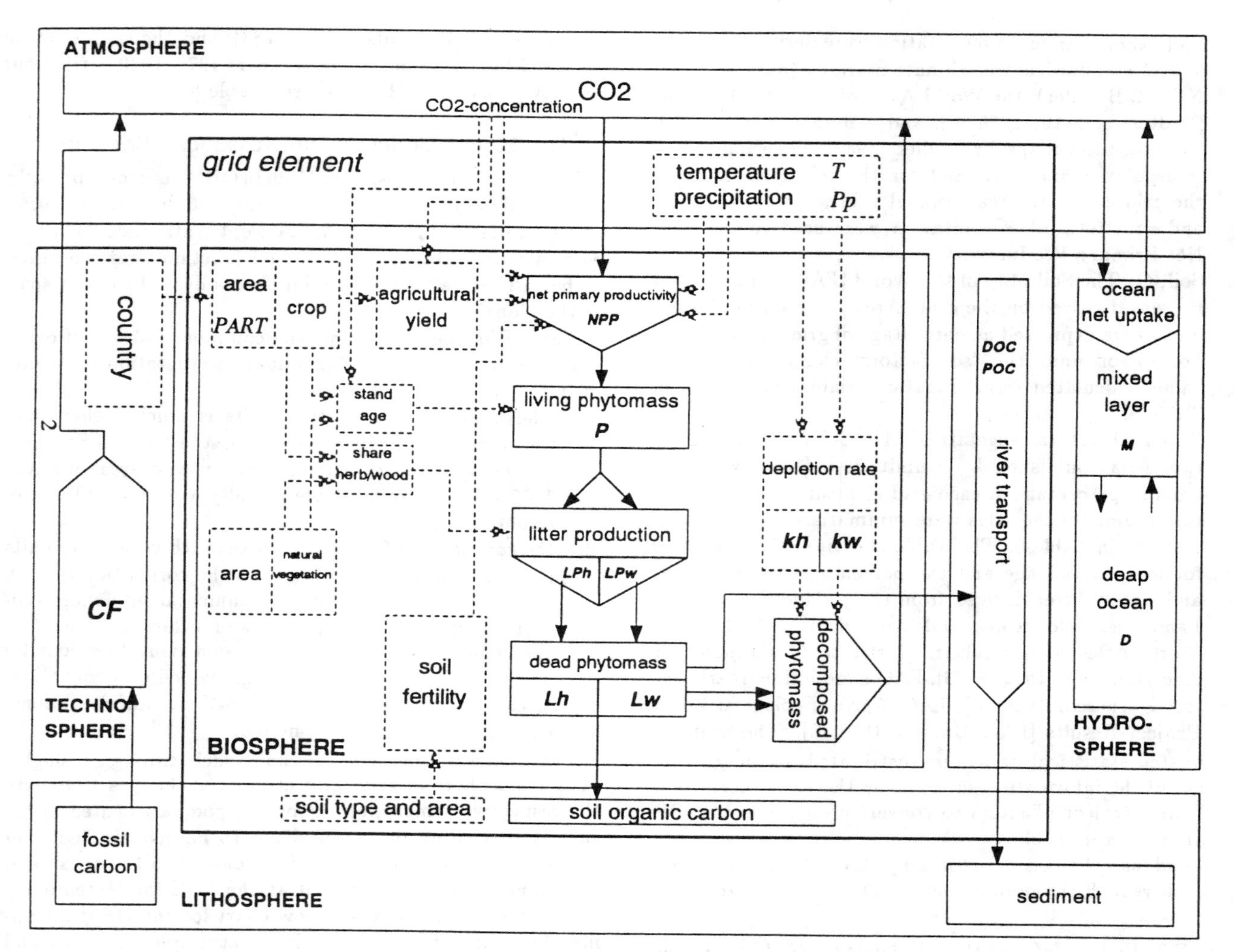

Fig. 2. Flow chart of the Osnabrück Biosphere Model with connections from and to Geosphere, Technosphere, Atmosphere and Hydrosphere. Central flow unit is Assimilation and Decomposition which varies quantitatively depending on enviromental constraints, utilizatian by men and intrinsic biosphere properties.

On the grid element level the usual systems analysis was carried out to determine the relevant carbon pools, fluxes, and control variables for the fluxes. The carbon pools are balanced by the fluxes. The system of differential equations is integrated using a standard forth-order Runge-Kutta method. It was attempted to replace the fluxes by functions of the pools. In this case, the obtained coefficients were expressed as depending on driving functions. If this was not possible, i.e. for the flux net primary productivity, then fluxes were derived from driving functions directly by means of empirical relationships. The procedure to calculate the annual carbon balance for each grid element is explained in the following paragraphs. Thirtythree equations are presented to assess uptake and release of carbon for each grid element into the atmosphere and for the transfer of carbon through rivers. The exchange between ocean and atmosphere and the input of carbon dioxide into the atmosphere by fossil fuel consumption were calculated as single figures per pool total and annum as specified earlier.

For the calculation of the global carbon flux and pool sizes a set of equations was developed which compute the model fluxes respectively the flux-coefficients grid element. The list of variables used in the description may be found in Table 4. The mass-balance of the model pools is carried out in the usual way, either by integrating the system of differential equations by using numeric routines, or by balancing the integrated fluxes explicitly at each time step, depending on the problem. The pool values necessary as initial conditions to start the model are computed using a fixed atmospheric CO_2 concentration. The atmosphere acts mathematically as an "unlimited" carbon source to fill the pools. These "pre-run" procedure requires considerable computing time

(2,000 model years) to get the large soil pools stable and to prevent drift in the subsequent model run. The following description of the OBM equations is taken nearly verbatim from Esser [1991].

Net Primary Productivity

The total flux net primary productivity is calculated as an annually integrated two-dimensional array according to equation (1). The productivity share of the potential natural vegetation is basically supplied by the equations (2) and (3) which correspond to the original MIAMI model [Lieth, 1975]. The result is consecutively modified by influence of the soil (second line of equation (1)). The soil factor array $F(m)$ is the weighted mean (4) of the individual soil factors of each soil type shown in Table 5 found in the respective grid element.

$$NPP(j,m) = \Bigg(\min[NPPT(j,m), NPPPp(j,m)] \cdot F(m) \cdot \frac{AV(j,m)}{AG(m)} + yield(j,m) \cdot FA(m) \cdot \frac{AA(j,m)}{AG(m)} \Bigg) \cdot FCO_2(j) \quad (1)$$

$$NPPT(j,m) = 3000/\{1 + \exp[1.315 - 0.119 \cdot T(j,m)]\} \quad (2)$$

$$NPPPp(j,m) = 3000 \cdot \{1 - \exp[-0.000664 \cdot Pp(j,m)]\} \quad (3)$$

$$F(m) = \frac{\sum_{o=1}^{O(m)} f(o) \cdot AS(m,o)}{\sum_{o=1}^{O(m)} AS(m,o)} \quad (4)$$

$$FCO_2(j,m) = A \cdot [1 - \exp\{-R \cdot (CO_2(j) - 80)\}] \quad (5)$$

$$A = 1 + \frac{F(m)}{4}$$

$$R = -\ln\left(\frac{A-1}{A}\right) \cdot \frac{1}{240}$$

The individual soil factors are empirical correction factors which relate the productivity measured on a given soil type to the MIAMI productivity. The values of $f(o)$ for the major FAO–Unesco soil units [FAO–Unesco, 1974 ff.] are found in Table 5.

The third line of equation (1) supplies the productivity share of agriculturally used areas of the grid element. The term depends linearly on the yields of the main field crops in the grid element. The factor array $FA(m)$ relates the total productivity of the crops to their yields. The variables *yield* and FA are again weighted means of the crop yields and their factors $fa(c)$ in the grid element. For the factor array $fa(c)$ see Table 7.

Finally, the term in line 3 of equation (1) is the CO_2 fertilization effect. In the OBM it is calculated from the actual

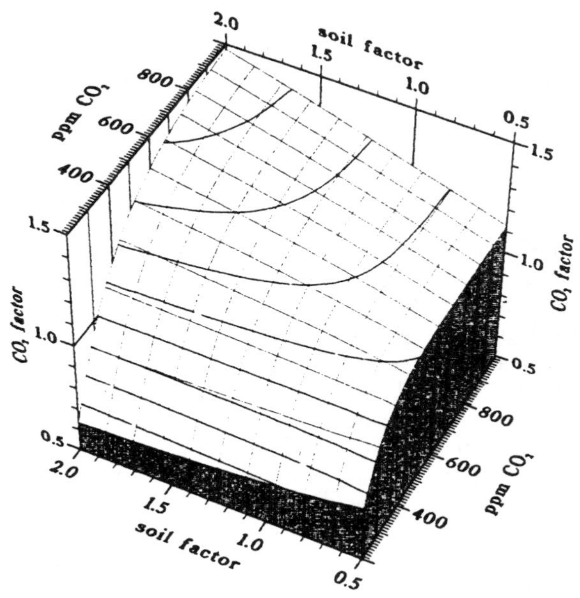

Fig. 3. Plot of function (5), $FCO_2(j,m) = f[CO_2(j), F(m)]$, to calculate the CO_2 fertilization factor on net primary productivity. The CO_2 compensation concentration for entire plants is assumed to be 80 $\mu l \cdot l^{-1}$ ($FCO_2 = 0$ irrespective of F(m)). For 320 $\mu l \cdot l^{-1}$ CO_2 concentration $FCO_2 = 1.0$, since it was assumed that the data used to calibrate the functions (2) and (3) were measured mainly during the 1960s when CO_2 was about 320 $\mu l \cdot l^{-1}$. The function achieves saturation at 1.125 for $F(m) = 0.5$, at 1.250 for $F(m) = 1.0$, and at 1.375 for $F(m) = 1.5$. For the present atmospheric CO_2 of 355 $\mu l \cdot l^{-1}$ the function yields the factors 1.034 for $F(m) = 0.5$, 1.052 for $F(m) = 1.0$, and 1.065 for $F(m) = 1.5$.

atmospheric CO_2 concentration and the soil fertility according to equation (5). The implementation of the fertilization effect assumes that the productivity of the potential natural vegetation as well as of agricultural crops is enhanced similarly. The existence of such a fertilization effect in nature has been controversial since the beginning of carbon cycle research. In terms of plant physiology and ecophysiology, the processes are quite clear: the CO_2 concentration acts in three ways:

1. Through a direct effect of the CO_2 partial pressure in the plant cell on the enzyme kinetics of the Ribulose–1,5-bisphosphate-carboxylase/oxygenase. The quantum-efficiency of photosynthesis is influenced directly by this effect.
2. Through an indirect effect on the transpiration of a plant through the stomatal resistance. The elevated external CO_2 level raises the intracellular partial pressure of CO_2. Plants may counteract by raising the stomatal resistance to keep the internal CO_2 partial pressure constant. The reduced transpiration per unit leaf area allows the plant to increase the leaf area or to extend the vegetation period. Since most of the global land areas having mean annual temperatures above 5 °C are

limited in water supply [Esser, 1987], this effect may be the most important on a global scale.

3. The lack of minerals limits the fertilization effect. The flux net primary productivity must always be accompanied by adequate fluxes of minerals from the soil, since the ratios carbon/minerals within plant organs are constant within narrow limits [Ingestad & Lund, 1986; Ingestad & Ågren, 1988]. Minor changes of the ratios probably occur in some plant species if grown at very high atmospheric CO_2 concentrations [Overdieck et al., 1991].

Equation (5) was calibrated against physiological data. Physiological and ecophysiological investigations of the fertilization effect have been carried out at many institutions [see Overdieck et al., 1991], but their value in determining the plant behavior in natural environments is limited. The function proposed here considers soil fertility, but the water interrelations are not considered explicitly. A plot of function (5) is shown in Figure 3.

In the model the productivity is partitioned into two separate fluxes: herbaceous and woody productivity, by use of a simple linear sharing factor H:

$$NPP(i) = H(i) \cdot NPP \tag{6}$$

The factors $H(i)$ are attributes of the vegetation units. In Table 6 the values for 31 vegetation formations used in the model are given.

Litter Production

The flux litter production is assumed to be proportional to the respective source pool phytomass. Separate fluxes for herbaceous and woody material are distinguished:

$$LP(i) = klp(i) \cdot P(i) \tag{7}$$

In order to derive the factors $klp(i)$ from the mean stand age of the plant material of the vegetation unit under consideration, an equation was used, which was originally developed to calculate phytomass from net primary productivity and stand age [Esser, 1984]:

$$P = 0.59181 \cdot NPP \cdot AGE^{0.79216} \tag{8}$$

Reformulated for the biomass compartments and extended by the biomass share factors $H(i)$ this equation reads:

$$P(i) = 0.59181 \cdot \frac{NPP(i)}{H(i)} \cdot AGE(i)^{0.79216} \tag{9}$$

If we assume mature stands, the mass balance equation for phytomass may be set to zero:

$$\frac{dP(i)}{dt} = NPP(i,t) - LP(i,t) = 0 \tag{10}$$

LP is replaced by the term of equation (7):

$$NPP(i,t) - klp(i) \cdot P(i,t) = 0 \tag{11}$$

and inserting (9):

$$NPP(i,t) - klp(i) \cdot 0.59181 \cdot \frac{NPP(i,t)}{H(i)} \cdot AGE(i)^{0.79216} = 0 \tag{12}$$

Divided by $NPP(i,t)$ and solved for $klp(i)$:

$$klp(i) = \frac{H(i)}{0.59181 \cdot AGE(i)^{0.79216}} \tag{13}$$

At maturity, the results of equations (9) and after integration (10) are similar. The stand ages for woody material of the 31 vegetation formations are found in Table 6. For herbaceous material, a mean stand age of 1 year is assumed generally, except for evergreen rain forests (1.2 years) and for needle–leaved evergreen forests (2 years).

Litter Depletion

It is assumed, that the decomposed amount of litter is proportional to the litter pool:

$$LD(t) = \sum_i -k(i) \cdot L(i,t) \quad i = woody, herbaceous \tag{14}$$

The value of the coefficient $k(i)$ depends on the composition of the decomposed material (in the OBM woody and herbaceous materials are distinguished), and on the elements of the environmental vector (the OBM considers temperature and precipitation). $k(i)$ depends exponentially on the temperature while the relation to precipitation is a maximum function. The analysis of the data and derivation of the functions is described in detail by Esser et al. [1982] and Esser & Lieth [1988]. The set of equations to predict the coefficients was calibrated by data listed in Esser [1986]. The functions are (with $i = h, w$):

$$kd(h,j,m) = \min[k_T(h,j,m), k_{Pp}(h,j,m)] \tag{15}$$

$$k_T(h,j,m) = 0.1063 \cdot \exp[0.0926(T(j,m) + 6.41)] + 0.2365 \tag{16}$$

$$k_{Pp}(h,j,m) = \left(\frac{0.4436}{0.0215 + \exp[4.2 - 0.0053 \cdot Pp(j,m)]} + 5.944\right) \cdot \left(\frac{0.094}{0.7 + \exp[0.0023 \cdot Pp(j,m) - 5.05]} + 0.076\right) \cdot (1 - \exp[-0.001 \cdot Pp(j,m)]) \tag{17}$$

$$kd(w,j,m) = \min[k_T(w,j,m), k_{Pp}(w,j,m)] \tag{18}$$

$$k_T(w,j,m) = 0.037 \cdot \exp[0.0522(T(j,m) + 31.63)] - 0.0348 \tag{19}$$

$$k_{Pp}(w,j,m) = \left(\frac{0.1927}{0.021 + \exp[8.53 - 0.0095 \cdot Pp(j,m)]} + 4.9352\right) \cdot \left(\frac{0.126}{1.51 + \exp[0.003 \cdot Pp(j,m) - 4.65]} + 0.05\right) \cdot (1 - \exp[-0.001 \cdot Pp(j,m)]) \quad (20)$$

The precipitation–driven maximum functions (17) and (20) consist of three brackets: The term in the first bracket controls the ascend, the term in the second bracket the descend of the function. The third term forces the function to pass through the zero point.

Soil organic carbon production

It is assumed that the polyphenolic compounds in the litter (lignins) contribute to the soil organic carbon. Thus soil organic carbon production is part of the flux litter production:

$$SOCP(i,j,m) = ksocp(i) \cdot LP(i,j,m) \quad (21)$$

The sharing factor $ksocp(i)$, which represents the lignin content of the respective material, is set to 0.30 for woody material and 0.11 for herbaceous material. In the OBM it does not depend on the grid element or the model year.

Soil Organic Carbon Depletion

It is assumed that lignin compounds have depletion coefficients $ksoc$ which amount to 1% of the coefficients of fresh herbaceous litter as given by the equations (15)–(17):

$$ksoc(j,m) = 0.01 \cdot kd(h,j,m) \quad (22)$$

Leaching of Dissolved and Particulate Organic Carbon (DOC and POC)

Dissolved and particulate organic carbon commonly occur as organic carbon freights in each body of fresh water. It could be shown by Esser & Kohlmaier [1990], who used data which were acquired by Degens et al. [1982, 1983, 1985], that the DOC and POC freights of a river mainly depend on its water discharge. In contrast, the correlations of transported DOC and POC with the extent of agricultural areas or their change in the watershed of the river were zero or even slightly negative. The authors established two equations to derive DOC and POC from the discharge of water from a m^2 of the watershed of a river:

$$DOC = 0.0064 \cdot DIS \quad (23)$$

$$POC = 0.0022 \cdot DIS \quad (24)$$

Since in many regions with high precipitation discharge correlates with precipitation, DIS may be replaced by mean annual precipitation.

Land–Use Changes and Deforestation

The direct human influence on vegetation acts upon the variables $yield(j,m)$ and relative agricultural area $AA(j,m)/AG(m)$ in equation (1), and the stand age $AGE(i)$ in equations (13) and (9), respectively.

During the period modelled, the area $AA(j,m)$ of a grid element under agricultural use may increase by deforestation, or decrease by reforestation. The OBM uses historical data for the years up to 1980 and a scenario for the decades after 1980 to obtain the relative agricultural area $PART(j,m) = AA(j,m)/AG(m)$ of each grid element in each model year. The land–use submodel is described in detail by Esser [1989]. Here only a short overview can be given.

Land–use changes 1860–1980. The basic land use information originates from the World atlas of Agriculture [Instituto Geographico de Agostini, 1969, 1971, 1973], which was digitized on the grid of the model. Five principal land–use classes where distinguished including arable land, permanent crops, meadows and pastures, forests, and rough grazing land. The land–use categories were digitized in terms of percentages of grid element area, which resulted in a finer areal resolution than suggested be the grid.

The areas digitized from the World Atlas of Agriculture were assumed to be valid for the year 1970. In order to model the changes within the period 1860–1980, the land–use areas given by the atlas are modified by a factor matrix $KORR(j,m)$:

$$PART(j,m) = PART(1970,m) \cdot KORR(j,m) \quad (25)$$

The factor matrix $KORR(j,m)$ was calculated by use of a data base published by Richards et al. [1983]. Since data were only available for the years 1860, 1920, 1930, 1950, 1961–1965, 1978, we interpolated the values for the years between in order to achieve at a complete sequence of $KORR$ factors. To test this approach, we began to evaluate Landsat images for land–use changes [Esser & Lieth, 1986]. 934 scenes of South America were evaluated to find out the changes of the areas influenced by humans between the beginning 1970s and the beginning 1980s. In Table 8 the results are compared with the OBM modeled changes based on the Richards data base.

Land–use development after 1980. The land–use scenario pertaining after the year 1980 must not be understood as an attempt to predict future developments, since scenarios are less predictions than logical assumptions about future developments. The basic concept is to derive $PART(j,m);\ j = 1981,\ldots,2100$ from a logistic function for each grid element:

$$PART(j,m) = PART(1980,m) + \frac{PART_{max} - PART(1980,m)}{1 + \exp[c_1(m) - c_2(m)]} \quad (26)$$

The parameters $c_1(m)$, $c_2(m)$ determine slope and turning-point (i.e. year of maximum clearing rate) of the function.

$PART_{max}$ is a user-defined upper limit of the relative agricultural area of the grid element. The year of the turning point of equation (26) is given by:

$$j_{TR}(m) = \frac{c_1(m)}{c_2(m)} \qquad (27)$$

c_2 is a user-supplied value. If it is set to 0.2, the time necessary to clear the grid element in the limits $0.05 \leq PART(j,m) \leq 0.95$ is 30 years. In order to obtain c_1 the year j_{TR} must be determined by use of the global scenario functions. On a global level, the two-step calculation includes:

1. ranking the grid elements according to their individual "clearing probability".
2. determination of the number $\Delta NT(j)$ of grid elements which have their turning point of function (26) in the model year j by use of the global logistic function.

The number of turning points occurring in the time span $j-(j-1)$ is:

$$NT(j) = \frac{2433 + NT_0}{1 + \left(\frac{2433+NT_0}{NT_0} - 1\right)\exp[-\beta(j-(j-1))]} \qquad (28)$$

This number of grid elements is taken successively from the ranked sequence of grid elements and their $c_1(m)$ are calculated using equation (27) with $j_{TR}(m) = j$.

Grid elements are ranked according to their "clearing probability", which is defined as the product of four individual probabilities:

$$WR(m) = WU(m) \cdot WP(m) \cdot WS(m) \cdot WL(m) \qquad (29)$$

The four individual probabilities are based on the following assumptions:

1. Clearing will occur earlier the more agricultural areas were present in the surrounding grid elements in 1970:

$$WU(m) = \frac{1}{N} \cdot \sum_{n=1}^{N} \left(\varrho(n) \cdot \frac{1}{K(n)} \cdot \sum_{k=1}^{K(n)} PART(1970,k) \right) \qquad (30)$$

n	order of circles of grid elements surrounding element m. $n = 1,\ldots,N$. The author used $N=5$.
$k(n)$	current number of grid element in circle of nth order. $k(n) = 1,\ldots,K(n)$.
$\varrho(n)$	weighting factor; $\varrho(n) = \frac{1}{2^{n-1}}$

2. The higher the natural productivity $NPP(1970,m)$ the earlier clearing occurs:

$$WP(m) = \frac{NPP(1970,m) - NPP_{lim}}{NPP_{max} - NPP_{lim}} \qquad (31)$$

NPP_{max}	the highest value of the matrix $NPP(1970,m)$
NPP_{lim}	The minimum natural productivity to allow economic use of the area, in the scenario set to either 0 or 500 [g·m^{-2}·yr^{-1}]

3. The higher the soil fertility expressed through the soil factor array $F(m)$ (see equation (4) and Table 5), the earlier clearing occurs:

$$WS(m) = \frac{F(m) - F_{min}}{F_{max} - F_{min}} \qquad (32)$$

F_{max}	highest value of the array $F(m)$
F_{min}	lowest value of the array $F(m)$

4. The steeper the clearing increased in the respective grid element in the period 1950–1980, the earlier it is continued after 1980:

$$WL(m) = \frac{KORR(1980,m)}{KORR(1950,m)} \cdot \frac{1}{WL_{max}} \qquad (33)$$

WL_{max}	the maximum value of the quotient-array of (33)
$KORR$	see equation (25).

Model Results

The first encouraging result of our modelling was our ability to simulate correctly the increase of CO_2 in the atmosphere between 1860 to the present [see Figure 4]. The predicted increase of the atmospheric CO_2 concentration in the period 1958 to present meets the results of the measurements at the Mauna Loa observatory [Keeling 1986]. It is necessary here to repeat again that those results must not be taken as absolute validations of the model. The model equations for the fluxes include assumptions which could not be proved on a large scale. Since the modelled net influence on the atmosphere is the balance of several large fluxes which largely compensate at annual time scales (i. e., fertilization effect and deforestation), minor changes of the assumptions may change the model results.

Therefore, this simulation was criticized for not being validated enough. We are therefore still engaged in further attempts to validate importent sections of our model.

The important questions concerning the global carbon cycle of the terrestrial biosphere include:

1. What is the importance of the CO_2 fertilization effect on a global scale and how is it geographically distributed.
2. How contribute the tropical deforestation and other anthropogenic influences on vegetation to the carbon balance.

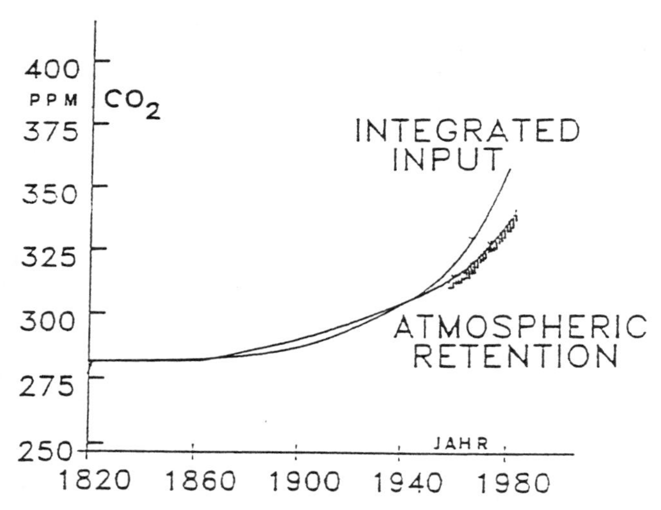

Fig. 4. The course of annual CO_2 increase in the atmosphere since the last century, simulated with the Osnabrück Biosphere Model and the MPI Hamburg Ocean Model by Maier-Reimer. Figure taken from Lieth [1988].

3. How important is the climate—carbon cycle feedback. In the past years several results of the OBM which refer to those problems have been published [Esser, 1986, 1987, 1990]. In summary the authors can draw some fundamental conclusions.

After deforestation the productivity of field crops is generally much lower than the productivity of the replaced natural vegetation. Table 9 shows that the productivity of agricultural crops is almost generally as low as 10–20% of the natural productivity. Only industrialized countries which maximized the crop yields per unit area of their fields without considering expenses achieve 100% or more of the natural productivity. Thus deforestation does not only reduce the phytomass (through reduction of the stand–age) but also net primary productivity. As a consequence, the input fluxes

TABLE 9. Comparison of the productivities of natural vegetation and agricultural crops for some tropical and extratropical countries. The productivities of the agricultural crops were calculated from the yields (FAO Production Yearbooks) by use of conversion factors given by Aselmann & Lieth [1983]. Values mean dry matter (for carbon multiply by 0.45). From Esser [1990].

country	agricultural productivity $[g \cdot m^{-2} \cdot yr^{-1}]$	natural productivity $[g \cdot m^{-2} \cdot yr^{-1}]$	ratio agric./natural
Zaire	180	1960	0.10
Kenya	350	1300	0.13
Niger	150	890	0.17
Kampuchea	310	1800	0.17
Bolivia	280	1500	0.19
Brazil	310	1620	0.19
Spain	510	750	0.68
FR Germany	1130	1190	0.95
Belgium, Luxemburg	1290	1210	1.07

into the litter and soil organic carbon pools are reduced. Since the depletion coefficients are basically unchanged, the pools are reduced. On the other side, reforestation does not restore the production of litter and soil organic carbon immediately, since immature forests produce less litter. Thus reforestation of old fallow land, which has a productivity flux similar to that of the regular natural vegetation of the site, may even reduce the soil and litter carbon pools during the first decades.

The CO_2 fertilization effect increases the net primary productivity. As a consequence, the phytomass of mature stands grows if the stand age is unchanged. There are no experimental results available to support the assumption, that the stand age is not changed. If we assume, that the stand age of mature stands is not changed, a carbon sink is induced in the growing phytomass as long as the atmospheric CO_2 is rising. Subsequently, the productions of litter and soil organic carbon are also rising and thus the target pools. Deforestation reduces the fertilization–induced sink in the phytomass pool (via the reduced stand age) as well as in the litter and soil organic carbon pools (via the reduced productivity).

Those processes are interacting. They may be deconvoluted by use of a model like the OBM. Results of such a deconvolution are shown in Table 10. Column (a) demonstrates the regional importance of the fertilization effect. It is highest in the humid tropics where net primary productivity and phytomass are high. Moist subtropical or temperate forests fall behind the tropics. In the boreal coniferous forest regions (i.e. Siberia, Canada, Alaska, Northern Europe) the fertilization effect hardly exceeds 20 g carbon per year and m^2, frequently it is much lower. This is a consequence of the low productivity and phytomass in those regions. The zonal sums of the phytomass change caused by the fertilization effect are listed in Table 10 for the reference year 1980. In total, 1.25 Gt of carbon were additionally sequestered in the global phytomass, 0.76 Gt or 60% between 20° south and 20° north. In the same table column (b–a) land use changes including deforestation are balanced against the fertilization effect. It is obvious that in regions with high deforestation rates net losses from the phytomass pool occur, although the fertilization effect alone may be high. An example for that is southeastern Brazil. In the southeastern United States carbon fixation by reforestation is more important than the fertilization effect. Globally, the 1.25 $Gt \cdot yr^{-1}$ of carbon sequestered by the fertilization effect are reduced to 0.45 $Gt \cdot yr^{-1}$, which means that 0.8 $Gt \cdot yr^{-1}$ were emitted from the living phytomass by land–use changes and deforestation globally. In the latitudinal belt 30° south to 30° north, the losses due to deforestation were 0.9 $Gt \cdot yr^{-1}$. In contrast, in the belt 30° to 50° north a net fixation of carbon of 0.2 Gt annually occurs probably due to reforestation (table 10 $b-a$). Therefore, the total global losses due to deforestation (northern reforestation not taken in account) amount to 1.1 $Gt \cdot yr^{-1}$. In a recent paper Houghton [1991] gives a range for the flux from changes in land use of 0.6 to 2.6 Gt C for

TABLE 10. Zonal integrals of the fertilization effect (a), the phytomass change due to fertilization effect and land–use changes together (b), and the total biospheric balance including phytomass, litter, and soil (c). The difference $(b - a)$ is the net effect of deforestation, reforestation, and land–use changes alone. Values are in 10^6 tons of carbon per year for zones of 10° latitude. Negative values are losses from the biosphere. The reference year is 1980.

hemisphere	latid. zone	fertilization	phytomass change	deforestation, reforestation	biosphere total
		(a)	(b)	$(b - a)$	(c)
north	70–80	1.1	1.0	−0.1	5.6
	60–70	40.4	34.9	−5.5	64.2
	50–60	80.7	62.6	−18.1	10.6
	40–50	102.5	171.1	68.6	143.9
	30–40	67.5	178.1	110.6	116.9
	20–30	80.5	−113.6	−194.1	−50.8
	10–20	82.3	−95.4	−177.7	−71.5
	0–10	233.9	79.5	−154.4	144.7
south	10–0	295.0	180.5	−114.5	215.0
	20–10	147.7	67.5	−80.2	115.4
	30–20	90.5	−102.3	−192.8	−57.3
	40–30	19.4	−20.9	−40.3	−31.0
	50–40	9.3	1.8	−7.5	1.6
	60–50	1.1	0.9	−0.2	1.9
total net biosph. sink		1251.9	445.7		608.2
total net biosph. source				−806.2	

1980, although he stated in a personal communication that he subjectively considers 1.0 to 2.0 Gt C more likely. The OBM result is on the low side of this range.

If the secondary effects of CO_2 fertilization and land–use changes on soil organic carbon and litter are also considered (see Table 10 column (c)), then the global biospheric carbon balance is raised to 0.6 Gt·yr^{-1} fixation in 1980. It is interesting to compare the situations in the eastern United States and in western Europe. Obviously, considerable losses from soils and litter occur in the U.S., while the phytomass is a strong sink. This is probably because reforestation and low productive agriculture are present in the same grid elements. In western Europe, the phytomass is hardly increasing, while the soils and litter are sinks due to the increasing productivity of the agriculture in the EC. In east Asia and parts of the tropics, soils and litter are sinks. In general, the net effects on litter and soil are small (≈ 0.2 Gt·yr^{-1} in 1980).

Thus, we have the strange situation that although tropical deforestation was already a problem at the beginning of the 1980s, the fertilization effect together with reforestation over–compensated the losses. The terrestrial biosphere has been a sink for atmospheric CO_2.

Tans et al. [1990] compared observed atmospheric concentrations of CO_2 with boundary layer concentrations over the oceans, which were calculated with the transport fields generated by a general circulation model for specified source–sink distributions. In the model, the observed north–south atmospheric concentration gradient could be only maintained if sinks of CO_2 were greater in the extratropical northern than in the southern hemisphere. The authors concluded that there must be a terrestrial sink at temperate latitudes of the northern hemisphere to balance the carbon budget and to match the north–south gradient of atmospheric CO_2. We can compare their suggestion with the results of the OBM in this paper as listed in Table 10 column (c). The extratropical northern hemisphere (> 20° N) is a net sink of about 0.3 Gt C annually, while the similar southern latitudes are a source of about 0.08 Gt C. This northern hemispheric sink given by the OBM is considerably lower than the respective value estimated by Tans et al. [1990] of 2.0–2.7 Gt·yr^{-1}. It can not be excluded at present, that the northern hemispheric sink might be larger than calculated by the OBM since this model does not yet consider the possible fertilization effect of anthropogenic emissions to the air besides CO_2, i.e. NO_x compounds.

Prior to Tans et al. [1990], Keeling et al. [1989] came to the conclusion, that the northern hemispheric sink necessary to explain the observed meridional atmospheric CO_2 gradient is only about 1 Gt·yr^{-1} C. They assumed the sink to be most likely mainly oceanic due to the observed $\delta\, ^{13}C$ profiles.

According to the OBM, the terrestrial biosphere has not always behaved as a sink for carbon. Time series of the pools calculated by the OBM for the period 1860–1980 (Table 11) show that until 1950 the phytomass decreased due to deforestation. The fertilization effect was too small to compensate for the total amount of carbon losses.

The litter and soil organic carbon show minor changes during the period 1860–1980. Therefore they probably contributed little to the carbon sequestering in the biosphere. Schlesinger [1990] comes to a similar conclusion in his recent paper in Nature. The results for the year 1980, which have been discussed in detail, suggest that the reduction of soil organic carbon which dominated since about 1900 has recently stopped.

In total, the terrestrial biosphere has been a small source for carbon (≈ 26 Gt) between 1860 and 1980. Further back in time the fertilization effect was certainly negligible, while

TABLE 11. Development of the global sums of the major carbon pools of the terrestrial biosphere and the atmosphere in the period 1860–1980. Results of a model run with standard climate (no changes). The terrestrial pools are stated in Gt C, the atmosphere in $\mu l \cdot l^{-1}$. From Esser [1990].

year	pools					accum. fluxes	
	atmo-sphere $[\mu l \cdot l^{-1}]$	phytomass nat.	phytomass agri.	litter	organic C in soil [GtC]	deforest.	fertil. effect
1860	285.0	668	1.6	91	1536	−0	0
1870	286.7	663	1.7	92	1537	−7	2
1880	288.6	658	1.9	92	1538	−15	5
1890	290.7	654	2.1	92	1539	−22	8
1900	293.1	650	2.2	92	1539	−30	12
1910	296.1	646	2.4	92	1538	−37	15
1920	299.4	643	2.5	92	1538	−44	19
1930	303.2	639	2.6	92	1538	−54	25
1940	307.0	636	2.8	92	1537	−63	31
1950	311.6	633	2.9	92	1536	−73	38
1960	317.6	634	3.0	92	1535	−81	47
1970	326.5	637	3.1	92	1533	−89	58
1980	338.8	644	3.2	92	1531	−96	73

the deforestation rate already reached high values. Deconvolution of atmospheric CO_2 data including ice core data suggests a total net biospheric CO_2 source of about 93 Gt C in the period 1740–1980 [Keeling et al. 1989]. This value is in good agreement with the OBM results as listed in Table 11, if we assume, that the net biospheric losses from deforestation prior to 1860 were in the same order of magnitude for each decade as in 1860–1870.

TABLE 12. Some model results based on three scenarios in the period 1860–1980. The atmospheric CO_2 concentration values given by the model are compared with measured data. Standard climate means the data set mentioned on page 6, temperature increase relates to a change of the mean annual temperature coupled to the atmospheric CO_2 to yield +0.8 °C in the period 1860–1980, no CO_2 fertilization assumes that NPP is uninfluenced by the atmospheric CO_2 level. Values are net changes in Gt carbon. Negative prefix indicates losses of the respective pool to the atmosphere. The CO_2 values are in $\mu l \cdot l^{-1}$. Mauna Loa means ΔCO_2 for the period 1958–1980 at the Mauna Loa observatory, Hawaii [Keeling, 1986]. Ice core value after Friedli et al. [1986]. The ocean flux was calculated using a box diffusion model [Oeschger et al., 1975].

	standard climate	temperature increase	no CO_2 fertilization
		$[\mu l \cdot l^{-1}$ (ppmv)]	
CO_2 1860	285	283	261
antarctic ice cores 1860		283 ± 3	
ΔCO_2 1958–1980	+22.6	+23.3	+37.4
Mauna Loa 1958–1980		+23.2	
changes 1860–1980		[Gt C]	
phytomass	−23	−20	−83
litter	+1	−0.5	−5
soil organic C	−5	−12	−26
ocean	+76	+78	+112
atmosphere	+114	+117	+164
fossil source		−163	

The results discussed so far were based on a standard climate data set with constant temperatures since 1860. It is probably more realistic to consider the global temperature rise since 1860. This would allow to investigate the temperature feedback effects. In this scenario the author used a temperature raise of 0.8 °C for the period 1860–1980, which is slightly higher than he observed (0.4–0.6 °C). The results are included in Table 12. The depletion fluxes of soil organic carbon and litter are enhanced while the losses from phytomass are 3 Gt less. This is due to the higher fertilization effect as consequence of the steeper CO_2 increase in the atmosphere. Although the effects are small, it nevertheless may be stated that there is a positive feedback of temperature changes.

The results in Table 12 also point out that it is probably unrealistic to ignore the CO_2 fertilization effect. If a fertilization effect is ignored it is necessary to assume an atmospheric CO_2 concentration of 261 $\mu l \cdot l^{-1}$ in 1860 to get the right value of 339 $\mu l \cdot l^{-1}$ for the year 1980. In contrast, the results of ice core analysis suggest 283 ±3 ppm for 1860 [Friedli et al., 1986], which is in excellent agreement with model results which consider CO_2 fertilization and the temperature rise since 1860. For this scenario the CO_2 concentration of the atmosphere in the period of the Mauna Loa records (1958–1980) increased by 23.3 $\mu l \cdot l^{-1}$. The Mauna Loa records indicate an increase of 23.2 $\mu l \cdot l^{-1}$. This is not an absolute argument for the existance of a fertilization effect, since any other sink with a similar time behavior including an under–estimated oceanic uptake could have balanced the atmospheric CO_2. Further model studies are necessary which include seasonal cycles and the stable isotope fractionation. Especially a correct simulation of the seasonal atmospheric CO_2 variation would be very helpful for the validation of our model.

Acknowledgments The authors wish to thank Charles D. Keeling, Scripps Institution of Oceanography, La Jolla (California), R. A. Houghton, The Woods Hole Research Center, Woods Hole (Massachusetts), William H. Schlesinger, Duke University, Durham (North Carolina), and Pieter Tans, NOAA, Boulder (Colorado), for careful reviews and helpful comments on earlier manuscript. We thank the German Science Foundation (DFG), the German Federal Environmental Agency (UBA), the German Federal Ministry of Research and Technology (BMFT), and the Commission of the European Communities (CEC) for financial supported of our work. I also wish to thank all colleagues in the German Climate Program and in the IGBP who promoted and encouraged the work by their assistance and critical discussion. Finally we thank Prof. Dr. Hantel for the invitation to present our model at the Hann Symposium in Vienna. We thank also Mrs. McCallum for preparing this manuscript and Mr. J. Berlekamp to proofread and copyedit the MS, and prepare it for publication.

REFERENCES

Aselmann, I., Lieth, H., 1983, The implementation of agricultural productivity into existing global models of primary productivity. In: Degens, Kempe, Soliman (eds.),Transport of carbon and minerals in major world rivers, Part 2, *Mitt. Geolog.-Paläontolog. Inst. Univ. Hamburg, SCOPE/UNEP Sonderband, Heft 55*, 107–118.

Degens, E. T. (ed.), 1982, Transport of Carbon and Minerals in the Major World Rivers, Part 1. *Mitt. Geol.-Paläont. Inst. Univ. Hamburg, SCOPE/UNEP Sonderband 52.*

Degens, E. T., Kempe, S., and Soliman, H. (eds.), 1983, Transport of Carbon and Minerals in Major World Rivers, Part 2. *Mitt. Geol.-Paläont. Inst. Univ. Hamburg, SCOPE/UNEP Sonderband 55.*

Degens, E. T., Kempe, S., and Herrera, R. (eds.), 1985, Transport of Carbon and Minerals in Major World Rivers, Part 3. *Mitt. Geol.-Paläont. Inst. Univ. Hamburg, SCOPE/UNEP Sonderband 58.*

Esser, G., 1984, The significance of biospheric carbon pools and fluxes for the atmospheric CO_2: A proposed model structure. *Progress in Biometeorology 3*, 253–294.

Esser, G., 1986, The carbon budget of the biosphere — structure and preliminary results of the Osnabrück Biosphere Model (in German with extended English summary). *Veröff. Naturf. Ges. zu Emden von 1814, New Series Vol. 7*, 160 pp. and 27 Figures.

Esser, G., 1987, Sensitivity of global carbon pools and fluxes to human and potential climatic impacts. *Tellus 39B*, 245–260.

Esser, G., 1989, Global land-use changes from 1860 to 1980 and future projections to 2500. *Eclogical Modelling 44*, 307–316.

Esser, G., 1990, Modelling global terrestrial sources and sinks of CO_2 with special reference to soil organic matter. In: Bouwman, A. F. (ed.), *Soils and the Greenhouse Effect*, Chapter 10., 247–261, John Wiley & Sons, Chichester · New York · Brisbane · Toronto · Singapore.

Esser, G., 1991, Osnabrück Biosphere Model: structure, construction, results. In: Esser, G., Overdieck, D. (eds), 1991, *Modern Ecology: Basic and Applied Aspects*, 679–709, Elsevier, Amsterdam · London · New York · Tokyo.

Esser, G., Aselmann, I., Lieth, H., 1982, Modelling the carbon reservoir in the system compartment "litter". In: Degens, Kempe (eds.), Transport of carbon and minerals in major world rivers, Part 1, *Mitt. Geolog.-Paläontolog. Inst. Univ. Hamburg, SCOPE/UNEP Sonderband, Heft 52*, 39–58.

Esser, G., Kohlmaier, G., 1990, Modelling terrestrial Sources of Nitrogen, Phosphorus, Sulfur, and organic Carbon to Rivers. In: Degens, Kempe, Richey (eds.), *Biogeochemistry of Maior World Rivers, SCOPE 42*, Chapter 14, John Wiley & Sons, Chichester · New York · Brisbane · Toronto · Singapore.

Esser, G., Lieth, H., 1986, Evaluation of climate relevant land surface characteristics from remote sensing. *Proc. ISLCP Conference, Rome, ESA Publ. SP-248* (May 1986), 205–211.

Esser, G., Lieth, H., 1988, Decomposition in tropical rain forests compared with other parts of the world. In: Lieth, Werger (eds.), *Tropical rain forest ecosystems, Ecosystems of the World Vol. 14B*, Elsevier Amsterdam.

Esser, G., Overdieck, D. (eds), 1991, *Modern Ecology: Basic and Applied Aspects*, Elsevier, Amsterdam · London · New York · Tokyo.

FAO-Unesco, 1978 ff., Report on the Agro-Ecological Zones Project. *World Soil Resources Report 48/1 ff.*, Food and Agricultural Organization of the U. N., Rome.

FAO-Unesco, 1980 ff., Production yearbooks, Vol. 33 ff. *FAO Statistics Series No. 28 ff.* Food and Agricultural Organization of the U. N., Rome.

Friedli, H., Lötscher, H., Oeschger, H., Siegenthaler, U., Stauffer, B., 1986, Ice core record of the $^{13}C/^{12}C$ ratio of atmospheric CO_2 in the past two centuries. *Nature 324*, 237–238.

Guderian, R. (ed.), 1985, Air pollution by photochemical oxidants. *Ecological Studies 52*, Springer-Verlag Berlin · Heidelberg · New York · Tokyo.

Houghton, R. A., 1991, Tropical deforestation and atmospheric carbon dioxide. *Climatic Change* (in press).

Ingestad, T., Lund, A.-B., 1986, Theory and techniques for steady-state mineral nutrition and growth of plants. *Scand. J. For. Res. 1*, 439–453.

Ingestad, T., Ågren, G. I., 1988, Nutrient uptake and allocation at steady-state nutrition. *Physiol. Plant. 72*, 450–459.

Instituto Geographico de Agostini, 1969, 1971, 1973, *World Atlas of Agriculture*, Novara (Italy).

Keeling, C. D., 1986, Atmospheric CO_2 concentrations — Mauna Loa Observatory, Hawaii 1958-1986. NDP-001/R1 Carbon Dioxide Information Centre, Oak Ridge, Tennessee (regularly updated).

Keeling, C. D.; Bacastow, R. B.; Carter, A. F.; Piper, S. C.; Whorf, T. P.; Heimann, M.; Mook, W. G.; Roeloffzen, H., 1989, A three-dimensional model of atmospheric CO_2 transport based on observed winds: 1. Analysis of observational data. In: Peterson, D. H. (ed.), Aspects of climate variability in the Pacific and the western Americas. *Geophysical Monograph 55*, 165–235, American Geophysical Union.

Keeling, C. D.; Piper, S. C.; Heimann, M., 1989, A three-dimensional model of atmospheric CO_2 transport based on observed winds: 4. Mean annual gradients and interannual variations. In: Peterson, D. H. (ed.), Aspects of climate variability in the Pacific and the western Americas. *Geophysical Monograph 55*, 305–363, American Geophysical Union.

Lieth, H., 1972, Modelling the primary productivity of the world. *Nature and Resources 8 (2)]*, 5–10, UNESCO, Paris.

Lieth, H., 1975, Historical survey of primary productivity research. In: Lieth, Whittaker (eds.), *Primary productivity of the biosphere, Ecological Studies 14*, 7–16, Springer-Verlag, New York, Heidelberg, Berlin.

Lieth, H., 1975, Modeling the primary productivity of the world. In: Lieth, Whittaker (eds.), *Primary productivity of the biosphere, Ecological Studies 14*, 237–283, Springer-Verlag, New York, Heidelberg, Berlin.

Lieth, H., 1978, Patterns of primary production in the biosphere. *Benchmark papers in Ecology 8*, Dowden, Hutchinson and Ross, Stroudsburg Penn.

Lieth, H., 1978, Vegetation and CO_2 changes. In: Williams (ed), *Proceedings of the IIASA Workshop*, 103–110, Pergamon press, Oxford, New York.

Lieth, H., 1984, Net primary production deduced with the Hamburg Model from climate change predictions with the GCMs for elevated CO_2 Scenarios. In: Lieth, Fantechi, Schnitzler (eds), *Interaction between Climate and Biosphere, Progress in*

Biometereology, Vol 3, 335–343, Swets & Zeitlinger, Lisse (The Netherlands).

Lieth, H., 1985, Biomass pools and primary productivity of natural and managed ecosystem types in a global perspective.*INTECOL Bulletin 1985 11*, 5–7.

Lieth, H., 1988, Kopplung des Klimas mit der Biosphäre. *VDI Berichte 703*, 151–179.

Lieth, H.; Esser, G., 1982, Zur Modellierung der Beziehung zwischen globaler Netto–Primärproduktivität und Umweltfaktoren. III. Arbeitstagung Umweltbiophysik der DDR vom 25. bis 29. März 1981 in Templin. In: Unger, Schuh (eds), *Umweltstress, Wiss. Beiträge 1982/35 (P 17)*, 303–321, Halle (Saale).

Lieth, H.; Esser, G., 1985, The attempt to simulate the global carbon flux from 1860 to 1981 using the Osnabrück Biosphere Model. *Mitt. Geol.–Paläontolog. Inst. Univ. Hamburg, SCOPE/UNEP Sonderband 58*, 137–144.

Limbert, D.W.S., 1991, Weather events of 1990 and their consequences. *WMO Bulletin Vol. 40 (4)*, 328–351.

Maier–Reimer, E.; Hasselmann, K., 1987, Transport and storage of CO_2 in the ocean — an inorganic ocean–circulation carbon cycle model. *Climate Dynamics 2*, 63–90.

Müller, M. J., 1982, Selected climatic data for a global set of standard stations for vegetation science. In: Lieth (ed.), *Tasks for vegetation science, Vol. 5*, Dr. W. Junk Publ., The Hague, Boston, London.

NCAR, National Center for Atmospheric Research, Data tape documentations TD–9645 and TD–9618. Boulder, Co., U.S.A.

Oeschger, H.; Siegenthaler, U.; Schotterer, U.; Gugelmann, A., 1975, A box diffusion model to study the carbon dioxide exchange in nature. *Tellus 27*, 168–192.

Ramanathan, V., 1987, Climate–chemical interactions and effects of changing atmospheric trace gases. *J. Geophys. Res. 25*, 1441–1482.

Richards, J. F., Olson, J. S., Rotty, R. M., 1983, Development of a data base for carbon dioxide releases resulting from conversion of land to agricultural uses. Institute for energy analysis, Oak Ridge Ass. Universities, ORAU/IEA–82–10(M); ORNL/TM–8801.

Rotty, R., 1983, Distribution of and changes in industrial carbon dioxide production. *J. Geophys. Res. 88 (C2)*, 1301–1308.

Schlesinger, W. H., 1990, Evidence from chronosequence studies for a low carbon–storage potential of soils. *Nature 348*, 232–234.

Schmithüsen, J., 1976, Atlas for Biogeography. Meyers Grosser Physischer Weltatlas 3, Bibl. Inst. Mannheim, Wien Zürich.

Tans, P. P.; Fung, I. Y.; Takahashi, T., 1990, Observational constraints on the global atmospheric CO_2 budget. *Science 247*, 1431–1438.

FAO–Unesco, 1974 ff., Soil Map of the World, 1:5,000,000. Vol. I–X, Unesco Paris.

Walter, H., Lieth, H., 1960 ff., World Atlas of Climate Diagrams. Gustav Fischer Verlag Jena, GDR.

Wuebbles, D. J., Edmonds, J., 1988, A Primer on Greenhouse Gases. Report for United States Dept. of Energy TR040, DOE/NBB–0083.

G. Esser, Department of Botany, University of Giessen, Heinrich–Buff–Ring 38, 6300 Giessen, FRG.

H. Lieth, Department of Biology, University of Osnabrück, P.O. Box 4469, 4500 Osnabrück, FRG.

Diurnal Variations in the Water Vapor Budget Components over the Midwestern United States in Summer 1979

ABRAHAM ZANGVIL[1], DIANE H. PORTIS, AND PETER J. LAMB[2]

Cooperative Institute for Mesoscale Meteorological Studies, The University of Oklahoma
Norman, Oklahoma

The water vapor budget for summer 1979 is studied for a rectangular area (1300 x 750 km) centered over east central Illinois that extends across most of the Midwestern United States (surface to 300 hPa). The budget is calculated for consecutive 12 hour periods between 00 and 12 UT, 12 and 00 UT etc. Area averaged evapotranspiration is obtained as the residual of the water vapor budget equation. A pronounced diurnal variation is found in the moisture flux divergence, evapotranspiration, time change of total precipitable water, and to a lesser extent in precipitation. The vertical structure of the diurnal variation in some moisture budget-related quantities is also examined on the basis of differences between 00 and 12 UT observations. The vertical p-velocity exhibits diurnal variations from the surface to 300 hPa with a maximum amplitude near 700 hPa, while the specific humidity has a main diurnal maximum near 850 hPa and a secondary maximum near 500 hPa. The diurnal variations of the vertical moisture advection peak at 800 hPa; those of the horizontal divergence (in presence of moisture) peak at 900 hPa; while those in the horizontal moisture advection are much weaker and confined to the layer below 850 hPa. By using an estimated value for the ratio between the night-time evapotranspiration and the total daily evapotranspiration over the region, it was possible to obtain an improved average water vapor budget for the 12 day-time hours and the 12 night time hours. Also, it was possible to get an estimate of the amplitude and phase of the moisture flux divergence first harmonic (3.2 mm day^{-1} and time of maximum near 1400 CST, respectively). By further assuming that the time of maximum evaporation is near 1230 CST it was possible to estimate the amplitude and phase of the time change of precipitable water first harmonic (3.5 mm day^{-1} and time of maximum near 1100 CST, respectively). The main source of moisture (in the moisture budget equation) for the day-time precipitation is local evapotranspiration. However, for large rainfall amounts, the vertical moist advection also plays an important role. For the night-time precipitation two sources can be identified: depletion of stored precipitable water and moist vertical advection.

INTRODUCTION

The atmospheric water vapor budget of a given region is an extremely complex function of the meteorological-hydrological processes in the region and of the soil characteristics and type of land use. Any change in one of these factors, artificial or natural, may affect the relationships between the moisture budget terms. The study of these relationships on different time scales may yield important insight into the hydrological cycle of the

[1] On leave from the Jacob Blaustein Institute for Desert Research, Ben Gurion University of the Negev, Sede Boker 84990, Israel.

[2] Also in School of Meteorology, University of Oklahoma, Norman, OK .

Interactions Between Global Climate Subsystems, The Legacy of Hann
Geophysical Monograph 75, IUGG Volume 15

region and the atmospheric processes involved. For this reason, the diurnal variations in temperature, wind, and pressure have been well documented for many regions. Very obvious also are the diurnal variations in sensible and latent heat fluxes. For the United States, the diurnal variation in precipitation has been analysed by several authors [e.g., Wallace, 1975; Landin and Bosart, 1985] and the diurnal variation in the water vapor flux divergence has been reported by Rasmusson [1966, 1967]. Furthermore, Bonner [1968] studied the diurnal variations in the low level jet over the Great Plains, while the diurnal oscillations of the central United States windfield have been documented by Hering and Borden [1962] and Reiter and Tang [1984].

Most of the above work on diurnal variations deals with only one meteorological variable. No attempt has yet been made to study the simultaneous diurnal variations of all the water vapor budget components over a given region using observational data. One of the reasons for this is the lack of upper-air data with a fine time resolution. An attempt to study the diurnal variation of some of the water vapor budget terms for the southern U.S. Great Plains has been made by Hao and Bosart [1987]. In addition, Randall et al. [1991] studied the diurnal variations of the water vapor budget components over the eastern half of North America in experiments conducted using the UCLA/GLA GCM.

It is the intention of this paper to show that, even using the existing coarse resolution (i.e., twice-daily) upper-air data, some meaningful information can be obtained about the amplitude and phase of the diurnal variations in the water vapor budget components and the interrelations between them. We focus on the Midwestern United States for May-August 1979, which experienced pronounced interweek rainfall fluctuations [Peppler and Lamb, 1989, Table 1]. Because of this strong intraseasonal variability, the results obtained are unlikely to be biased towards any extreme and should be representative of a range of May-August periods.

THEORY AND COMPUTATIONAL PROCEDURES

Theory

Following Yanai et al. [1973], the water vapor budget per unit mass of air can be written as

$$\frac{\partial q}{\partial t} + V \cdot \nabla q + \omega \frac{\partial q}{\partial p} = e - c \tag{1}$$

where q is the specific humidity, V is the horizontal wind vector, w is the vertical p-velocity, and e and c are the cloud evaporation and condensation rates per unit mass, respectively. By using the mass continuity equation, Eq. (1) can be expressed as

$$\frac{\partial q}{\partial t} + \nabla \cdot (qV) + \frac{\partial}{\partial p}(q\omega) = e - c \tag{2}$$

Vertical integration of Equations (1) and (2) with respect to pressure, and horizontal averaging over the study area, yields

$$\underbrace{\frac{1}{g}\frac{\partial}{\partial t}\int_{P_T}^{P_S} \bar{q}\, dp}_{dPW} + \underbrace{\frac{1}{g}\int_{P_T}^{P_S} \bar{V} \cdot \overline{\nabla q}\, dp}_{HA} + \underbrace{\frac{1}{g}\int_{P_T}^{P_S} \bar{\omega}\frac{\partial \bar{q}}{\partial p}\, dp}_{VA} = E - P \tag{3}$$

and

$$\underbrace{\frac{1}{g}\frac{\partial}{\partial t}\int_{P_T}^{P_S} \bar{q}\, dp}_{dPW} + \underbrace{\frac{1}{g}\int_{P_T}^{P_S} \bar{V} \cdot \overline{\nabla q}\, dp}_{HA} + \underbrace{\frac{1}{g}\int_{P_T}^{P_S} \bar{q}\, \overline{\nabla \cdot V}\, dp}_{HD} = E - P \tag{4}$$

where E and P are respectively the surface evapotranspiration and precipitation rates, P_T and P_S are respectively the pressures at the upper integration limit and the earth's surface, and an overbar denotes horizontal averaging. In deriving Eqs. (3) and (4) from Eqs. (1) and (2), the horizontal divergence of the horizontal eddy flux of moisture has been neglected following Yanai et al. [1973], McBride [1981], Hao and Bosart [1987], and many others. In the forthcoming analysis, dPW, HA, VA, and HD will be used to represent the terms that are so labelled in Eqs. (3) and (4) above.

Computations

The area for which the study is made is a 1300 x 750 km rectangle centered over east central Illinois (Fig. 1). The basic data set consists of U.S. National Weather Service (NWS) rawinsonde observations (50 mb vertical resolution) for 00 and 12 UT (May-August 1979) interpolated to a 190 km grid. The basic computations performed for this area were:

1. 12 hr area-averaged precipitation totals for 00 to 12 UT (1800 to 0600 CST, subsequently defined as "night") and 12 to 00 UT (0600 to 1800 CST, subsequently defined as "day") derived from hourly precipitation measurements at 590 stations.

2. 12 hr area-averaged vertically-integrated (surface to 300 hPa) moisture flux divergence including the separate HD and HA terms. These were obtained using a

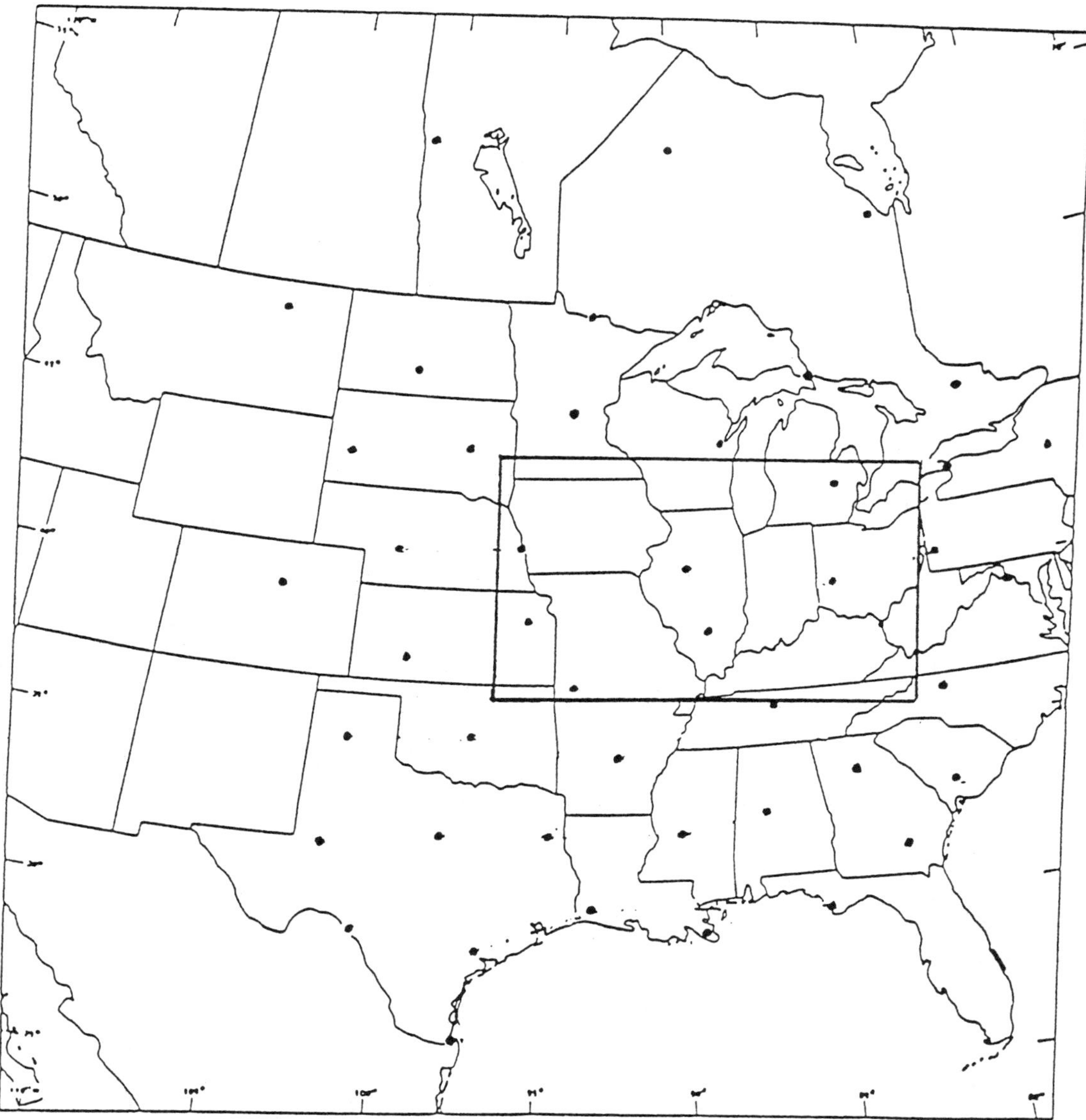

Fig. 1. Orientation map showing the rectangular region over which the study was performed. Dots mark the original rawinsonde stations used.

standard finite difference method from a Cressman-type objective analysis of rawinsonde data for the stations in Fig. 1 to a 190 km grid, for the aforementioned "day" and "night" periods based on rawinsonde data at 00 and 12 UT.

3. 12 hr area-averaged precipitable water (surface to 300 hPa) differences for 00-12 and 12-00 UT. These were obtained from calculations for the rawinsonde stations shown in Fig. 1, which were interpolated to a 190 km grid via Akima's bivariate scheme for irregular data.

4. Area averaged kinematic vertical velocity estimates for 00 and 12 UT, which were obtained by use of the modified-Barnes [1964, 1973] objective analysis method of Achtemeier [1986]. This calculates divergence fields by center-differencing objectively interpolated (190 km grid) wind fields, where the interpolation involves a Fourier integral representation of the observed wind field (consistent with Barnes) within a locally varying radius of influence for a fixed number of stations (key modification). See also Portis and Lamb [1988]. The vertical velocity was computed

using surface to 100 hPa data. To guarantee vertically integrated non-divergence, the divergence profiles were subjected to the well known O'Brien [1970] mass budget/profile form correction scheme.

5. 12 hr evapotranspiration estimates were obtained for the above "day" and "night" periods as residuals of Eq. (4).

DIURNAL VARIATIONS IN UPPER-AIR QUANTITIES DERIVED FOR 00 AND 12 UT OBSERVATION TIMES

Before discussing the analysis of the water vapor budget components in Eqs. (3) and (4) for the above half-day periods, it is desirable to examine the diurnal variations in some of the area-averaged vertically integrated measured and/or derived basic quantities available for 00 UT and 12 UT. We first consider the moisture flux divergence (MFD, i.e. HA + HD) and the precipitable water (PW). First, a simple mean of these quantities at 00 and 12 UT was calculated. The results are:

MFD: 12 UT = -1.04 mm day^{-1}
00 UT = 2.24 mm day^{-1}

PW: 12 UT = 27.82 mm
00 UT = 28.94 mm

Clearly, the MFD values for 00 and 12 UT are markedly different. This difference is statistically significant at the 0.005 level, using a standard t test. Such diurnal variations of the MFD for this region have also been documented by Rasmusson [1966] and others, based on rawinsonde data for 00 and 12 UT. When subjected to a standard t test, the difference between the PW means for 12 and 00 UT is not statistically significant. This still does not rule out the existence of a diurnal variation in PW, since the large standard deviation in PW (used in the t-test) may originate in oscillations with periods longer than 24 hours. To establish that the PW and MFD differences between 00 and 12 UT do not originate from some spurious random noise, we performed the following analyses.

Filtering

The original 00-12-00 etc. UT time series of MFD and PW were first smoothed using a three-point filter with weights of 0.25, 0.50, and 0.25. Then, the smoothed values were subtracted from the original time series. The result of this standard operation is to create a high-pass filter which reduces the amplitudes of the low frequency variation to zero as the frequency approaches zero, while retaining the full amplitude of the first diurnal harmonic. No change of phase is effected by this symmetric filter. The response (G) of this filter as a function of frequency (f) is given in Table 1. The resulting high-pass filtered time series for MFD (surface to 700 hPa) and PW (surface to 300 hPa) are shown in Figs. 2 and 3, respectively. These show clearly that the differences between the MFD and PW values for 00 and 12 UT arise from diurnal variations in these variables, which characterize most of the 1979 summer days.

TABLE 1. Response of the high pass filter (see text) as a function of frequency, f (day^{-1}).

f	G
0.0	0.0
0.2	0.095
0.4	0.345
0.6	0.655
0.8	0.905
1.0	1.0

Vertical Structure of the Diurnal Variations

To obtain some insight into the vertical structure of the diurnal variations of MFD and PW, we generated vertical profiles of the differences between the 12 and 00 UT 1979 summer averages of high-pass filtered times series of several of the moisture budget-related quantities in Eqs. (3) and (4). This used the same high-pass filter as in the preceeding Subsection. The differences between the 12 and 00 UT means have been tested for significance using a t test. The results are presented in Table 2.

Note that HA experiences small diurnal variations compared to the other quantities. On the basis of Table 2, it is concluded that the diurnal variation in HA has an amplitude of less than 0.5 of the VA (all levels) or HD (beneath 500 hPa) terms. Also, it is interesting that the specific humidity has a maximum diurnal variation near 850 hPa and not closer to the surface. However, the amplitude of the diurnal variation of HA is largest near 900 hPa, whereas the amplitude of the diurnal variation of VA has a maximum near 800 hPa. The vertical velocity exhibits a highly significant diurnal variation throughout the troposphere with a maximum amplitude near 700 hPa.

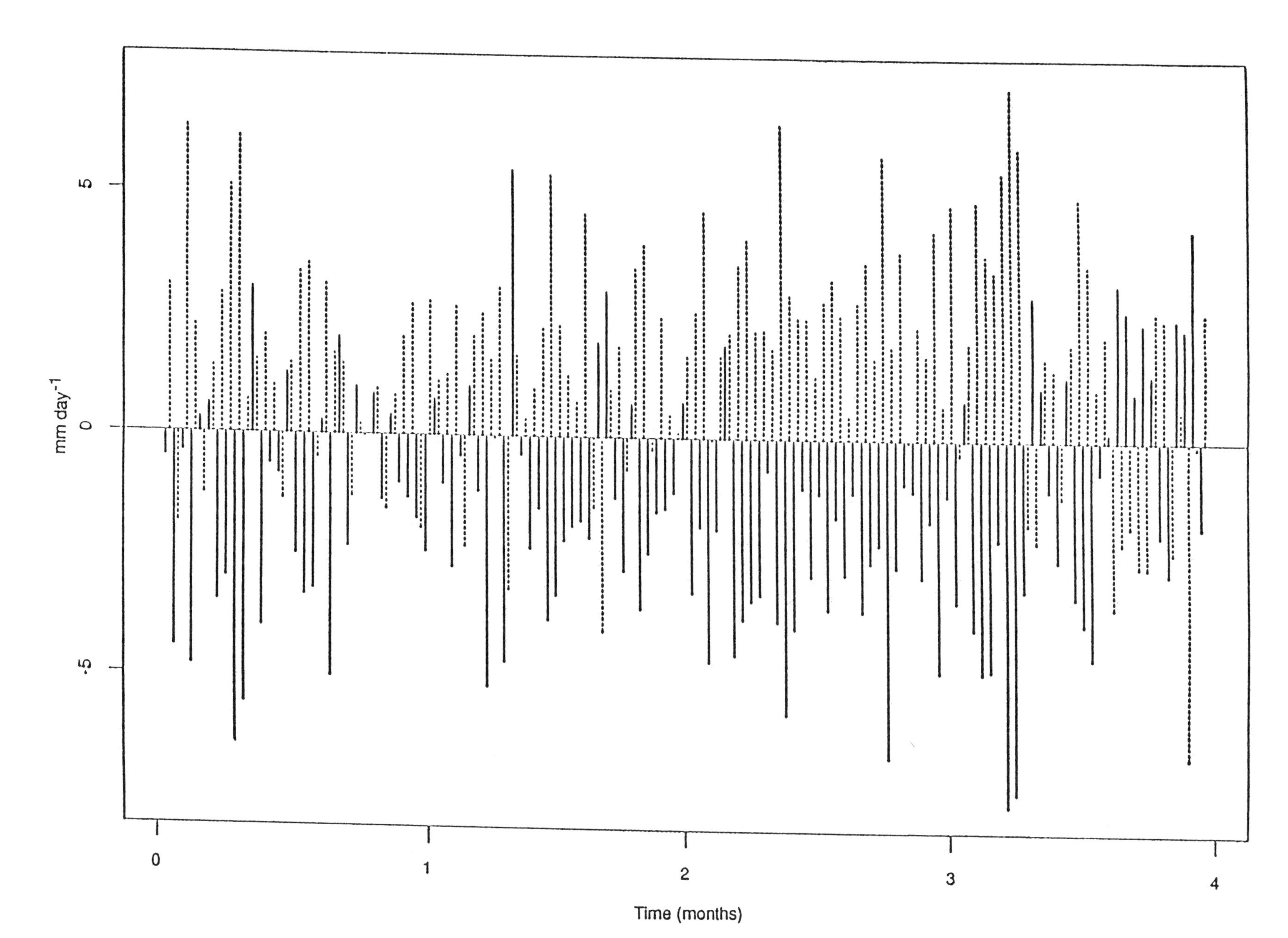

Fig. 2. Time series of the diurnal component of MFD (surface to 700 hPa). Dashed lines are 00 UT values and solid lines are values at 12 UT (see text). The times on the abscissa are denoted in months from the beginning of May to the end of August 1979.

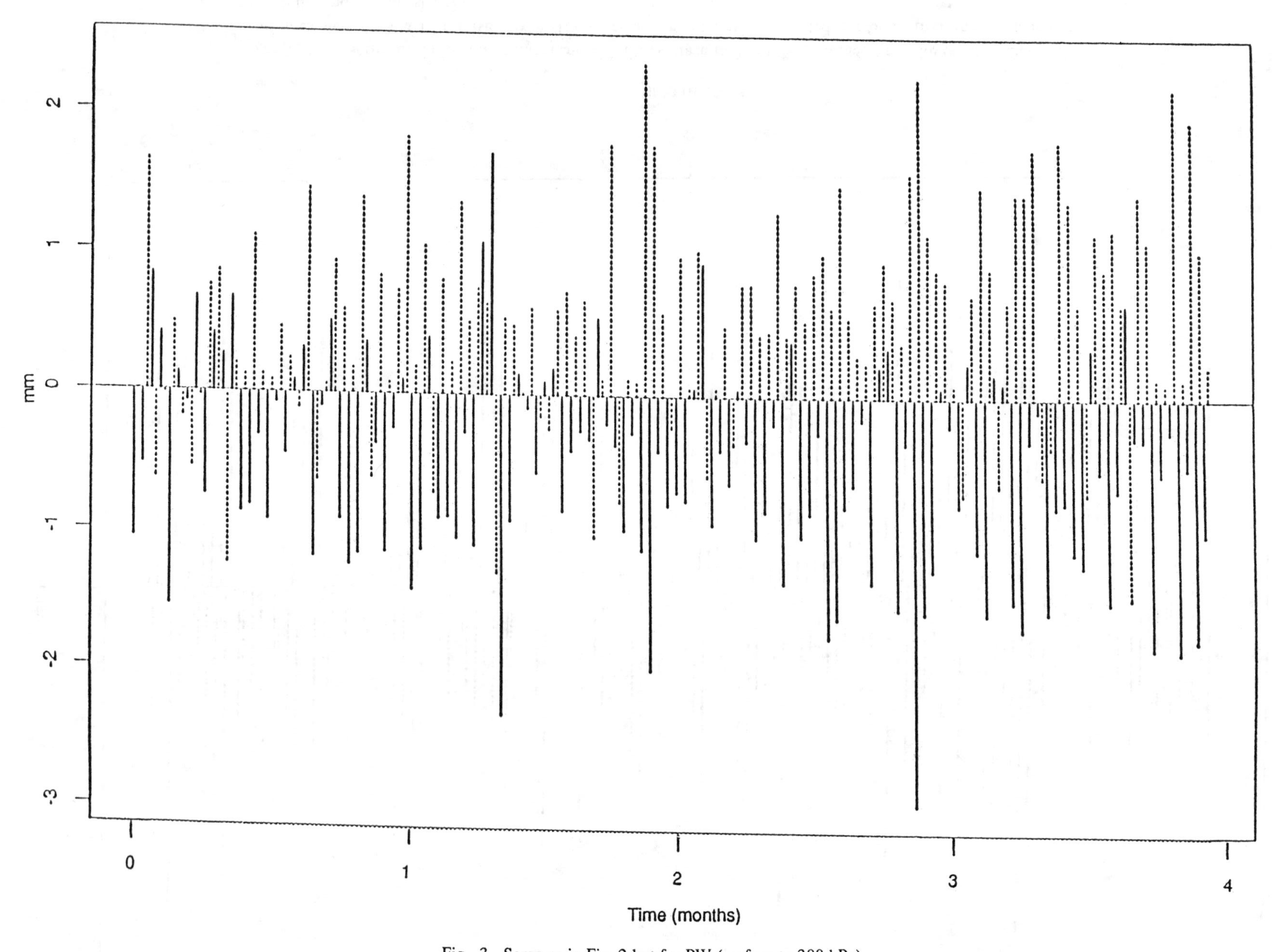

Fig. 3. Same as in Fig. 2 but for PW (surface to 300 hPa).

TABLE 2. Time mean differences (12 minus 00 UT) of high-pass filtered quantities in Eqs. (3) and (4) as a function of pressure (hPa). The dimensions of $\bar{q}\,\overline{\nabla \cdot V}$, $\bar{\omega}\, \partial\bar{q}/\partial p$, and $\bar{V} \cdot \overline{\nabla q}$ are 10^{-8} kg wv (kg air s)$^{-1}$; those for $\bar{q}$ are 10^{-4} kg wv (kg air)$^{-1}$; and $\bar{\omega}$ is given in 10^{-2} Pa s^{-1}. The columns denoted by "SIG" show the statistical significance levels of the differences in the immediately preceeding columns. N means significance level under 0.10, H means significance level above 0.005, and numbers give intermediate significance values.

Pressure	$\bar{q}\,\overline{\nabla \cdot V}$	SIG	$\bar{\omega}\frac{\partial\bar{q}}{\partial p}$	SIG	$\bar{V}\cdot\overline{\nabla q}$	SIG	$\bar{q}$	SIG	$\bar{\omega}$	SIG
350	0.04	H	-0.08	H	-0.03	0.050	-0.05	N	-2.59	H
400	0.05	H	-0.15	H	-0.05	0.050	-0.30	H	-3.03	H
450	0.03	.100	-0.22	H	-0.07	0.050	-0.53	H	-3.20	H
500	0.03	N	-0.31	H	-0.01	N	-0.94	H	-3.17	H
550	0.06	.100	-0.29	H	0.02	N	-0.58	0.010	-3.15	H
600	0.13	H	-0.30	H	-0.06	N	-0.45	0.050	-3.33	H
650	0.07	.100	-0.46	H	-0.03	N	-0.05	N	-3.59	H
700	-0.04	N	-0.62	H	-0.16	0.025	0.14	N	-3.70	H
750	-0.28	H	-1.01	H	-0.12	0.050	-0.63	0.025	-3.60	H
800	-0.83	H	-1.24	H	0.07	N	-3.71	H	-3.19	H
850	-1.58	H	-0.84	H	0.22	H	-6.61	H	-2.37	H
900	-1.95	H	-0.22	H	-0.27	H	-5.43	H	-1.27	H
950	-1.22	H	-0.05	H	-0.58	H	-0.96	0.010	-0.36	H

Diurnal Variations in the 12 hr Water Vapor Budget Components

From the standpoint of digital filtering, the precipitation, MFD, and dPW have different key characteristics. Precipitation is a mean accumulation for 12 hr periods (not a running mean). It is therefore possible, depending on the phase of the 12 hr period with respect to the phase of an existing diurnal variation, that this averaging process retains part (or all) of the amplitude of the diurnal precipitation variation. For the MFD, on the other hand, running means of two consecutive values are used and this eliminates the diurnal variation from the 12 hr averaged MFD values. Nevertheless, it is known from the previous analysis that a pronounced diurnal variation actually exists in the original observation-time MFD.

When obtaining the dPW storage term for the moisture budget, simple differences of 00 UT and 12 UT precipitable water values were used. This differencing operation acts on PW as a digital filter that essentially retains the diurnal variation but doubles its amplitude. In this case, the phase of the basic 00 and 12 UT observations with respect to an existing diurnal variation is of crucial importance. Fortunately, the 00 and 12 UT observations are made at 1800 and 0600 CST, respectively. These two times constitute a natural separation between day and night. In the following, the budget averaged from 00 to 12 UT will be considered "night" and the budget averaged from 12 to 00 UT will be termed "day". The average "day" and "night" water vapor budgets for the summer of 1979 are shown in Table 3.

A pronounced diurnal variation in dPW is evident, while the well known diurnal variation in MFD (given by HA and HD) has been eliminated by the averaging process.

TABLE 3. Mean and vertically integrated water vapor budgets (terms as defined previously). The first row for each half-day gives the budget terms in mm (12hr)$^{-1}$. The bracketed (unbracketed) numbers in the second rows express the water vapor source (sink) budget terms on a percentage basis.

	P	E	HA	HD	dPW
Day	1.61	3.07	.70	-0.39	1.16
	46.5	[88.7]	20.2	[11.3]	33.3
Night	1.85	1.18	.69	-0.39	-0.97
	72.8	[46.5]	27.2	[15.3]	[38.2]

Despite this deficiency, a marked diurnal variation is obtained for the E, calculated as the residual of Eq. (4). It is obvious that this calculated diurnal variation in E mainly resulted from the diurnal variation in dPW. It may be assumed that, since the diurnal variation in MFD is filtered out, the full amplitude of the diurnal variation in E was not obtained. We will return this point in the following Section. A t test on the "day"-"night" differences of dPW, E, and P showed those for dPW and E to be significant at the 0.005 level, while the difference in P is not statistically significant.

It is evident from Table 3 that, during the "day", E (especially) and HD are moisture sources for P, dPW, and HA. At "night", E alone cannot support the observed precipitation, and so the depletion of precipitable water in addition to the horizontal moisture convergence may act as sources for the observed precipitation. From the percentages in the square brackets, it is clear that during the "day" E is the main moisture source, while at "night" the role of stored atmospheric water vapor is almost as important. It is also interesting to note that, of the total moisture available during the "day", 46.5% is used to produce precipitation, 20.2% for dry advection (HA > 0), and 33.3% is stored in the atmosphere. At "night", on the other hand, a much larger percentage (72.8%) is used to produce precipitation.

When the water vapor budgets for the "day" and "night" periods are stratified according to rainfall intensity, we can obtain some more interesting information about the behavior of the budget (Table 4). First, the importance of E as a moisture source decreases with increasing precipitation, and is smaller during the "night" than during the "day". Second, although the percentage of the available night-time moisture contributed by dPW is about the same for all rainfall categories, in absolute terms this source increases with increasing precipitation. Finally, the percentage of available water vapor used to produce precipitation is larger during the "night" than during the "day", and increases with increasing precipitation.

TABLE 4. "Day" and "night" water vapor budgets stratified and averaged for individual half-day rainfall amounts. The first row for each half-day gives the budget terms in mm $(12\ hr)^{-1}$. The bracketed (unbracketed) numbers in the second rows express the water vapor source (sink) budget terms on a percentage basis.

	Precipitation Categories mm $(12\ hr)^{-1}$	P	E	HA	HD	dPW	No. of Cases
Day	0-1	0.42	2.75	0.20	0.62	1.51	55
		15.3	[100.0]	7.3	22.5	54.9	
Night	0-1	0.41	0.94	0.53	0.46	-0.46	49
		29.3	[67.1]	37.8	32.9	[32.9]	
Day	1-4	2.06	3.17	1.22	-0.87	0.76	56
		51.0	[78.5]	30.2	[21.5]	18.8	
Night	1-4	2.23	1.11	0.67	-0.70	-1.09	61
		76.9	[38.3]	23.1	[24.1]	[37.6]	
Day	4<	5.22	4.18	0.58	-3.01	1.39	11
		72.6	[58.1]	8.1	[41.9]	19.3	
Night	4<	5.74	2.50	1.48	-2.24	-2.48	12
		79.5	[34.6]	20.5	[31.0]	[34.4]	

From Table 3, a weak nocturnal maximum in precipitation is evident. However, it should be mentioned that the diurnal variation of precipitation is not homogeneous over our research area. Wallace [1975] found that over the northwestern portion of our area a nocturnal maximum of precipitation exists, while over the southeastern portion a late afternoon maximum is predominant. Many explanations have been offered for the observed nocturnal maximum in precipitation over some parts of the Mid-West. Wallace [1975] lists two main physical processes capable of producing diurnal variations in the boundary layer convergence that is a major contributor to this rainfall: (i) a uniform diurnal heating cycle in regions of sloping terrain [Holton, 1967; Lettau, 1967]; and (ii) changes in frictional drag associated with the diurnal variation of static stability within the planetary boundary layer [Blackadar, 1957]. One may argue on the basis of the above discussion that, in addition to the explanations listed above, that the ability of the atmosphere to utilize a higher percentage of the available moisture at night to generate precipitation (Table 4) may be important for nocturnal precipitation, and not only for the regions of nocturnal maximum. However, from moisture budget considerations alone, it is not possible to determine the cause of a nocturnal maximum in precipitation. More work is needed to elaborate this point.

AMPLITUDE AND PHASE OF FIRST DIURNAL HARMONIC IN MOISTURE FLUX DIVERGENCE AND TIME CHANGE OF PRECIPITABLE WATER

General remarks

Although no direct evapotranspiration measurements exist for our research area, it appears that our atmospheric moisture budget-based estimate of 24 hr evapotranspiration (~4.25 mm) is reasonable. The absence of direct evapotranspiration measurements was one of the reasons for undertaking this moisture budget calculation. Our above result is consistent with earlier moisture budget calculations for the same general area by Rasmusson [1971]. Furthermore, Kunkel [1992] recently applied his soil moisture model [Kunkel, 1990] over much of our research area to calculate the evapotranspiration for July 1979, with the result of 4.32 mm day^{-1} being consistent with our finding for the entire 1979 summer.

However, as mentioned in the preceeding Section, our partitioning of the calculated evapotranspiration between "day" and "night" is probably not so accurate. This is because average values of the MFD for 00 and 12 UT are used for both "day" and "night". In an attempt to correct this situation, we introduced a more realistic "day"-"night" evapotranspiration partition to replace the value of E in Table 3. The substitution of these values into Eq. (4) will yield, as residuals, more realistic "day" and "night" values of MFD (= HA+HD). In doing this, we capitalized on the relatively large amount of available information on the surface energy balance over regions of the world having a similar climate to that of the Midwest, and also for the Midwest itself. This information is found in Sellers [1965], Griffiths and Driscoll [1982], Oke [1987], Sorbjan [1989], and Kunkel [1992], among others. From these references, it is clear that most of the summer evapotranspiration in midlatitude humid climates occurs during the "day". From energy balance data published in those sources, we concluded that the ratio (R) of evapotranspiration occuring during the "night" to the daily total is very low -- in the vicinity of 0.02-0.08.

This assumption provided the basis for our computations of the amplitude and phase of the first diurnal harmonic of the MFD. Furthermore, in order to calculate the amplitude and phase of dPW, an assumption about the amplitude and phase of the first diurnal component of evapotranspiration had to be made. This also rested on the above published information on the diurnal variations of the surface energy balance. We assumed the maximum evapotranspiration to occur at about 1230 CST. We further assumed the diurnal variation of evaportranspiration to consist mainly of the first diurnal harmonic, S1. With these assumptions, the ratio R determines the amplitude of S1. That determines the amplitude and phase of the dPW S1 as well.

Calculation of Amplitude and Phase

For the purpose of demonstration, and following the procedure outlined above, the calculated 24 hr evapotranspiration has been partitioned as follows: Day = 3.95 mm and Night = 0.3 mm (ratio, R, of 0.071; see literature cited in the above Subsection). Then, the MFD for the day and night can be calculated as residuals in the moisture budget equation. The result is shown in Table 5. It is noted that, by using this method, the partition of MFD between the horizontal advection and the horizontal divergence terms cannot be determined. From Table 5 it is clear that the role of evapotranspiration as a moisture source during the "night" has considerably diminished (compared to the case of the uncorrected moisture budget, Table 3), while its role as a source of moisture during the "day" has increased.

Using the information about the 12 hr mean MFD in Table 5 and about the instantaneous values of MFD at 00 UT and at 12 UT in Section 3, and assuming (somewhat

TABLE 5. "Corrected" moisture budget using R = 0.071 (see text). Dimensions are mm $(12\ \text{hr})^{-1}$.

	E	P	MFD	dPW
Day	3.95	1.61	1.18	1.16
Night	0.30	1.85	-0.58	-0.97

crudely) that most of the diurnal variation is in the first diurnal harmonic (S1), it is possible to calculate the amplitude and phase of this first harmonic. The result is: Amplitude=3.21 mm day^{-1}, and the MFD maximum occurs near 1402 CST.

Based on (i) the above knowledge of the MFD S1 amplitude and phase, (ii) an assumed (from discussion above) evapotranspiration S1 amplitude and phase (Table 6), and (iii) the calculated amplitude and phase information of the observed precipitation S1 (amplitude 0.38 mm, time of maximum 0000 CST), it is possible to calculate the amplitude and phase of the dPW S1. The result of this calculation gave an amplitude of 3.46 mm day^{-1} and a time of the maximum dPW at 1102 CST.

An analysis of the diurnal MFD variations in the UCLA/CSU GCM experiments performed by Randall et al. [1991] showed similar results for the eastern half of the United States -- a MFD amplitude of 2.38 mm day^{-1} with the maximum MFD at 1415 CST; and a dPW amplitude of 3.6 mm day^{-1} with the maximum dPW at 1048 CST. Since summer 1979 was highly variable in a meteorological sense, this general agreement may not be coincidental. This comparison with Randall et al. [1991] is made because that study contains the only other results available on the diurnal variation of the moisture budget components over North America. The good agreement between their model results and our empirical findings is encouraging. However, we realize that more work is needed on this important issue.

Since the above results appear to depend on the particular choice of the ratio (R) between night versus total daily evapotranspiration, it is desirable to establish the sensitivity of the calculated MFD and dPW amplitude and phase to R. The results of these calculations are presented in Table 6 for the 1979 study year, and show little sensitivity to R. Preliminary results (not shown) for other May-August periods (1975, 1976) confirm this finding. This suggests that a reasonable assumption about R yields good results for the above parameters.

TABLE 6. Summary of calculated amplitude (mm day^{-1}) and phase (local time of maximum, CST) of the first diurnal harmonic, S1 in the moisture budget terms for different ratios, R. The time of maximum E is 1230 CST (see preceeding Subsection).

R	E	MFD		dPW	
	Amp.	Amp.	Phase	Amp.	Phase
0.023	6.42	3.77	1342	3.44	1107
0.048	6.09	3.48	1352	3.45	1105
0.071	5.79	3.21	1402	3.46	1102
0.094	5.47	2.95	1414	3.47	1059

CONCLUSIONS

1. Analysis of the water vapor budget for a rectangular area (1300 x 750 km) centered over east central Illinois for May-August 1979 showed marked diurnal variations in all moisture budget components (except precipitation) and related quantities such as vertical velocity and precipitable water.

2. Computation of the evapotranspiration as a residual in the moisture budget equation showed strong diurnal variations, but over-estimated the "night" time values and under-estimated the "day" time values.

3. When the "day" and "night" moisture budget terms were computed by the use of an estimated diurnal variation of the evapotranspiration, it was possible to estimate the amplitude and phase of the first diurnal harmonic, S1, of the moisture flux divergence (3.2 mm day^{-1} and time of maximum near 1400 CST) and of the time change of precipitable water (3.5 mm day^{-1} and time of maximum near 1100 CST).

4. By separately considering the moisture budgets for the "day" and for the "night", it was established that the main source of moisture for the "day" precipitation is the local evapotranspiration. For large rainfall amounts the vertical moist advection also plays an important role.

5. For the "night" precipitation, two major moisture sources were identified: depletion of stored precipitable water and moist vertical advection.

Acknowledgments. This research was supported by NSF Grants ATM 85-20877 and ATM 89-08545. We benefitted from discussions with Kenneth Kunkel, Beth Reinke and Michael Richman, from the assistance of Scott Isard and Ginger Rollins, and from the cooperation of Professor Michael Hantel.

REFERENCES

Achtemeier, G.L., On the notion of varying influence radii for a successive correlations objective analyses. *Mon. Wea. Rev., 114*, 40-49, 1986.

Barnes, S.L., A technique for maximizing details in numerical weather map analysis. *J. Appl. Meteor., 3*, 396-409, 1964.

Barnes, S.L., Mesoscale objective map analysis using weighted time series observations. *NOAA Tech. Memo.* ERT NSSL-62, 60 pp. [NTIS COM-73-10781], 1973.

Blackadar, A.K., Boundary layer wind maxima and their significance for the growth of nocturnal inversions. *Bull. Amer. Meteor. Soc., 38*, 283-290, 1957.

Bonner, W.D., Climatology of the low level jet. *Mon. Wea. Rev., 96*, 833-850, 1968.

Griffths, J.F., and D.M. Driscoll, *Survey of Climatology.* Charles E. Merrill Publishing Co., Columbus, OH, 358 pp., 1982.

Hao, W., and L.F. Bosart, A moisture budget analysis of the protracted heat wave in the Southern Plains during the summer of 1980. *Wea. and Forecasting, 2*, 269-288, 1987.

Hering, W.S., and T.R. Borden, Jr., Diurnal variations in the summer wind field over the central United States. *J. Atmos. Sci., 19*, 81-86, 1962.

Holton, J.R., The diurnal boundary layer wind oscillation above sloping terrain. *Tellus, 19*, 199-205, 1967.

Kunkel, K.E., Operational soil moisture estimation for the Midwestern United States. *J. Appl. Meteor., 29*, 1158-1166, 1990.

Kunkel, K.E., Measurement and estimates of evaporation in Midwestern United States during the 1988 drought. *Proceedings of Workshop on 1988 U.S. Drought,* University of Maryland (in press), 1992.

Landin, M.G., and L.F. Bosart, Diurnal variations of precipitation in the northeastern United States. *Mon. Wea. Rev., 113*, 989-1014, 1985.

Lettau, H., Small to large scale features of the boundary layer structure over mountain slopes. *Proc. Symposium on Mountain Meteorology*, Dept. of Atmos. Sci., Colorado State University, 1967.

McBride, J.L., Analysis of diagnostic cloud mass flux models. *J. Atmos. Sci., 38*, 1977-1990, 1981.

O'Brien, J.J., Alternative solutions to the classical vertical velocity problem. *J. Appl. Meteor., 9*, 197-203, 1970.

Oke, T.R., *Boundary Layer Climates.* Methuen; London and New York, 435 pp, 1987.

Peppler, R.A., and P.J. Lamb, Tropospheric static stability and central North American growing season rainfall. *Mon. Wea. Rev., 117*, 1156-1180, 1987.

Portis, D.H., and P.J. Lamb, Estimation of large-scale vertical motion over central United States for summer. *Mon. Wea. Rev., 116*, 622-635, 1988.

Randall, D.A., Harshvardhan, and D.A. Dazlich, Diurnal variability of the hydrologic cycle in a General Circulation Model. *J. Atmos. Sci., 48*, 40-62, 1991.

Rasmusson, E.M., Diurnal variations in the summer water vapor transport over North America. *Water Resources Res., 2*, 469-477, 1966.

Rasmusson, E.M., Atmospheric water vapor transport and the water balance of North America: Part I. Characteristics of the water vapor flux field. *Mon. Wea. Rev., 95*, 403-426, 1967.

Rasmusson, E.M., A study of the hydrology of eastern North America using atmospheric vapor flux data. *Mon. Wea. Rev., 99*, 119-135, 1971.

Reiter, E.R., and M. Tang, Plateau effects on diurnal circulation patterns. *Mon. Wea. Rev., 112*, 638-651, 1984.

Sellers, W.D., *Physical Climatology*. The University of Chicago Press, 272 pp., 1966.

Sorbjan, Z., *Structure of the Atmospheric Boundary Layer.* Prentice Hall, New Jersey, 317 pp., 1989.

Wallace, J.M., Diurnal variation of precipitation and thunderstorm frequency over the conterminous United States. *Mon. Wea. Rev., 103*, 406-419, 1975.

Yanai, M., S. Esbensen, and J.H. Chu, Determination of bulk properties of tropical cloud clusters from large-scale heat and moisture budgets. *J. Atmos. Sci., 30*, 611-627, 1973.

A. Zangvil, D.H. Portis, P.J. Lamb, Cooperative Institute for Mesoscale Meteorological Studies, The University of Oklahoma, Norman, OK 73019.

Ocean Heat Transport Across 24°N Latitude

HARRY L. BRYDEN

Woods Hole Oceanographic Institution, Woods Hole, Massachusetts

Direct estimates of ocean heat transport indicate that the meridional circulation across 24°N carries 1.2 PW of heat northward in the Atlantic Ocean and 0.8 PW of heat northward in the Pacific Ocean. The total ocean heat transport across 24°N of 2.0 PW is substantially less than recent indirect estimates of ocean transport based on satellite radiation measurements and atmospheric energy transport estimates. Heat transports in the Atlantic and Pacific Oceans are accomplished by two different mechanisms. In the Atlantic, a deep vertical-meridional circulation cell with northward flowing warm surface waters and southward flowing cold deep waters effects the northward heat transport across 24°N. In the Pacific, a horizontal circulation cell in the upper waters with northward flowing warm waters on the western side of the ocean and southward flowing colder waters in the central and eastern parts of the ocean effect the northward heat transport across 24°N. Prospects for direct determination of the ocean heat transport as a function of latitude are described with emphasis on the need to identify the principal mechanisms of ocean heat transport in each ocean basin.

INTRODUCTION

The earth gains heat from the sun in the form of shortwave incoming radiation principally in the tropical regions, but radiates heat back to space in the form of outgoing longwave radiation nearly uniformly over all latitudes. Thus, there is a net radiational heating of the earth equatorward of about 35°latitude and a net radiational cooling poleward of 35°(Figure 1). Since there is little storage of heat by the earth, particularly averaged over an annual period, compared with the latitudinal imbalances, the atmosphere and ocean must transport heat poleward from the tropical regions to the polar regions in order to maintain the observed radiation balance at the top of the atmosphere. On the basis of recent satellite radiation measurements at the top of the atmosphere, Stephens, Campbell and Vonder Haar [1981] have concluded that about 5.5×10^{15} W (or 5.5 PetaWatts = 5.5 PW) of heat must be transported poleward across 35°latitude.

There has been a long and continuing controversy over whether the ocean or atmosphere effects the bulk of the required heat transport. Jung [1956] summarized the historical views on the relative roles of the air and the sea in transferring heat toward the poles. In recent years, emphasis has returned to the ocean's role in the required heat transport due to a series of papers by Oort and Vonder Haar [Vonder Haar and Oort, 1973; Oort and Vonder Haar, 1976; Oort and Peixoto, 1983; Carissimo, Oort and Vonder Haar, 1985] in which direct estimates of the atmospheric heat transport as a function of latitude are subtracted from the total heat transport required by the satellite radiation measurements to derive the ocean heat transport as a function of latitude (Figure 2). Such analyses suggest that the maximum ocean heat transport is larger than the maximum atmospheric heat transport and that the ocean transports the majority of the heat for latitudes equatorward of 30°. Thus, meteorologists now argue that the ocean effects most of the poleward heat transport required by the observed radiation balance.

On the other hand, oceanographers who calculate air–sea exchange of heat using bulk formula and then integrate the resulting ocean heat losses spatially to determine meridional ocean heat transport as a function of latitude have reported that the ocean carries very little heat poleward. For example, Budyko's [1974] classic study of the heat budget of the earth including bulk formula estimates of air–sea heat exchange yielded maximum ocean heat transports of 1.4 and 2 PW at latitudes 20°N and S respectively, far lower than the maximum ocean heat transport of nearly 4 PW estimated by Carissimo, Oort and Vonder Haar. In the most recent summary of indirect estimates of ocean heat transport based on bulk formula calculations of ocean heat loss to the atmosphere, Talley [1984] found a maximum ocean heat transport in the Northern Hemisphere of less than 1 PW and a maximum in the Southern Hemisphere of about 2 PW. In a reversal of the Oort and Vonder Haar approach, Talley subtracted the estimates of small ocean heat transport

Interactions Between Global Climate Subsystems, The Legacy of Hann
Geophysical Monograph 75, IUGG Volume 15

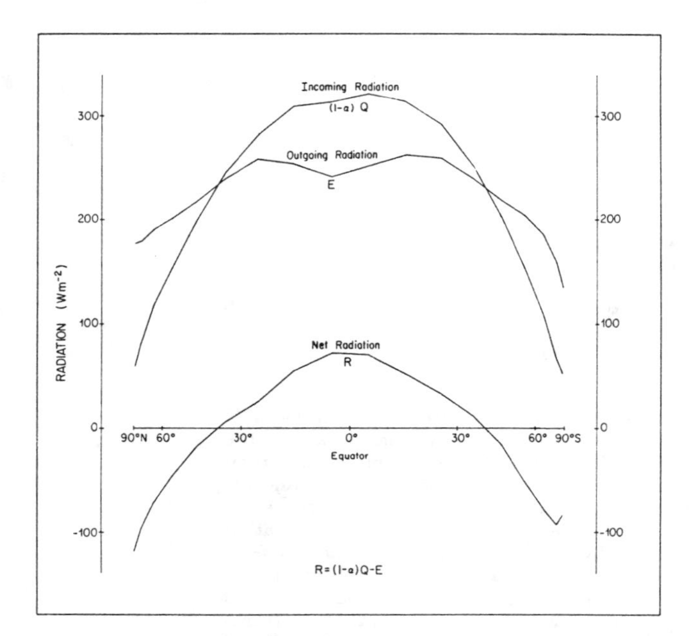

Fig. 1. Annual-averaged radiation at the top of the atmosphere (derived from tables in Stephens, Campbell and Vonder Haar, 1981). Net incoming short-wave radiation, $(1-\alpha)I$ where I is incoming radiation and α is the albedo, is concentrated in the tropical regions. Outgoing long-wave radiation, E, is more uniform with latitude. The net incoming radiation, $R=(1-\alpha)I-E$, then is positive for latitudes equatorward of about 35°and negative for latitudes poleward of 35°. A uniform bias in net incoming radiation of 9 W m^{-2} has been subtracted in order to ensure radiation balance over the globe.

based on bulk formula from the total heat transport required by the radiation balance to conclude that the atmosphere effects nearly all of the required poleward heat transport (Figure 3). Thus, the present situation is one in which meteorologists argue that the ocean is carrying most of the total meridional heat transport but oceanographers argue that the atmosphere is carrying nearly all of the total heat transport.

With increasing societal emphasis on measuring, predicting and planning for changes in global climate due either to natural variability or to anthropogenic effects, it is essential to understand how the present global climate is maintained. A basic question about the maintenance of the present climate is whether the ocean or the atmosphere carries the bulk of the poleward heat transport required to maintain the earth's radiation balance. Because of the difficulty in making measurements in the ocean, much of what we now know about ocean heat transport has been derived indirectly from atmospheric observations and analyses. There is now underway a concerted effort by oceanographers to make the necessary measurements so that the ocean heat transport can be directly determined for comparison with the atmospheric heat transport. In this paper, the initial results on direct estimates of the total ocean heat transport across 24°N are presented and compared with atmospheric heat transport and the total ocean+atmosphere heat transport required by the radiation balance across 24°N; the mechanisms of ocean heat transport in the Atlantic and Pacific Oceans across 24°N are contrasted; and the prospects for future direct estimates of ocean heat transport are outlined.

OCEAN HEAT TRANSPORT ACROSS 24°N

24°N is a good choice for estimating ocean heat transport because it marks the center of the subtropical gyre in both the Atlantic and Pacific Oceans. In each ocean at 24°N,

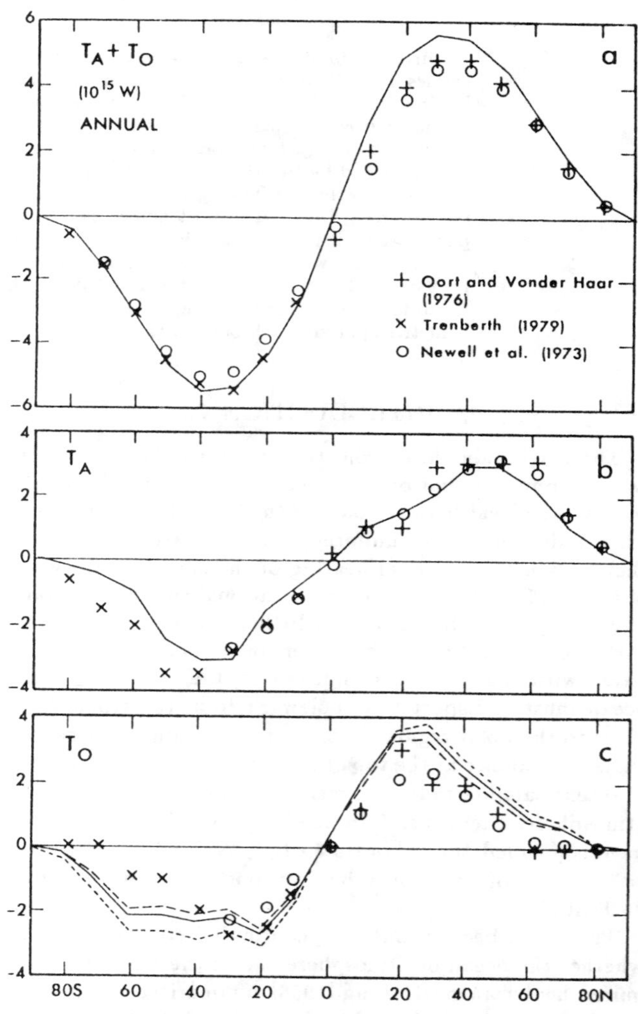

Fig. 2. Annual transport of energy (Carissimo, Oort and Vonder Haar, 1985, Figure 4): (a) total atmosphere+ocean energy transport determined from satellite radiation measurements in Figure 1 (Stephens, Campbell and Vonder Haar, 1981); (b) atmospheric transport determined from analysis of the rawinsonde network (Carissimo, Oort and Vonder Haar, 1985); and (c) oceanic transport determined as a residual between (a) and (b). In (c) the three curves for oceanic transport are based on different bias corrections to the satellite radiation measurements. Results obtained by some previous investigators are added for comparison.

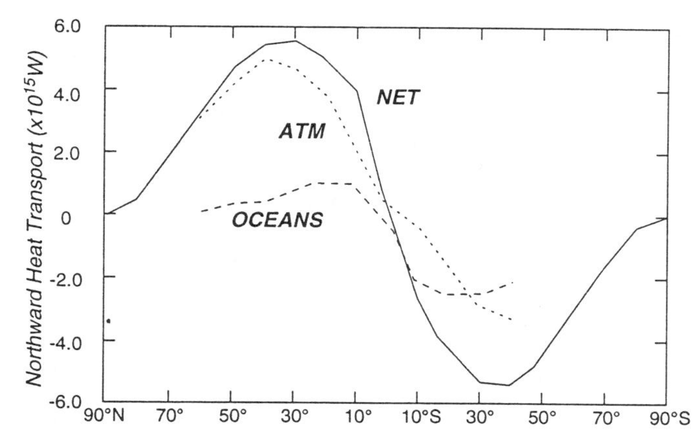

Fig. 3. Northward heat transport by the oceans, the atmosphere and the combined ocean+atmosphere (Talley, 1984, Figure 4). NET is the total atmosphere+ocean energy transport determined from satellite radiation measurements in Figure 1 (Stephens, Campbell and Vonder Haar, 1981). OCEANS is the ocean heat transport determined from integration of bulk formula calculations of air-sea heat exchange by Talley (1984). ATM is the energy transport by the atmosphere determined as a residual between NET and OCEANS.

there is a warm northward flowing western boundary current (Gulf Stream in the Atlantic and Kuroshio in the Pacific) that in popular views carries heat toward subpolar and polar regions. 24°N also marks the approximate boundary between ocean heat gain from the atmosphere in the tropics and ocean heat loss in the polar regions as revealed in bulk formula calculations of air-sea heat exchange (e.g. Figure 4 for the Atlantic). As a result, 24°N should be close to the latitude of maximum ocean heat transport. Furthermore, because there is essentially no ocean north of 24°N in the Indian Ocean, the Atlantic and Pacific heat transports make up the total global ocean heat transport across 24°N.

The ocean heat transport across 24°N in the Atlantic was first estimated about a decade ago to be 1.2 PW using a historical zonal transatlantic hydrographic section across 24°N and direct current measurements of the Gulf Stream flow through Florida Straits [Bryden and Hall, 1980; Roemmich, 1980; Wunsch, 1980]. These were the first modern direct estimates of ocean heat transport, although Bryan [1962] and Bennett [1978] had attempted similar calculations earlier that suffered from lack of information on the western boundary current transports.

A modern estimate of ocean heat transport across 24°N in the Pacific had to await the carrying out of a transoceanic hydrographic section, which occurred in 1985, and its subsequent analysis. From this section, Roemmich and McCallister [1989] and Bryden, Roemmich and Church [1991] have estimated that the Pacific Ocean circulation transports 0.8 PW of heat northward across 24°N. Combining the Atlantic and Pacific heat transports then yields a total ocean heat transport across 24°N of 2.0 PW northward.

The ocean heat transport of 2.0 PW is larger than Oort's estimate of the atmospheric energy transport across 24°N of 1.7 PW. In fact, according to Oort and Peixoto [1983], 1.0 PW of the atmospheric energy transport across 24°N is latent heat transport, which is actually a joint atmosphere-ocean process resulting from the northward flow of water vapor in the atmosphere compensated by the southward flow of freshwater in the ocean. Thus, the ocean does appear to accomplish more of the meridional heat transport than the atmosphere across 24°N. Adding the ocean and atmosphere heat transports, however, yields a total heat transport of only 3.7 PW across 24°N, far less than the 5.2 PW required by Stephens, Campbell and Vonder Haar's analysis of the earth's radiation budget.

Where is the missing 1.5 PW? Have the oceanographers missed it? Is it in the atmosphere? Or is there some bias in the satellite radiation measurements that might account for the discrepancy? Since this paper is written by an oceanographer, there will be no conclusion that the ocean could possibly be carrying an extra 1.5 PW. Oceanographers have generally stated the uncertainties in direct estimates of ocean heat transport to be ±0.3 PW [Hall and Roemmich, 1986]. Thus, it is reasonable to state that the total ocean heat transport across 24°N may be off by ±0.45 PW. But to make the ocean contribute even an extra 1 PW in meridional heat transport across 24 °N would require that oceanographers have entirely missed the equivalent of a Gulf Stream in their understanding of ocean circulation. Such myopia is unlikely.

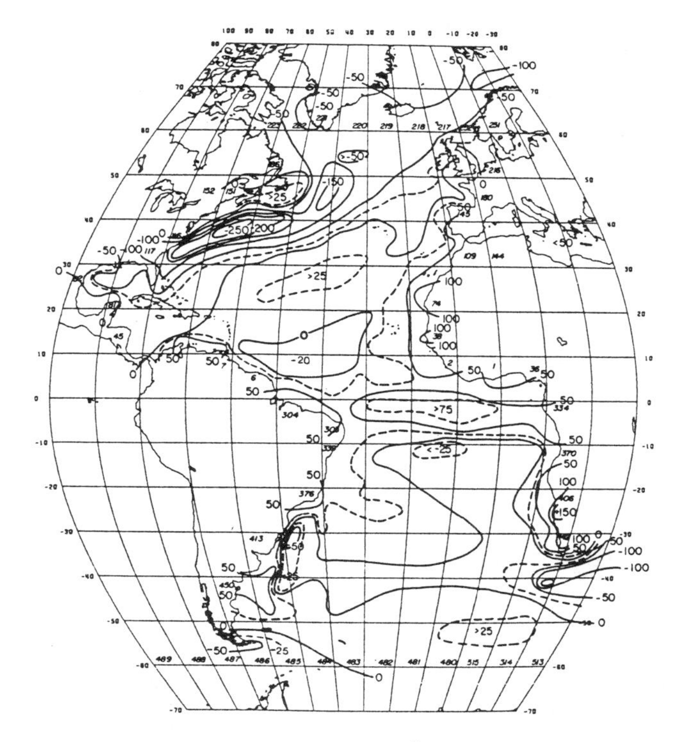

Fig. 4. Net annual heat gain (W m^{-2}) by the Atlantic Ocean as determined from bulk formula calculations of air–sea heat exchange (Bunker, 1988, Figure 4).

For the atmosphere, Masuda [1988] and Michaud and Derome [1991) have developed an alternative approach to estimating the energy transport by using state-of-the-art numerical model assimilations for the atmospheric variables. Masuda analyzed both the ECMWF (European Center for Medium-Range Weather Forecasting) and GFDL (Geophysical Fluid Dynamics Laboratory) global model assimilations for the FGGE year 1978-79 when the observational data was optimal, while Michaud and Derome emphasized the ECMWF assimilations for 1986, after various improvements particularly related to humidity parameterizations in the tropics had been made to the model. Both analyses avoid the obvious land bias in the observational network used by Oort [1983] by using numerical model forecasts to interpolate over data-sparse regions, but there is clearly a question whether the single FGGE year or the single year 1986 are representative of the long-term atmospheric circulation and heat transport contained in Oort's analysis of the 15-year period from 1958 to 1973. In any case, both Masuda and Michaud and Derome found that the atmosphere did carry about 1 PW more heat poleward at latitudes of 20°to 40°N than Oort had found in his analyses (Figure 5). The larger transport in each case is due to enhanced northward eddy sensible and latent heat transports. Michaud and Derome concluded that Oort's interpolation of the land-based rawinsonde observations over the oceans likely yields less developed structure in the transient eddies over the oceans than the model assimilations exhibit. Thus, it is possible that uncertainties in the atmosphere energy transport may be large enough to account for much of the discrepancy between direct estimates of the ocean+atmosphere energy transport and the total energy transport required by satellite radiation budget analyses. Both Masuda and Michaud and Derome, however, still estimate a northward ocean heat transport across 24°N of about 3.0 PW by subtracting their atmospheric transports from the satellite radiation budget requirements, and such a transport is still 1.0 PW, or 50%, larger than the direct estimate of ocean heat transport across 24°N of 2.0 PW.

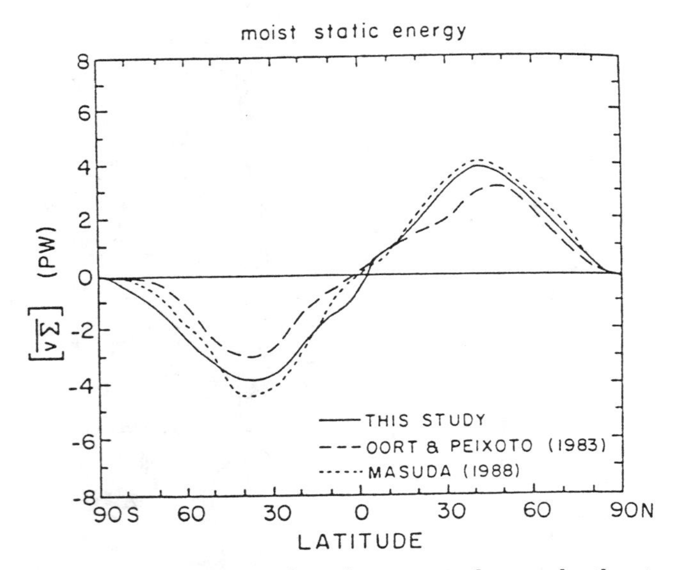

Fig. 5. Annual mean northward transport of energy by the atmosphere as a function of latitude (Michaud and Derome, 1991, Figure 5). The solid line gives the zonally and vertically integrated transport of moist static energy for the period December 1985 through November 1986. The long-dashed curve refers to the corresponding results of Oort and Peixoto (1983). The short-dashed curve refers to results obtained by Masuda (1988) using the FGGE-ECMWF main (old) analyses.

In terms of radiation budget analyses, it is fair to point out that the total direct estimate of the ocean+atmosphere heat transport of 3.7 PW across 24°N reported above is in reasonable agreement with pre-satellite radiation balance analyses which yielded about 4.0 PW across 24°N [e.g. as presented in Palmen and Newton, 1967]. Satellite radiation measurements, however, have changed our understanding of the radiation balance in many ways. For example, the solar constant has increased from 1352 to about 1365°W m^{-2} due to the new satellite radiation measurements uncontaminated by atmospheric effects [Barkstrom, Harrison and Lee, 1990]. The total meridional heat transport required by satellite radiation measurements, however, is very sensitive to small biases in both incoming and outgoing radiation measurements and how they are accounted for during analysis. Hastenrath [1982] and Hartmann, Ramanathan, Berroir and Hunt [1986] have presented various latitudinal distributions of meridional energy transport derived from different satellite systems and analysis procedures and the maximum northward energy transport varies from about 4.8 to 6.7 PW in the northern hemisphere. Hastenrath [1984] has argued that the differences may be due to real interannual variations, while Hartmann et al. [1986], although stating that "the reason for the difference is not known", imply that there may be some aliasing in sun-synchronous satellites that leads to excessively large meridional transports. Clearly, there is an uncertainty of order ±1 PW in the amount of total meridional heat transport required by the satellite radiation measurements.

Thus, the discrepancy between direct estimates of meridional transport in the ocean and atmosphere and the meridional energy transport required by satellite radiation budget analyses is likely due to the combination of an underestimate of the atmospheric energy transport, probably the result of the land bias in the observational network, and to small biases in satellite radiation measurements which can lead to large uncertainties in the required meridional energy transport. Because the ocean heat transport has only been directly estimated at a single latitude of 24°N, there is a clear need to expand our understanding of the latitudinal variation in the meridional ocean heat transport by making the necessary measurements so that the poleward ocean heat transport can be directly determined throughout the northern and southern hemispheres. Such direct estimates of ocean heat transport may also provide an effective means of diagnosing whether any small biases in satellite radiation

TABLE 1. NORTHWARD TRANSPORT ACROSS 24°N

Temperature Class	Atlantic Depth Range (m)	Atlantic Transport (Sv)	Pacific Transport (Sv)	Pacific Depth Range (m)
θ >17°C	0–290	8.9	10.6	0–190
12°< θ <17°C	290–610	2.5	-5.7	190–350
7°C< θ <12°C	610–950	6.5	-4.6	350–590
4°C< θ <7°C	950–1800	-2.3	0.0	590–960
θ <4°C	1800–Bottom	-15.6	-0.3	960–Bottom

θ is potential temperature and 1 Sv = 10^6 m^3 s^{-1}. Note how much deeper the isotherms are on average in the Atlantic, i.e. how much warmer the Atlantic is than the Pacific at a given depth.

are uniform with latitude or are confined to specific latitudinal zones.

MECHANISMS OF OCEAN HEAT TRANSPORT

Direct determinations of the meridional ocean heat transport have the added advantage that an understanding of the mechanisms of ocean heat transport is achieved in carrying out the estimates. The indirect estimates of ocean heat transport based either on bulk formula calculations of air–sea energy exchange or on the residual method whereby the atmospheric energy transport is subtracted from the radiation balance requirements do not identify the oceanic processes that are responsible for transporting heat. These indirect methods leave an impression that the ocean passively responds to changes in atmospheric conditions or radiational parameters, and this impression conflicts with general intuition that the ocean is an active participant in long-term climate changes. In this section, I would like to show how the mechanisms of northward ocean heat transport across 24°N are substantively different in the North Atlantic and Pacific Oceans, even though the amount of heat transport is approximately the same, and to suggest how the different mechanisms are related to present climate differences between the North Atlantic and North Pacific.

The northward heat transport of 1.2 PW across 24°N in the Atlantic is principally due to a deep vertical-meridional circulation cell. Warm surface and thermocline waters flow northward across 24°N, lose heat to the atmosphere as they make their way into the Norwegian and Labrador Seas where they are converted into deep water in late-winter formation events, and the deep water then returns southward across 24°N at very cold temperatures. Hall and Bryden [1982] tabulated the meridional transport across 24°N as a function of temperature to show that about 18 Sv (1 Sv = 1×10^6 $m^3s{-}1$) of water warmer that 12°C flows northward and a similar amount of water colder than 4°C returns southward across 24°N (Table 1). Surprisingly, they found that the large northward Gulf Stream flow through Florida Straits is actually colder than the smaller southward mid-ocean flow at all depths greater than 200 m. The northward heat transport then cannot be attributed to a horizontal circulation in which warm Gulf Stream waters flow northward, lose their heat and return southward at similar depth but at colder temperatures, as might have been anticipated. Instead the northward heat transport is due to a net northward flow of warm Gulf Stream waters in the thermocline that only return southward after they have been converted into cold deep water.

In contrast to the Atlantic, the northward heat transport of 0.8 PW across 24°N in the Pacific Ocean is due to a horizontal circulation in the upper waters above 800 m depth [Bryden, Roemmich and Church, 1991]. Warm waters flow northward across 24°N in the Kuroshio on the western side of the Pacific and in the wind-driven Ekman surface layer, lose heat in the subtropical and subpolar North Pacific, and return southward across 24°N in the central and eastern Pacific at colder temperatures at depths above 800 m. There are no sources of deep water in the North Pacific, so the deep circulation is sluggish. Only about 5 Sv of 1°C bottom water originating from the Antarctic flows northward across 24°N; it mixes and warms slightly from downward heat diffusion north of 24°N, and then returns southward across 24°N as Pacific Deep Water with a temperature of about 2°C. There is essentially no heat transport associated with the deep water circulation in the Pacific. Formally, the heat transport can be divided into two components: one due to the horizontal circulation cell at each depth and one due to the zonally averaged vertical-meridional circulation (Table 2). For the Pacific, half of the northward heat transport across 24°N is in each component. However, the heat transport carried by the horizontal circulation all occurs in the upper 800 m of the water column; and all but 0.02 PW of the heat transport in the vertical meridional circulation is due to the zonally averaged circulation above 800 m depth [Bryden, Roemmich and Church, 1991]. Thus, we can conclude that the northward heat transport across 24°N in the

TABLE 2. OCEAN HEAT TRANSPORT ACROSS 24°N

Heat Flux Component	Northward Heat Flux (10^{15} W) Atlantic	Pacific
Vertical-Meridional Overturning		
$\rho\, C_p \int < v(z) > < \theta(z) > \ L(z) dz$	1.28	0.38
Horizontal Circulation Cell		
$\rho\, C_p \int \int (v- < v >)\ (\theta- < \theta >)\ dxdz$	−0.06	0.38
Total	1.22	0.76

$L\ (z)$ is the width of the transoceanic section at each depth, z; v and θ are the meridional velocity and potential temperature; brackets, $< >$, denote a zonal, x, average; ρ is density and C_p is the specific heat of seawater.

Pacific is essentially associated with the circulation of the upper waters above 800 m depth.

In contrast to the Pacific, the northward heat transport in the Atlantic is due entirely to the zonally averaged vertical meridional circulation and the horizontal circulation actually contributes a small southward heat transport across 24°N (Table 2). Furthermore, in the Pacific there is no net meridional flow of cold, deep waters, in dramatic contrast with the Atlantic (Table 1). Thus, the two Northern Hemisphere oceans have two different mechanisms of transporting heat northward: the Atlantic mechanism is a deep vertical-meridional circulation cell that ultimately depends on deep water formation in the polar regions; but the Pacific mechanism is a horizontal circulation confined to the upper kilometer of the ocean that more traditionally depends on northward flow in a warmer western boundary current and southward flow in colder mid-ocean and eastern boundary currents. The North Atlantic Ocean might then be considered as a basin in which the thermohaline forcing determines the size of the vertical circulation cell and the associated meridional heat transport while the North Pacific Ocean is a basin in which the wind forcing determines the size of the horizontal circulation and the meridional heat transport.

The differences in circulation and heat transport mechanisms lead to marked differences in the climate of the North Atlantic and North Pacific Oceans. At subtropical and particularly at subpolar and polar latitudes the North Atlantic is much warmer than the the North Pacific (Figure 6). In both oceans, the warm surface waters extend to quite high latitudes in late winter, keeping the climate of the adjoining land more hospitable to human settlement, particularly on the western edges of the continents. For example, the Pacific surface temperature is still 5°C in February at the southern limit of Alaska near 55°N. Atlantic surface temperatures in February are even more impressive with the 10°C isotherm reaching beyond 50°N off Ireland and the 5°C isotherm extending nearly to 70°N off northern Norway. In terms of the above discussion of heat transport mechanisms, we can offer an explanation for these relatively high surface temperatures. In the Pacific, warm water carried by the Kuroshio along the western boundary off Japan makes its way northward and eastward across the Pacific losing heat to the atmosphere as it flows. Near the eastern boundary off North America, the west wind drift current splits at about 50°N with some of the residual warm water flowing into the Gulf of Alaska and the remainder turning southward to flow as a relatively cold eastern boundary current along the coast of California. In the Atlantic, the warm Gulf Stream water also flows eastward and northward across the Atlantic. The process of deep water formation requires a source of surface waters for conversion to deep water and hence the warm subpolar waters are drawn northward into the Norwegian-Greenland Sea, maintaining the high surface temperatures in the far northern Atlantic.

Do we understand the physical reasons underlying the difference between the North Atlantic and North Pacific circulations? There are arguments that no deep water can be formed in the North Pacific because the surface waters of the subpolar region are too fresh due to high rainfall [Warren, 1983]. There are arguments that deep water is formed in the North Atlantic because the Mediterranean outflow keeps the subpolar region salty enough to form deep water [Reid, 1979] or because the North Atlantic extends far into northern latitudes where large buoyancy loss by the ocean to the atmosphere can lead to dense bottom water formation. But I am not aware that any coupled atmosphere-ocean model could predict the difference between the two observed circulations in the North Atlantic and North Pacific. In particular, it is not clear how well present models used for climate prediction are able to distinguish between the two types of circulation and heat transport observed in the North Atlantic and North Pacific. Because climate models are generally adjusted to reproduce the present-day ocean circulation as initial conditions, no test of the model's ability to predict the inherent

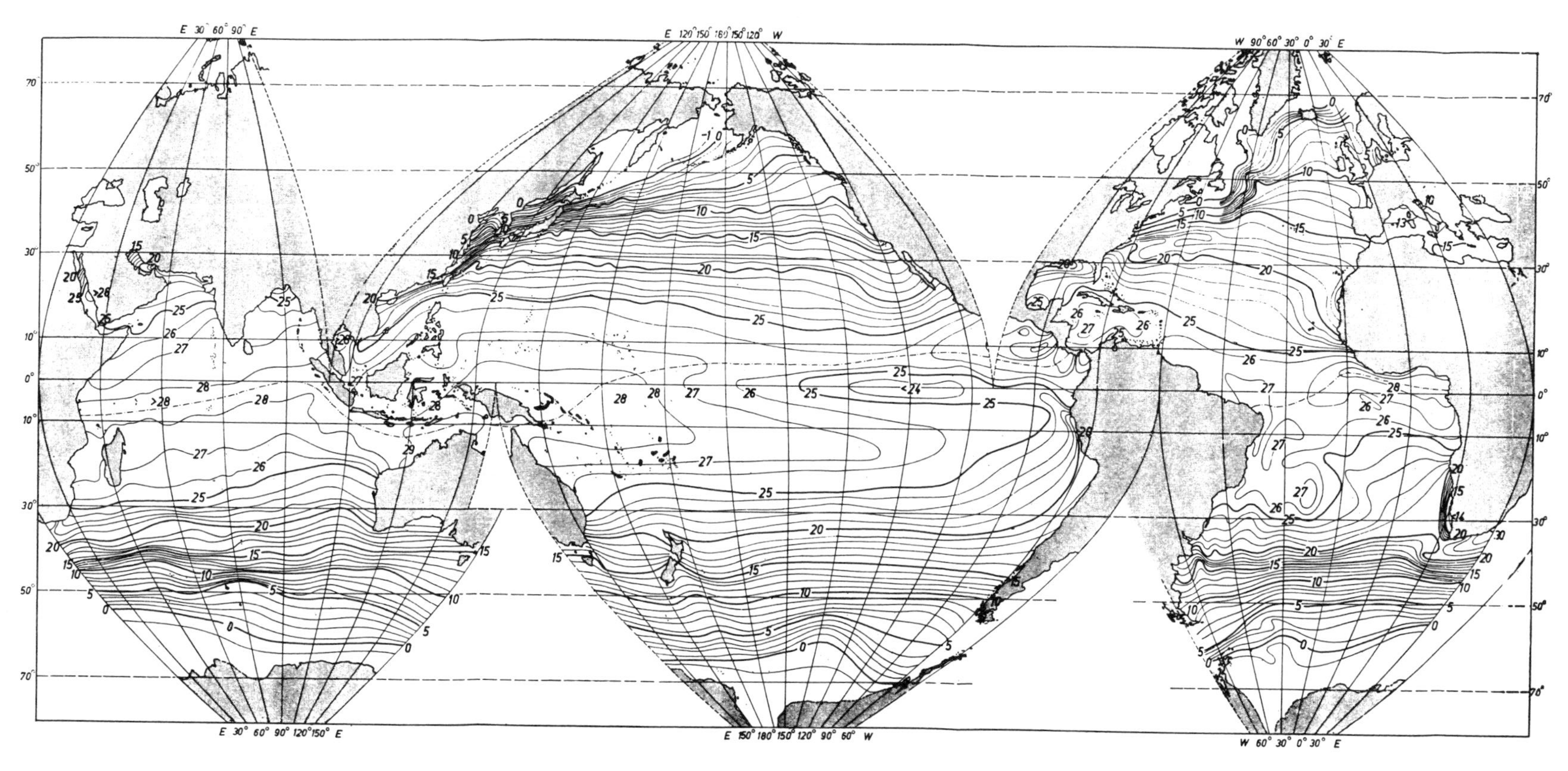

Fig. 6. Surface temperature (°C) of the world oceans for February (Defant, 1961, Plate 3A).

differences between the Atlantic and Pacific is made. In fact, large flux corrections are apparently needed in many coupled models to keep ocean temperatures from drifting quickly away from their observed values. Such flux corrections imply that the models are not simulating the ocean heat and freshwater transports in a realistic manner. An ability to predict such basic differences in ocean circulation as those observed in the North Atlantic and North Pacific Oceans would give confidence that a coupled atmosphere-ocean model might be suitable for predicting the direction of future climate change.

FUTURE PROSPECTS FOR OCEAN HEAT TRANSPORT DETERMINATIONS

There is a clear need to determine directly the meridional ocean heat transport across a number of latitudes in each ocean basin in order to derive a distribution of ocean heat transport as a function of latitude that is independent of satellite radiation analyses and atmospheric energy transport estimates. Instead of relying on indirect estimates of ocean heat transport based on a difference of two large numbers (as in Figure 2), oceanographers would like to make direct estimates of the ocean heat transport as a function of latitude in order to assess the ocean's role in maintaining the present global heat budget and to help diagnose problems in the atmospheric and radiation analyses.

24°N is the single latitude for which there are reliable direct estimates of the total meridional ocean heat transport. Earlier, Bennett 1978) attempted to estimate the total meridional ocean heat transport across 30°S in the Atlantic, Indian and Pacific Oceans, but we now consider that his estimates suffer from poor resolution of the western boundary currents in each ocean. Modern analyses for the meridional heat transport across 35 or 36°N have been carried out for both the Atlantic and Pacific Oceans, but there is a lack of agreement on the size of the heat transport in each basin. Roemmich and Wunsch [1985] estimated a northward heat transport of 0.8 PW across the 36°N transatlantic section, but Rintoul and Wunsch [1991] estimated 1.3 PW for the same section. The difference appears to be a question of how well the eddy heat transport is resolved both by the hydrographic section that runs close to and parallel to the eastward Gulf Stream flow and by the inverse analysis techniques. In the Pacific, Roemmich and McCallister [1989] estimated a southward heat transport of -0.1 PW across the 35°N transpacific section while McBean [1991] estimated a northward heat transport of 1.0 PW across the same hydrographic section but using an additional more detailed section across the Kuroshio western boundary current off Japan. McBean concluded that "35°N where the Kuroshio is full of meanders and the transport is near zonal" is a difficult latitude to make a reliable estimate of the meridional ocean heat transport. The uncertainties in estimating ocean heat transport across 35°N in the Atlantic and Pacific serve to emphasize the need for accurate measures of the western boundary current transport and for careful analyses of the eddy heat transport when determinations of ocean heat transport are carried out. In many respects, 24°N is a singular latitude for ocean heat transport estimates because of the relatively firm knowledge of western boundary current transports across 24°N and the relatively small eddy activity in the mid ocean at 24°N in both the Atlantic and Pacific.

The above comparison of the 24°N ocean heat transport with atmospheric transport and the total transport required by satellite radiation measurements brought out several problems in our present understanding of the global heat budget. Preliminary estimates of ocean heat transport across other latitudes suggest that there may be even more severe problems at other latitudes, particularly in the Southern Hemisphere. In the Southern Hemisphere, the total atmosphere+ocean heat transport required by the satellite radiation measurements appears to be similar to the Northern Hemisphere transport, that is the total heat transport is reasonably symmetric about the equator with a poleward heat transport maximum of about 6 PW near 35°S (Figure 2). Because there is so little land in the Southern Hemisphere, the rawinsonde network is sparse and hence the determination of atmospheric heat transport is much more uncertain. Therefore, direct estimates of ocean heat transport in the Southern Hemisphere can be quite valuable.

Available estimates of ocean heat transport indicate that the total ocean heat transport is small in the Southern Hemisphere. Bennett's [1978] original estimates surprisingly indicated a large (1 to 2 PW) equatorward ocean heat transport across 30°S. He found large equatorward heat transports in both the South Atlantic and South Indian Oceans, with a small and less well-determined poleward heat transport in the South Pacific. Recent evaluations of ocean heat transport have confirmed an equatorward heat transport in the South Atlantic of 0.25 to 0.8 PW across 30°S [Rintoul, 1991; Fu, 1981] and a small poleward heat transport of 0.18 PW across 28°S in the South Pacific [Wunsch, Hu and Grant, 1983], but have suggested that the ocean heat transport across 32°S in the Indian Ocean is poleward with a magnitude of 0.25 to 0.6 PW [Fu, 1986; Toole and Raymer, 1985]. Thus, the total ocean heat transport across 30°S appears to be small. Using the most recent estimates (presumably the most reliable) yields a total poleward heat transport across 30°S of only 0.2 PW [Rintoul, 1991; Wunsch, Hu and Grant, 1983; Fu, 1986]. If such results hold up in the analysis of the new measurements during the World Ocean Circulation Experiment (WOCE) specifically designed to determine ocean heat transport, they would imply that the atmospheric energy transport in the Southern Hemisphere must be very large. Clearly, direct estimates of ocean heat transport as a function of latitude are central for composing a basic understanding of how the present global heat balance is maintained.

In addition to determining ocean heat transport as a function of latitude, there is a need to understand the mechanisms of ocean heat transport in each ocean basin. We have seen that the mechanisms of ocean heat transport are very different in the North Atlantic and North Pacific Oceans, leading to differences in ocean climate. In each of the Southern Hemisphere Oceans, there is a substantial equatorward flow of Antarctic Bottom Water that can lead to large poleward heat fluxes, as we found in the North Atlantic. The Southern Hemisphere western boundary currents, which represent a second mechanism for effecting poleward ocean heat transport, are somewhat strange: the Brazil Current transport, long considered to be very small, has been recently reevaluated to achieve a transport of 80 Sv at 36°S [Zemba, 1991]; the East Australian Current was not evident in the 28°S transpacific section used by Wunsch et al. [1983] to determine South Pacific heat transport; only the Agulhas Current in the South Indian Ocean has appeared to be a normal western boundary current with a poleward transport of about 40 Sv across 30°S. For ocean heat transport determinations across 30°S, there is a clear need for western boundary current measurements in each basin as well as for the transoceanic hydrographic sections emphasized above.

Thus, we have only a rudimentary understanding of the meridional circulation across each of the South Atlantic, South Pacific and South Indian Oceans and of the mechanisms of meridional ocean heat transport in the Southern Hemisphere. How do we understand the apparent equatorward ocean heat transport in the South Atlantic? Is it just the result of the North Atlantic drawing in warm surface waters across the equator for ultimate conversion into deep waters? Or is it due to the balance in deep water transports between northward flowing Antarctic Bottom Water and southward flowing North Atlantic Deep Water so that there is no net meridional deep water transport? How can the South Pacific have such a small poleward heat transport when the vast tropical Pacific appears to gain so much heat from the atmosphere? Furthermore, there is an estimated equatorward flow of Antarctic Bottom Water in the South Pacific of 18 Sv across 28°S [Warren, 1981], similar to the deep equatorward flow in the North Atlantic, so how can the heat transport be so small? In the Indian Ocean, how can the sizeable equatorward transport of bottom waters be converted into warm surface waters over such a small basin that is closed north of 20°N? Is there a throughflow of warm waters from the Pacific to the Indian Ocean that effects a substantial poleward heat flux in the combined South Pacific and Indian Oceans [Gordon, 1986]? Thus, we need to evaluate the ocean heat transport and to understand the mechanisms of heat transport in each ocean basin, particularly for the Southern Hemisphere oceans, in order to gain a basic understanding of the ocean's contribution to maintaining the present global heat budget.

Fig. 7b. Transoceanic zonal hydrographic sections across the Pacific Ocean during the period 1985–1993. Solid lines denote sections planned for 1992–1993; dotted lines indicate station positions for sections made during 1985–1989.

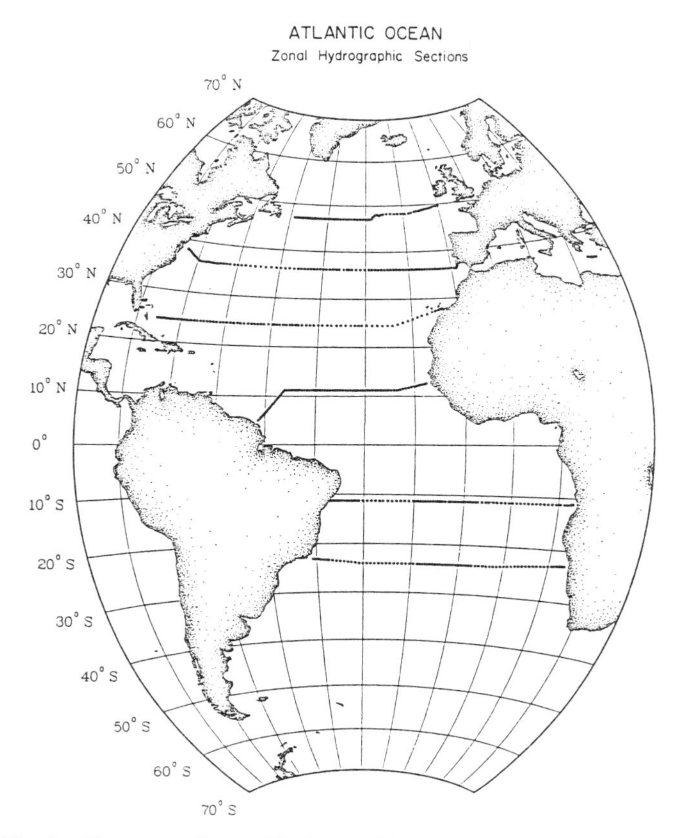

Fig. 7a. Transoceanic zonal hydrographic sections across the Atlantic Ocean during the 1980's.

To make estimates of ocean heat transport and to understand the mechanisms of heat transport, oceanographers have been carrying out a series of transoceanic zonal hydrographic sections over the past decade, from which modern

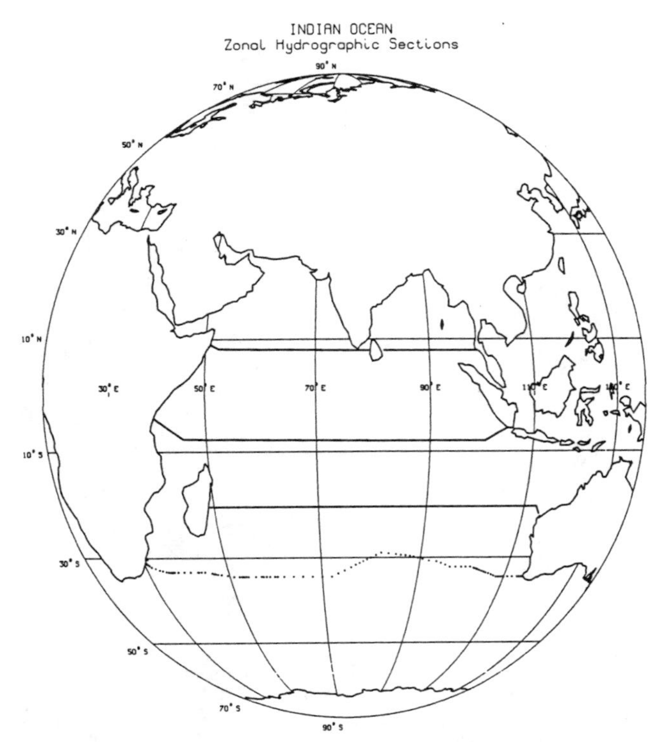

Fig. 7c. Transoceanic zonal hydrographic sections across the Indian Ocean during the period 1988–1995. Solid lines denote sections being planned as part of the World Ocean Circulation Experiment survey of the Indian Ocean for 1994–1995; dotted line indicate station positions for the section made during 1987.

estimates of ocean heat transport can be reliably made. In the Atlantic (Figure 7a), 6 zonal sections have been made in the 1980's across 48°N, 36°N, 24°N, 11°N, 12°S and 23°S. Analyses of these sections should result in a direct estimate of the distribution of meridional ocean heat transport as a function of latitude for the Atlantic Ocean. In the Pacific, 3 zonal sections have already been carried out across 47°N, 24°N, and 10°N during 1985-89, and two more are planned for 1992-93 across 17°S and 32°S (Figure 7b). Completion of these sections and their analyses should then result in direct estimates of meridional ocean heat transport as a function of latitude for the Pacific Ocean. Finally, one modern zonal section across 30°S in the Indian Ocean was carried out in 1988 and a survey of the Indian Ocean is now being planned for 1994–95 as part of WOCE that includes 3 more zonal sections across 8°N, 8°S and 20°S (Figure 7c). Completion of this survey and the analyses of these sections should then result in direct estimates of ocean heat transport as a function of latitude for the Indian Ocean. Thus, in about 5 years time, oceanographers should have direct estimates of ocean heat transport across various latitudes in each ocean basin and should understand the mechanisms of ocean heat transport in each basin. Such results will be a remarkable improvement on our present understanding of the ocean's role in maintaining the present global climate.

Acknowledgments. Support for this work has been provided by the National Science Foundation under Grant OCE-9101493. Critical comments by two anonymous reviewers were very helpful for clarifying the text. I thank Gordon McBean for giving me the opportunity to present this paper during the Hann Symposium at the IUGG in August, 1991. This is Contribution Number 7954 from Woods Hole Oceanographic Institution.

REFERENCES

Barkstrom, B. R., E. F. Harrison and R. B. Lee, III. Earth radiation budget experiment: Preliminary seasonal results, *EOS, 71* (9), February 27, 297,304–305, 1990.

Bennett, A. F., Poleward heat fluxes in Southern Hemisphere oceans, *Journal of Physical Oceanography, 8*, 785–798, 1978.

Bryan, K., Measurements of meridional heat transport by ocean currents, *Journal of Geophysical Research, 67*, 3403–3414, 1962.

Bryden, H. L., and M. M. Hall., Heat transport by currents across 25°N latitude in the Atlantic Ocean, *Science, 207*, 884–886, 1980.

Bryden, H. L., D. H. Roemmich and J. A. Church, Ocean heat transport across 24°N in the Pacific, *Deep-Sea Research, 38*, 297–324, 1991.

Budyko, M. I., *Climate and Life*, Academic Press, New York, 508 p, 1974.

Bunker, A. F., Surface energy fluxes of the South Atlantic Ocean, *Monthly Weather Review, 116*, 809–823, 1988.

Carissimo, B. C., A. H. Oort and T. H. Vonder Haar, Estimating the meridional energy transports in the atmosphere and ocean, *Journal of Physical Oceanography, 15*, 82–91, 1985.

Defant, A., *Physical Oceanography*, Volume I, Pergamon, Oxford, 729 p, 1961.

Fu, L.-L., The general circulation and meridional heat transport of the subtropical South Atlantic determined by inverse methods, *Journal of Physical Oceanography, 11*, 1171–1193, 1981.

Fu, L.-L., Mass, heat and freshwater fluxes in the South Indian Ocean, *Journal of Physical Oceanography, 16*, 1683–1693, 1986.

Gordon, A. L., Interocean exchange of thermocline water, *Journal of Geophysical Research, 91*, 5037–5046, 1986.

Hall, M. M., and H. L. Bryden, Direct estimates and mechanisms of ocean heat transport, *Deep-Sea Research, 29*, 339–359, 1982.

Hall, M., and D. Roemmich, Heat flux investigations within the World Ocean Circulation Experiment, in WOCE *Discussions of Physical Processes*, U.S. WOCE Planning Report 5, U.S. Planning Office for WOCE, College Station, Texas, 143 p, 1986.

Hartmann, D. L., V. Ramanathan, A. Berroir and G. E. Hunt, Earth radiation budget data and climate research, *Reviews of Geophysics, 24*, 439–468, 1986.

Hastenrath, S., On meridional heat transports in the World Ocean, *Journal of Physical Oceanography, 12*, 922– 927, 1982.

Hastenrath, S., On the interannual variability of poleward transport and storage of heat in the ocean-atmosphere system, *Archives for Meteorology, Geophysics and Bioclimatology, A33*, 1–10, 1984.

Jung, G. H., Energy transport by air and sea, Contribution 69 of the Department of Oceanography and Meteorology, Texas A&M University, College Station, Texas), 19 p, 1956.

Masuda, K., Meridional heat transport by the atmosphere and the ocean: analysis of FGGE data, *Tellus, 40A*, 285– 302, 1988.

McBean, G. A. Estimation of the Pacific Ocean meridional heat flux at 35°N, *Atmosphere-Ocean, 29*, 576–595, 1991.

Michaud, R., and J. Derome, On the mean meridional transport of energy in the atmosphere and oceans as derived from six years of ECMWF analyses, *Tellus, 43A*, 1–14, 1991.

Newell, R. E., J. W. Kidson, D. G. Vincent and G. J. Boer, *The*

General Circulation of the Tropical Atmosphere, Volume 1, M.I.T. Press, 258 p, 1972.
Oort, A. H., Global atmospheric circulation statistics, 1958–1973, NOAA Professional Paper No. 14, 180 p, 1983.
Oort, A. H., and J. P. Peixoto, Global angular momentum and energy balance requirements from observations, *Advances in Geophysics, 25*, 355–490, 1983.
Oort, A. H., and T. H. Vonder Haar, On the observed annual cycle in the ocean-atmosphere heat balance over the Northern Hemisphere, *Journal of Physical Oceanography, 6*, 781–800, 1976.
Palmen, E., and C. W. Newton, *Atmospheric Circulation Systems*, Academic Press, New York, 603 p, 1969.
Reid, J. L., On the contribution of the Mediterranean Sea outflow to the Norwegian-Greenland Sea, *Deep-Sea Research, 26*, 1199–1223, 1979.
Rintoul, S. R., South Atlantic interbasin exchange, *Journal of Geophysical Research, 96*, 2675–2692, 1991.
Rintoul, S. R., and C. Wunsch, Mass, heat, oxygen and nutrient fluxes and budgets in the North Atlantic Ocean, *Deep-Sea Research, 38* (Supplement), S355–S377, 1991.
Roemmich, D., Estimation of meridional heat flux in the North Atlantic by inverse methods, *Journal of Physical Oceanography, 10*, 1972–1983, 1980.
Roemmich, D., and T. McCallister, Large scale circulation of the North Pacific Ocean, *Progress in Oceanography, 22*, 171–204, 1989.
Roemmich, D., and C. Wunsch, Two transatlantic sections: meridional circulation and heat flux in the subtropical North Atlantic Ocean, *Deep-Sea Research, 32*, 619–664, 1985.
Stephens, G. L., G. C. Campbell and T. H. Vonder Haar, Earth radiation budgets, *Journal of Geophysical Research, 86*, 9739–9760, 1981.
Talley, L. D., Meridional heat transport in the Pacific Ocean, *Journal of Physical Oceanography, 14*, 231–241, 1984.
Toole, J. M., and M. E. Raymer, Heat and fresh water budgets of the Indian Ocean-revisited, *Deep-Sea Research, 32*, 917–928, 1985.
Trenberth, K. E., Mean annual poleward energy transports by the oceans in the Southern Hemisphere. *Dynamics of Atmospheres and Oceans, 4*, 57–64, 1979.
Vonder Haar, T. H., and A. H. Oort, New estimate of annual poleward energy transport by Northern Hemisphere oceans, *Journal of Physical Oceanography, 3*, 169–172, 1973.
Warren, B. A., Deep Circulation of the World Ocean, in *Evolution of Physical Oceanography*, edited by B. A. Warren and C. Wunsch, M.I.T. Press, Cambridge, 6–41, 1981.
Warren, B. A., Why is no deep water formed in the North Pacific?, *Journal of Marine Research, 41*, 327–347, 1983.
Wunsch, C., Meridional heat flux of the North Atlantic Ocean, *Proceedings of the National Academy of Sciences, U.S.A., 77*, 5043–5047, 1980.
Wunsch, C., D. Hu and B. Grant, Mass, heat, salt and nutrient fluxes in the South Pacific Ocean, *Journal of Physical Oceanography, 13*, 725–753, 1983.
Zemba, J. C., The structure and transport of the Brazil Current between 27°and 36°South, Ph.D. Thesis, Massachusetts Institute of Technology-Woods Hole Oceanographic Institution Joint Program, WHOI-91-37, 160 p, 1991.

H.L. Bryden, Woods Hole Oceanographic Institution, Woods Hole, Massachusetts.

The Role of the Oceans in the Global Water Cycle

RAYMOND W. SCHMITT AND SUSAN E. WIJFFELS

Woods Hole Oceanographic Institution, Woods Hole, Massachusetts

An assessment is made of the contribution of the oceans to the global water cycle. Because of the multiply-connected nature of the world ocean, it is not possible to infer the transport of water by the oceans from surface forcing (E-P) and runoff alone. Direct ocean measurements are required to provide the constants of integration for summations of surface flux divergence in order to determine transports. The only inter-ocean transport known with any confidence is that through Bering Straits; it allows us to estimate the freshwater and salt transports in the North Pacific and Atlantic basins. An ocean transport picture emerges that is surprisingly different from that proposed by Baumgartner and Reichel [1975], who assumed zero freshwater transport across the Atlantic equator. Given the great uncertainties in evaporation and precipitation estimates over the oceans, direct ocean transport measurements provide important constraints on the global water cycle and must be more fully utilized in future research programs.

INTRODUCTION

The global water cycle is of obvious first order importance to the redistribution of energy, the maintenance of climate and the survival of life on the planet. The oceans represent the primary reservoir of water, containing 97% of the Earth's supply; the atmosphere holds only 0.001%. Yet there is poor understanding of the ocean's role in the world water budget, to the extent that fundamental misconceptions persist in the literature. Here we attempt to clarify the function of the ocean in the global water cycle.

The atmosphere is capable of transporting water (as vapor) in any direction on the planet, but the oceans are constrained by the continents. If the land were simply connected, then knowledge of the net surface flux of water from the difference between evaporation and precipitation plus river runoff, could be sufficient to characterize the ocean transports. However, the continents are not simply connected and multiple pathways exit between ocean basins. In some cases the flows between ocean basins exceed atmospheric transports by more than two orders of magnitude, indicating that proper understanding of oceanic transports is essential to development of a global water budget.

In the following, we outline the fundamentals of ocean freshwater transports as in Wijffels et al. [1992] (Section 2) and apply presently available estimates to the North Atlantic basin and the global ocean (Sections 3 and 4). In addition, we note the importance of constraining the freshwater budget with direct ocean measurements (Section 5). Indeed, given the difficulty of estimating evaporation and precipation over the sea, there may be no better way to monitor the global water cycle.

Interactions Between Global Climate Subsystems, The Legacy of Hann
Geophysical Monograph 75, IUGG Volume 15

OCEAN FRESHWATER TRANSPORTS

Following Wijffels et al. [1992] we want to more clearly define the ocean contribution to global freshwater fluxes. One point requiring special attention is clarification of what is meant by "freshwater" when discussing its transport by a saline ocean. Often what is meant is a freshwater anomaly relative to some reference salinity, or the divergence of water flux in a control volume with varying salinity on its boundaries. Such definitions may be appropriate in regional studies. However, when discussing the global ocean it must be appreciated that there is no universal reference salinity. The mean salinity of the ocean is maintained as a balance between input by rivers and sea-bed mineralization processes, but it is believed to have varied with changing ice volume on climatic time scales. Salinities are modified by atmospheric and riverine water sources, and can be used to develop an understanding of ocean freshwater transports. The oceans carry both water and salt and the only logically consistent treatment of their budgets must separately address their fluxes.

Our definition of freshwater flux will use the total water content of seawater, thus freeing us from any implicit assumptions about the salt budget. That is, 96.5% of seawater is freshwater, if the salinity is 35 psu. This allows for the development of a consistent discussion of transports in a multiply connected ocean.

The ocean exchanges freshwater with the atmosphere via evaporation and precipitation at the surface of the ocean and runoff from the continents at its boundaries. Because of this exchange, the steady-state mass divergence in oceanic volumes will not be precisely zero. For an oceanic basin, the changes in the zonally integrated meridional transport are related to the surface fluxes as follows (ignoring small scale diffusion):

$$\frac{d}{dy} \int \int \rho v dx dz = \int F\ (x, y)\ dx \tag{1}$$

where x, y and z are the zonal, meridional and vertical coordinates respectively, v the meridional oceanic velocity and ρ the density of seawater. Here $F\ (x, y)$ is the net gain of mass at the ocean surface due to precipitation, evaporation and land runoff, P, E and R respectively. Hence $F\ (x, y) = P - E + R$.

Because there is no significant atmospheric pathway for salt, the long term divergence of salt in an oceanic volume in a steady state must always be zero. This does not imply that there is no salt flux across oceanic sections, only that for any given volume of ocean the ingoing and outgoing fluxes are exactly equal, that is:

$$\frac{d}{dy} \int \int \rho v S dx dz = 0 \tag{2}$$

where S is the salinity of seawater. The freshwater balance is the difference between (1) and (2):

$$\frac{d}{dy} \int \int \rho v(1 - S) dx dz = \int F\ (x, y)\ dx \tag{3}$$

In (1) through (3) the left hand side denotes the divergence of the meridional fluxes in the ocean. Hence the surface mass exchange is the difference between the incoming and outgoing advective fluxes. The oceanic advective fluxes may themselves dwarf the air–sea exchanges, making the surface fluxes appear negligible, a small difference between large numbers. However, one should keep two things in mind: (1) the surface fluxes drive the thermohaline circulation; and (2) the surface freshwater flux has implications for the salt budget of oceanic volumes.

It is useful to note the relationships between the freshwater flux and salt flux for flows of varying complexity. That is, we can separate the velocity and salinity fields into section averages and spatially varying components and let the over-bar represent the section average (or equally well, a time average) such that:

$$\overline{S} = \frac{\int S(x, z) dx dz}{\int dx dz}, \quad S'\ (x, z) = S(x, z) - \overline{S} \tag{4}$$

Similarly for $\overline{v}$ and v'. Neglecting density variations, the salt flux across a section can be written:

$$\rho\ (\overline{v}\overline{S} + \overline{v'S'}) \tag{5}$$

and the freshwater flux is:

$$\rho\ (\overline{v} - \overline{v}\overline{S} - \overline{v'S'}) \tag{6}$$

We see that the freshwater flux is equal to the negative salt flux only in the case where $\overline{v} = 0$. This may apply to the two layer flows found in straits connecting enclosed basins (and only where there is no net mass gain or loss in the basin due to runoff and evaporation). For the through-flows connecting the major ocean basins and for sections across the basins the more general expression must be used. Note that $\rho\ \overline{v'S'}$ as defined here, represents the salt flux across a section due to all scales of flow beyond uniform advection, such as the large scale horizontal and vertical circulations associated with the wind-driven gyres and thermohaline circulations, and also includes mesoscale eddy fluxes.

It is important to emphasize that a knowledge of the surface forcing function $F(x, y)$ is not sufficient to determine the net fluxes in the ocean because of its multiply connected nature. That is, the surface gain or loss of water only represents the mass divergence within a region, not the total mass transport itself. In order to establish the fluxes within basins it is necessary to know the interbasin exchanges as well. The interbasin flows can be thought of as non-divergent fluxes, at least for salt. Baumgartner and Reichel [1975] constructed a schematic of ocean freshwater transports based on zero ocean transport across the Atlantic and Indian Ocean equators. These assumptions (perhaps motivated by a desire for hemispheric symmetry) are quite improbable from an oceanographic standpoint. It happens that the resulting transport estimate for Bering Straits corresponds to the freshwater flux divergence of this through-flow within the Artic itself, and thus their assumption seemed reasonable to later workers [Broecker et al., 1990]. While it is the flux divergence that is of dynamical significance, the total flux must be considered in order to properly assess divergences at latitudes further south in the Atlantic, since the reference salinity will differ from that chosen for the Arctic. Because the distillation processes of evaporation and freezing can separate water and salt in any ocean basin, it is necessary to account for the total flux of both components of seawater in order to develop consistent global budgets. Thus, information on interbasin exchanges is essential before a consistent picture of ocean freshwater fluxes can be developed.

INTERBASIN EXCHANGES

Unfortunately, we have rather imprecise knowledge of the largest interbasin exchanges. The Antarctic Circumpolar Current (ACC) is by far the largest with a zonal volume transport of order $100 \times 10^6\ m^3\ s^{-1}$. Maximum atmospheric transports are less than 1% of this, well below any uncertainty level we might ascribe to ocean estimates of the ACC transport. Thus, no significant statements about the global water cycle seem possible, at least in terms of southern hemisphere zonal variations. (Meridional transports of water are of greater consequence for the global water cycle in any case.)

Smaller but even less well known is the through-flow between the Pacific and Indian oceans in the Indonesian archipelago. Variously estimated to be between 1 and 20×10^6 m^3 s^{-1}, [Gordon, 1986; Fieux et al., 1991] it will dominate the fresh water and salt budgets of the Indian and South Pacific oceans, where net E-P-R is less than $.5\times10^6$ m^3 s^{-1}.

Smallest and best known is the flow through Bering Straits from the North Pacific to the Arctic Basin. The Pacific stands 50 to 60 cm higher than the Atlantic because of its lower salinity, due in part to the excess of precipitation over evaporation in the North Pacific. Much of this excess derives from evaporation in the Atlantic and atmospheric transport across Central America. The Pacific - Arctic head difference drives the flow through Bering Straits, though local winds strongly modulate the flow (even to the point of occasional reversal). Coachman and Aagaard [1988] have estimated this transport to be 0.8 ($\pm$0.1)$\times10^6$ m^3 s^{-1} with an average salinity of 32.5 psu from a one year current meter array and proxy data over a longer period. This relatively "accurate" number allows us to sketch a preliminary picture of the freshwater and salt fluxes for the North Pacific and Atlantic basins, given available estimates of evaporation, precipitation and runoff.

For the present purposes, we have used compilations of global E, P, and R by Baumgartner and Reichel [1975], and estimates of E-P for the North Atlantic by Schmitt et al. [1989] to illustrate the patterns in ocean freshwater transports. The surface flux data were summed in 5 degree latitude bands and integrated from the north using the Bering Straits transport and Baumgartner and Reichel's summation for the Arctic. The resulting global water transports according to Wijffels et al. [1992] are reproduced in Figure 1. For contrast, the very different Baumgartner and Reichel scheme is given in Figure 2 . The major feature of the new plan is that the North Pacific freshwater excess is largely transported to the Atlantic via the Bering Straits and the Arctic, not through the southern ocean, as has previously been supposed.

One issue of great concern in attempting to develop a picture of ocean water transports is the accuracy with which surface exchange of water can be estimated. Precipitation

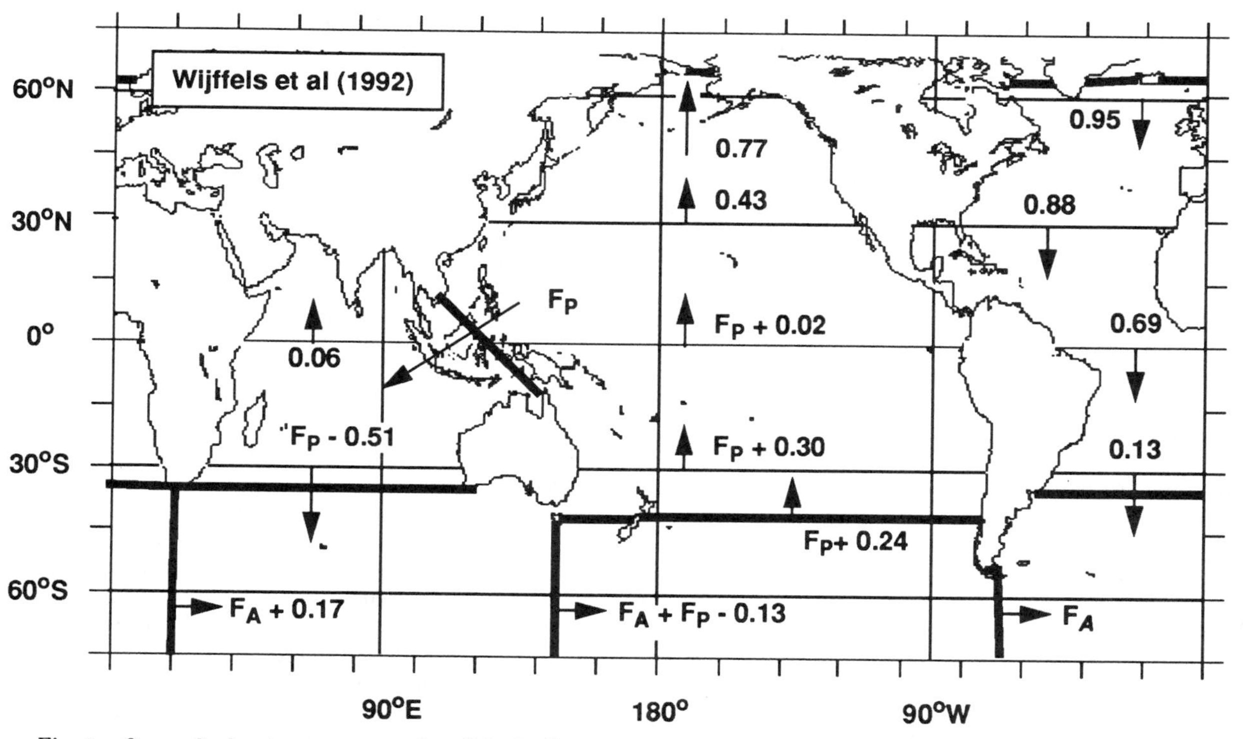

Fig. 1. Ocean freshwater transports ($\times10^9$ kg/s^{-1}) in the multiply connected ocean, according to Wijffels *et al.*, (1992). The fluxes into or out-of individual volumes from precipitation, evaporation and run-off can be summed to give the transport through a trans-ocean section, but an integration constant is required to establish the absolute fluxes. The flow through Bering Straits has been used to establish the transports in the North Pacific and Atlantic basins. Here F_P and F_A refer to the freshwater fluxes of the Pacific-Indian throughflow and the Antarctic Circumpolar Current in Drake Passage, which are poorly known at present.

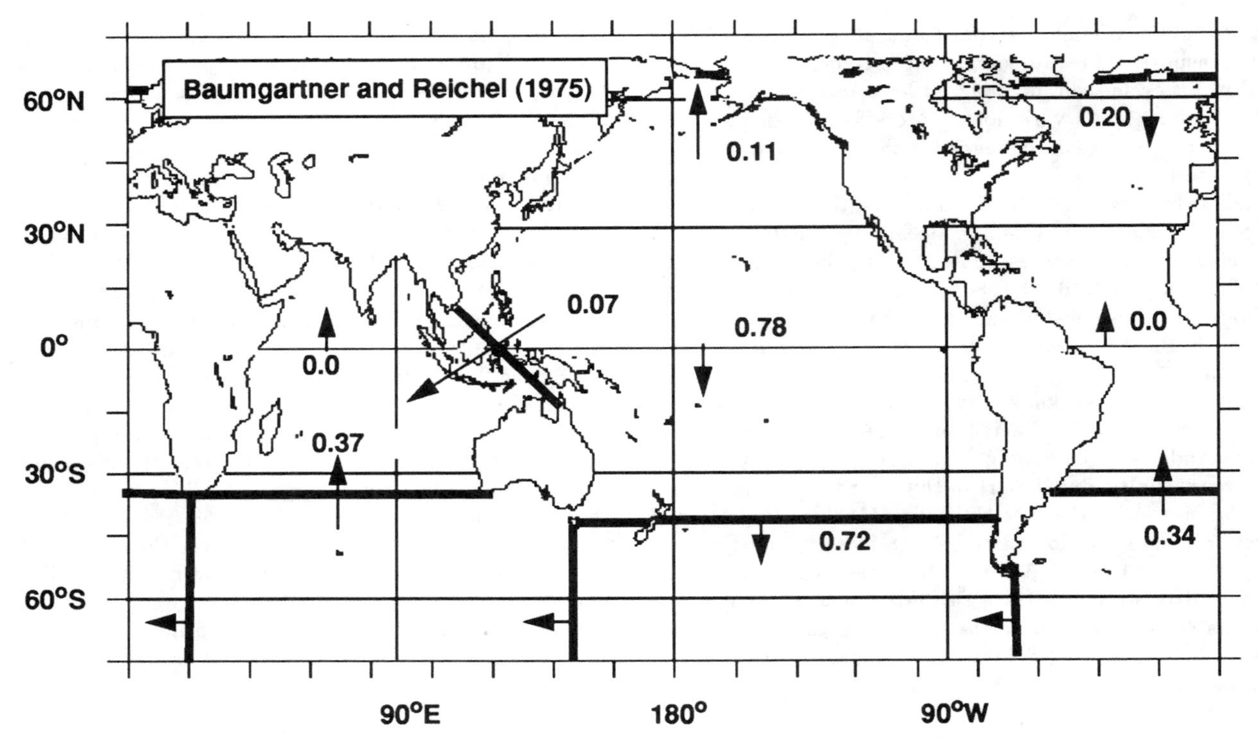

Fig. 2. The transport of freshwater in the ocean according to Baumgartner and Reichel (1975) ($\times 10^9$ kg s^{-1}). They assumed zero transport across the equator of the Atlantic and Indian Oceans in order to develop a global picture, rather than measured interbasin transport. This leads to a dramatically different pattern of ocean fluxes.

measurements at sea are almost nonexistent and satellite rainfall estimation techniques are in their infancy. Recent climatologies rely on weather reports from ships and empirical regression against coastal and island data [Dorman and Bourke, 1981]. Evaporation is estimated with "bulk" formula requiring wind speed, humidity and air and sea temperatures, as well as exchange coefficients which still generate debate. For the moment, all we can do is contrast available climatological estimates. By summing E-P data over the North Atlantic in 5 degree latitude bands to get the meridional transport we can easily see where the major discrepancies arise. Figure 3 displays the freshwater transport in the North Atlantic using three different data sets. The Schmitt et al. [1989] summations which use the evaporation estimates of Bunker [1976] and the precipitation estimates of Dorman and Bourke [1981], have more net water loss than Baumgartner and Reichel's in the subtropics. The Isemer and Hasse [1987] recomputation of the surface flux estimates of Bunker [1976] has even more evaporation, despite a downward adjustment of the exchange coefficient, because of an upward shift in the Beaufort wind scale at low to moderate wind speeds. The great variability reflects the uncertainties in at-sea precipitation estimates and use of bulk formula for evaporation rates. Bunker used a higher exchange coefficient in part to compensate for under-sampling of high wind events by ships' avoidance of severe weather. Isemer and Hasse used an estimate of the ocean heat flux at 25°N to adjust exchange coefficients. Such differences in data treatment lead to transport discrepancies of up to 5×10^9 kg/s^{-1} at the equator. This difference is over 2 1/2 times the flow of the Amazon, by far the largest of all rivers! Thus, great uncertainties exist in our understanding of ocean freshwater forcing and transports, and we must seek new approaches to this challenging and climatically significant problem.

THE OCEAN-ATMOSPHERE SYSTEM

As in Wijffels et al., [1992], we can sum the northward freshwater transports in each basin to get the global meridional flux carried in the ocean as a function of latitude. This can be compared with the estimated flux of water (as vapor) in the atmosphere compiled by Peixoto and Oort [1983] from rawinsonde profiles and models. This is displayed in Figure 4. A reassuring complementarity of the flux carried by the ocean and atmosphere can be seen. The ocean (atmosphere) has maximal (minimal) northward transport at about 15°N and 40°S and minimal (maximal) transport at 40°N and 10°S. For the most part ocean and atmospheric fluxes balance each other; the meridional flux carried by rivers is negligible, given the uncertainties of the available estimates. For comparison, the flux carried by the Mississippi, one of the largest north-south rivers, is shown as well.

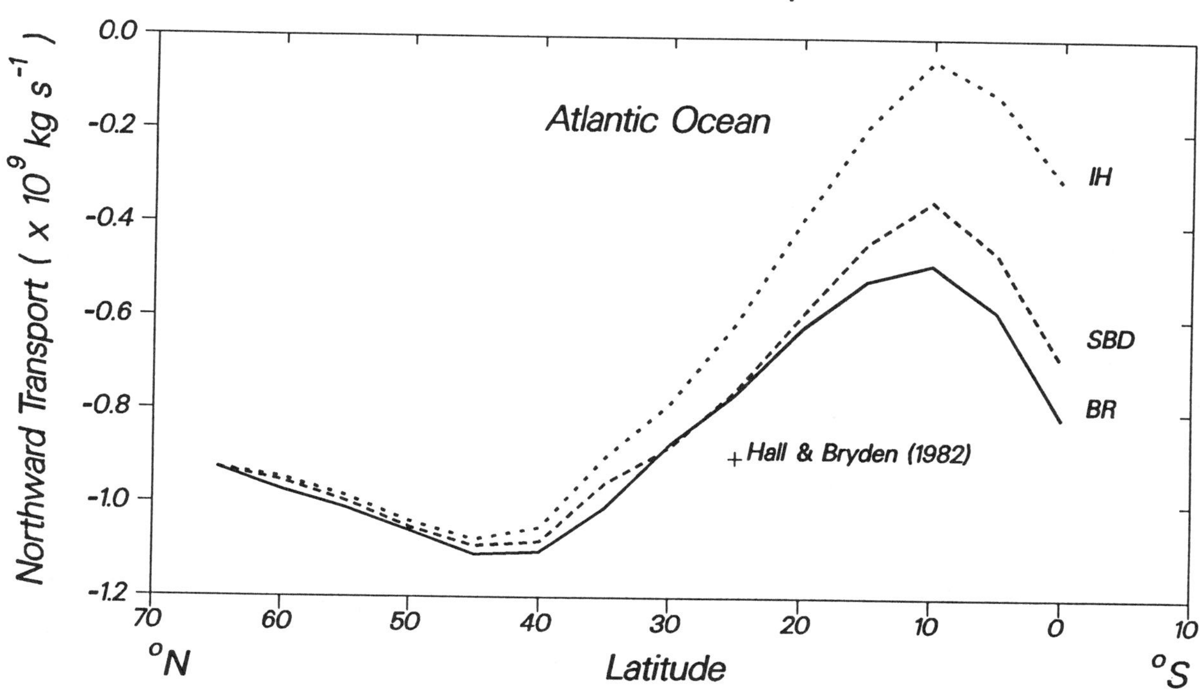

Fig. 3. Northward transport ($\times 10^9$ kg s^{-1}) of freshwater as a function of latitude for the North Atlantic Ocean. The measured flow through Bering Straits has been used with estimates of surface fluxes from: Baumgartner and Reichel (1975,——), Schmitt *et al.* (1989,- - -) and the combination of Dorman and Bourke (1981) and Isemer and Hasse (1987,·····). Also shown is a direct ocean estimate derived from the analysis of Hall and Bryden (1982) and the Bering Straits transport estimates of Coachman and Aagaard (1988).

It carries less than 2% of the flux of the North Atlantic at its discharge latitude. It is perhaps ironic that the best known component of the hydrologic cycle and the one of most direct importance to civilization (river flow) is a minor portion of the global water budget.

It is also worth noting that the combined ocean-atmosphere water transport system accounts for a significant fraction of the poleward heat flux required by the radiation budget of the Earth. The water transported poleward as vapor by the atmosphere and returned to the tropics as condensate by the ocean represents a substantial portion of the total heat budget. Wijffels *et al.* estimate that a latent heat flux of around 1.5×10^{15} watts can be attributed to the water cycle in midlatitudes, which represents a significant fraction of the heat flux usually ascribed to the atmosphere and 25 to 35% of the total heat flux of the ocean- atmosphere system. (Much of the rest of the flux is carried by the sensible heat content of the ocean itself.) By monitoring the heat flux by the ocean and its contribution to the return leg of the hydrologic cycle it would be possible to account for two major pathways for heat transport on the planet (the sensible heat and potential energy fluxes in the atmosphere being the largest remaining pathways). Thus, improvement of our knowledge of the patterns and processes of ocean heat and water transports is fundamental to our understanding of the Earth's energy system.

DIRECT OCEAN ESTIMATES

One possibility for significantly advancing our insight into the global heat and hydrologic cycles is the use of direct oceanographic estimates of heat and salt fluxes in mid-basin as well as narrow straits. From zonal oceanographic sections spanning an ocean basin we can compute geostrophic velocities, and evaluate near-surface Ekman transports from wind stress estimates. Monitoring of western boundary current transports and properties allows computation of the meridional salt flux, the $\overline{v'S'}$ term of Section 2. Hall and Bryden [1982] have performed such a calculation for 24°N in the Atlantic. The zonally averaged meridional velocity and salinity profiles show good correlation, (Figure 5) indicating that much of the salt transport is accomplished by the basic overturning cell of the North Atlantic circulation. Incorporating the most recent Bering Straits transport and using equations 1-6, the Hall and Bryden freshwater flux estimate is found to be 0.9×10^6 m^3/s southward, substantially lower than all three estimates of air-sea exchanges (Figure 3). The higher evaporations of the Isemer and Hasse compilation

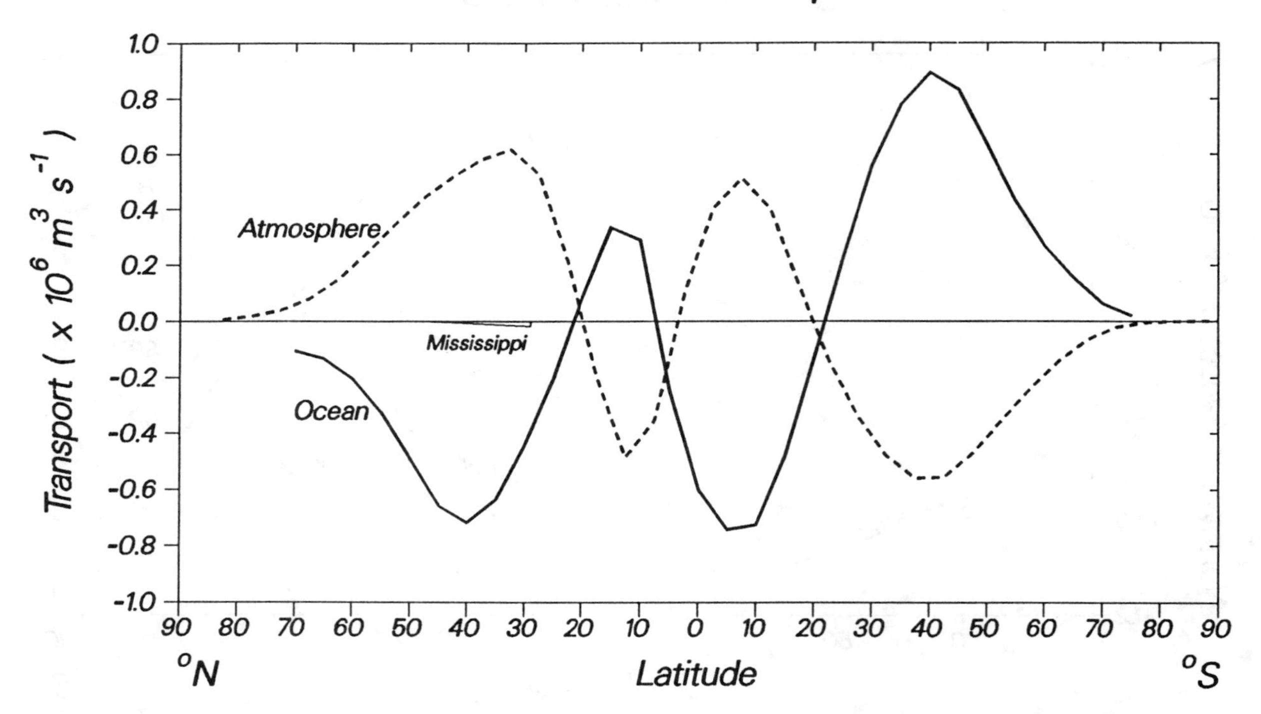

Fig. 4. Northward transport ($\times 10^6$ m^3s^{-1}) of freshwater as a function of latitude in the ocean (solid line) and the atmosphere (dashed line). Ocean transport at each latitude is the sum of the individual ocean transports in Figure 1. Atmospheric transport is derived from Peixoto and Oort's (1983, Table I) water vapor flux divergence values. The transport of the Mississippi River is also shown.

show the most discrepancy, suggesting that either reduced evaporation or increased precipitation at higher latitudes is required.

It is also possible to evaluate the meridional freshwater transport at 24°N in the Pacific from the measurements of Bryden et al. [1991]. Their estimate of salt flux due to the $\overline{v'S'}$ term can be combined with the observed Bering Straits transport to yield a value of northward water transport based only on oceanographic measurements. We obtain a value of 0.59×10^6 m^3/s. When differenced with the southward Atlantic transport of 0.9×10^6 m^3/s, a net southward transport of 0.31×10^6 m^3/s for the oceans at 24°N is obtained. This compares favorably with the integration of ocean surface E-P estimates given in Figure 5. Southward river transport at 24°N sums to approximately 10% of this value or 0.03×10^6 m^3/s (from data of Milliman and Meade [1983]). The ocean and river sum of 0.34×10^6 m^3/s is in good agreement with the Peixoto and Oort [1983] values for northward water transport by the atmosphere at this latitude, and corresponds to nearly a petawatt ($=10^{15}$ watts) in latent heat transport. This suggests that the "missing petawatt", discussed by Bryden [1992], might not be discovered in unresolved latent heat fluxes in the atmosphere, since present oceanographic and meteorological data are in general agreement on the amplitude of the hydrologic cycle, at least for this one latitude.

Twenty-four north is a latitude of intermediate freshwater flux, between the minima and maxima at 45°N and 10°N. These patterns emerge because of the net precipitation north of 45, the net evaporation between 10 and 45, and the rain band under the intertropical convergence zone. As previously noted, the climatologies are in poorest agreement in low latitudes. An oceanographic flux estimate at 10°N would be especially valuable for ascertaining the merits of the various estimates. The oceanic flux divergence between 10°N and 24°N could be a decisive factor in determining the best method for estimating evaporation under the trade winds.

However, a significant uncertainty in the oceanic estimates is the magnitude of seasonal and interannual variability. The upper ocean response to the seasonal cycle in winds is particularly important, as the Ekman transport can be a significant portion of the total flux. For example, at 24°N the Ekman transport is about 5 Sv, about one sixth of the Gulf Stream flow in Florida Straits. This is rather modest compared to other latitudes and its seasonal variation is less than 10%. The tropics are more problematic, as the Ekman transport can be much larger and show seasonal variability

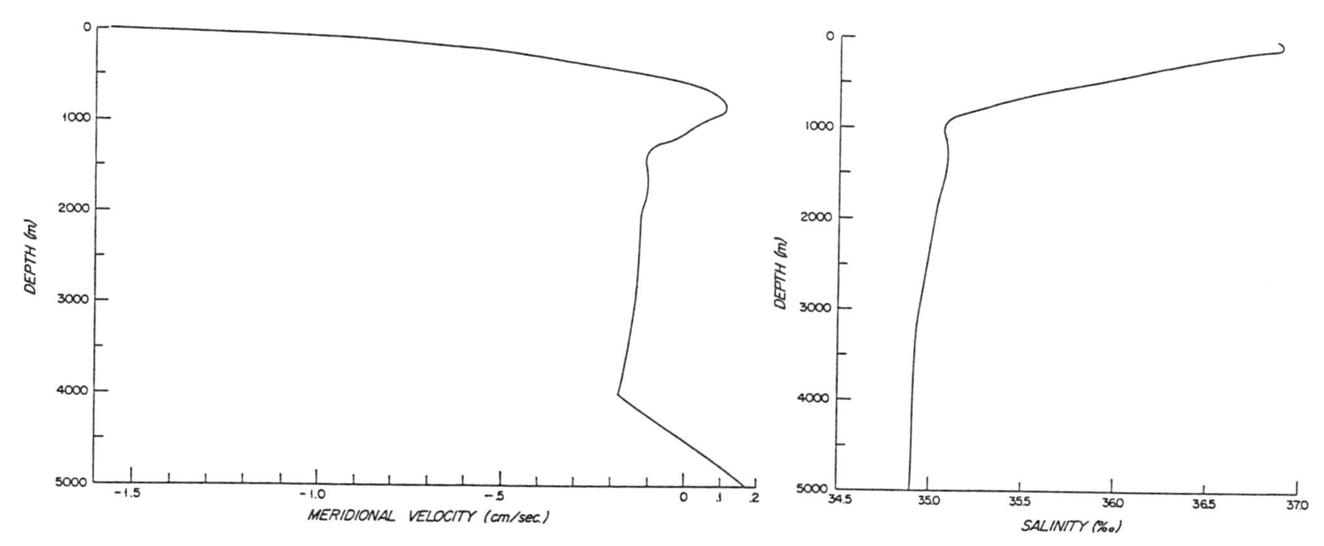

Fig. 5. The zonally averaged meridional velocity (a) and salinity (b) in the North Atlantic at 24°N, according to Hall and Bryden (1982). The profiles display correlation and can be used be used to compute the salt flux within the ocean, knowledge of which is necessary for assesment of the freshwater budget.

sufficient to change the sign of the poleward fluxes. There is also very little data on the interannual variation in transports computed from zonal sections. The 24°N section has been occupied in 1957 [Hall and Bryden, 1982], again in 1981 with higher resolution [Roemmich and Wunsch, 1985] and will be occupied again in 1992. Small but significant differences in abyssal temperatures were found between the 1981 and 1957 sections, indicating that a steady state assumption for the ocean heat and salt storage is not strictly accurate. Much more observational work needs to be done before we begin to resolve the temporal variability of oceanic flows.

SUMMARY

The oceans are the primary reservoir of water on Earth and are largely responsible for the redistribution of water transported by the atmosphere. The combined ocean-atmosphere hydrologic cycle accounts for a significant fraction of the poleward heat flux on the planet, and much of the rest is carried by the sensible heat content of the ocean itself. Thus, much could be learned about the global heat and water cycles by an improved understanding of the magnitude and mechanisms of ocean transports. In order to achieve this we require better assessments of air–sea fluxes and a comprehensive suite of direct ocean measurements.

The urgency of such studies is becoming increasingly clear. It is evident that the freshwater cycle, in addition to being a substantial component of the ocean-atmosphere heat engine, is also a controlling factor for the ocean thermohaline circulation itself. The paleoclimatic record contains dramatic evidence of strong changes in ocean deep water formation and circulation in response to the discharge of glacial melt water [Broecker, 1987; Keigwin et al., 1991]. High latitude freshwater fluxes can "cap-off" the ocean, creating a stable, low salinity surface layer that prevents deep convection, greatly decreasing heat transfer to the atmosphere and the formation of bottom water. Such processes seem to be occurring on decadal time scales as well, according to recent data describing an upper-ocean low salinity anomaly which was advected around the subpolar gyre of the North Atlantic during the '60s and '70s [Dickson et al., 1988]. By acting as a valve on the intensity of the large overturning cell by which near-surface warm water is transported poleward and deep cold water transported equatorward, high latitude freshwater fluxes play a critical role in climate regulation. Recent models of the ocean circulation [Joyce, 1991; Weaver, Sarachik and Marotzke, 1991] display extreme sensitivity to freshwater forcing, with strong changes in the mode of ocean overturning arising from small changes in surface fluxes. Unfortunately, we have little confidence in estimates of surface fluxes, especially in poorly sampled polar regions. Even in the subtropics there is great need to improve freshwater flux data, since the resulting effect on seawater density can dominate the net surface buoyancy flux there [Schmitt et al., 1989].

New satellite techniques, further development of climatologies and continued improvement of meteorological models may all contribute to better estimates of net freshwater flux over the ocean. However, the direct measurement of transports within the ocean will provide an important quantitative check on such estimates and allow evaluation of the largest terms in the planetary heat budget. The monitoring of flows in certain key straits could also provide vital information on the workings of the global hydrologic cycle. In addition, such work will help to define the dynamical processes within the ocean which ultimately control climate.

Acknowledgments. R. Schmitt was supported by grants from the Atlantic Climate Change Program of NOAA (NA16RC0521-01), the National Science Foundation (OCE 89-11053 and the Office of Naval Research (N00014-89-J-1073) and S. Wijffels was supported by a NASA Graduate Student Fellowship in Global Change Research. Ann Spencer and Veta Green assisted with manuscript preparation.

REFERENCES

Baumgartner, A., and E. Reichel, *The World Water Balance.* Elsevier, New York, 179 p, 1975.

Broecker, W. S., Unpleasant surprises in the greenhouse? *Nature*, **328**, 123–126, 1987.

Broecker, W. S., T-H. Peng, J. Jouzel, G. Russell, The magnitude of global freshwater transports of importance to ocean circulation. *Climate Dynamics*, **4**, 73–79, 1990.

Bryden, H. L., Ocean heat transport across 24°N latitude. Hann Symposium Volume (this issue), 1992.

Bryden, H. L., D. H. Roemmich and J. A. Church, Ocean Heat transport across 24°in the Pacific. *Deep Sea Research*, **38**, (3), 297–324, 1991.

Bunker, A. F., Computations of surface energy flux and annual air–sea interaction cycles of the North Atlantic Ocean. *Monthly Weather Review*, **104**, 1122–1140, 1976.

Coachman, L. K., and K. Aagaard, Transports through Bering Strait: annual and interannual variability. *Journal of Geophysical Research*, **93**, 15,535–15,539. 1988.

Dickson, R.R.,J. Meincke, S.Malmberg and A. Lee, The "Great Salinity Anomaly" in the northern North Atlantic 1968–1982. *Progress in Oceanography*, **20**, 103–151, 1988.

Dorman, C. E., and R. H. Bourke, Precipitation over the Atlantic Ocean, 30°S to 70°N. *Monthly Weather Review*, **109**, 554–563, 1981.

Fieux, M. Andrié, C., Kartavtseff, A. Delecluse, P., Molcard, R., Ilahude, A. G., Mantisi, F., Swallow, J., Measurements within the Pacific-Indian throughflow region: Preliminary results of the Jade Cruise. Paper presented at XX General Assembly of IUGG, Session PS–02, 1991.

Gordon, A. L., Interocean exchange of thermocline water. *Journal of Geophysical Research*, **91**, 5037–5046, 1986.

Hall, M. M., and H. L. Bryden, Direct estimates and mechanisms of ocean heat transport. *Deep-Sea Research*, **29**, 339–359, 1982.

Isemer, H.J. and L. Hasse, *The Bunker Climate Atlas of the North Atlantic Ocean*, **2**, Springer-Verlag, 252 pp, 1987.

Joyce, T.J., Thermohaline catastrophe in a simple four-box model of the ocean climate. *Journal of Geophysical Research*, **96**, (C11), 20,393–20,402, 1991.

Keigwin, L., G.A. Jones, S. J. Lehman, Deglacial meltwater discharge, North Atlantic deep circulation, and abrupt climate change. *Journal of Geophysical Research*, **96**,(C9), 16,811–16,826, 1991.

Milliman, J. D. and R. H. Meade, World-wide delivery of river sediment to the oceans. *Journal of Geology*, **91**, (1), 1–21, 1983.

Peixoto, J. P., and A. H. Oort, The atmospheric branch of the hydrological cycle and climate. Variations in the Global Water Budget, edited by A. Street - Perrott *et al.*, Reidel, 5–65, 1983.

Roemmich D. and C. Wunsch, Two transatlantic sections: meridional circulation and heat flux in the subtropical North Atlantic Ocean. *Deep-Sea Research*, **32**, 619–664, 1985.

Schmitt, R. W., P. S. Bogden and C. E. Dorman, Evaporation minus precipitation and density fluxes for the North Atlantic, 1989. *Journal of Physical Oceanography*, **19**, 1208–1221, 1989.

Weaver, A.J., E.S. Sarachik and J. Marotzke, Freshwater flux forcing of decadal and interdecadal oceanic variability. *Nature*, **353**, (6347), 836–838, 1991.

Wijffels, S.E., R.W. Schmitt, H.L. Bryden and A. Stigebrandt, Transport of freshwater by the oceans. *Journal of Physical Oceanography*, **22**, (2), 155–162, 1992.

Raymond W. Schmitt and Susan E. Wijffels, Woods Hole Oceanographic Institution, Woods Hole, MA 02543.

Carbon Dioxide and Nitrous Oxide in the Arabian Sea

S.W.A. NAQVI, R. SEN GUPTA AND M. DILEEP KUMAR

National Institute of Oceanography, Dona Paula, Goa, India

This review concentrates on some aspects of biogeochemical cycling of carbon and nitrogen in the Arabian Sea with emphasis on the atmospheric fluxes of CO_2 and N_2O. The Arabian Sea appears to serve as a net sink for combined nitrogen of ~3 x 10^{13} gN y^{-1}. Inflow of deep and bottom waters should make up this deficit providing a net carbon input of ~1.75 x 10^{15} g y^{-1}. This appears to sustain high atmospheric fluxes of CO_2 observed particularly during the summer monsoon. The distribution of dissolved organic carbon (DOC) reveals significant gradients corresponding to changes in total carbon dioxide (TCO_2), oxygen consumption and nitrate deficits, suggesting an important role for DOC in the oceanic biogeochemical cycles. The results on the activity of the respiratory electron transport system reinforce this view. Steady northward increases in TCO_2 are observed at all depths due to regeneration from biogenic debris which also seems to increase from west to east. Large regional variations occur in the regeneration of soft tissue and skeletal material. The Arabian Sea appears to be a significant source for atmospheric N_2O. The N_2O distribution suggests substantial losses to reducing zones, implying a very rapid turnover of N_2O within the region.

INTRODUCTION

The Arabian Sea is one of the most important areas of the ocean in regard to the processes which affect the atmospheric composition. First, it houses some of the ocean's most extensive upwelling sites, where deep waters, rich in carbon dioxide (CO_2) and nitrous oxide (N_2O), are brought to the surface where these gases can escape to the atmosphere. At the same time, upwelling also results in very high primary production rates in the area [Babenerd and Krey, 1974; Qasim, 1977, 1982; Naqvi, 1991a]; thus, substantial amounts of carbon are pumped back to the deep sea with the sinking particles and as suspended and dissolved organic carbon (DOC). Secondly, the region also experiences a severe depletion of dissolved oxygen at mid-depth (~150-1,200 m) [Wyrtki, 1971]. This leads to the development of reducing conditions within a large body of intermediate water especially in the northeast [Sen Gupta and Naqvi, 1984; Naqvi et al., 1992]. As under these conditions the oxidized nitrogen compounds are utilized by the facultative bacteria for the oxidation of organic matter (denitrification), the cycling of N_2O is quite different, and its turnover much more rapid, than in the more common oxidizing environments [Codispoti and Christensen, 1985].

The waters within the oxygen-deficient layer are renewed rapidly, in spite of the fact that the Arabian Sea is surrounded by landmasses on three sides, the estimates of renewal times varying from 10 months [Naqvi and Shailaja, 1992] to 11 years [Olson et al., 1992]. This is manifested in pronounced short-term temporal variations in chemical composition of the intermediate waters, particularly along the eastern boundary [Naqvi et al., 1990]. These observations imply time-variable fluxes of the greenhouse gases CO_2 and N_2O across various boundaries (air-sea, sediment-water and oxic-anoxic). In spite of the obvious importance of the Arabian Sea in contributing to the atmospheric inputs of these gases, adequate attention has not been paid so far to quantify their fluxes across the air-sea interface and to determine their monsoon-related seasonality. We present here a brief review of some recent works on CO_2 system and N_2O cycling in the Arabian Sea.

CARBON AND NITROGEN BUDGETS

There have been some attempts to estimate the magnitudes of various sources and sinks for carbon and combined nitrogen in the Arabian Sea [Somasundar et al., 1990; Naqvi et al., 1992]. However, these budgets should be considered as tentative as some of the important terms such as the exchanges with the Indian Ocean through the southern boundary are not properly constrained especially for carbon. The nitrogen budget of Naqvi et al. [1992], reproduced in Table 1, incorporates several important improvements over the effort of Somasundar et al. [1990]. For example, it takes into account inputs from the atmosphere and losses through sedimentary denitrification and as N_2O (this will be addressed in detail later) not considered by Somasundar et al. [1990]. Also, it does not consider any change due to biological production as the nitrogen exported out of the euphotic zone with the sinking matter is supplied by the upward flux. Finally, the loss due to denitrification

Interactions Between Global Climate Subsystems, The Legacy of Hann
Geophysical Monograph 75, IUGG Volume 15

TABLE 1. Carbon and nitrogen budgets of the Arabian Sea (Fluxes in Tg y^{-1} modified from Somasundar et al. [1990] and Naqvi et al. [1992])

Source/Sink	Carbon	Nitrogen
SOURCES		
River runoff	2.3	0.1
Persian Gulf		0.3
Red Sea		0.8
Nitrogen fixation		1.5
Atmospheric inputs	?	0.7
TOTAL INPUTS (A)	2.3	3.4
SINK		
Persian Gulf	11.7	
Red Sea	42.8	
Sedimentary & water column denitrification		31.0
N_2O losses to the atmosphere and to reducing zones		2.5
Burial	1.7	0.2
Flux to atmosphere	43.0	
TOTAL LOSSES (B)	99.2	33.7
NET INPUT (through deep and bottom water i.e. B-A)	96.9 (?) see text	30.3

(flux to the atmosphere = 2.05 Tg N y^{-1}) considered by Somasundar et al. [1990] is much smaller than the rate of denitrification (~30 Tg N y^{-1}) [Naqvi, 1987; Naqvi and Shailaja, 1992]. Consequently, the budget of Naqvi et al. [1992], unlike that of Somasundar et al. [1990], suggests the Arabian Sea to be a net sink for combined nitrogen of substantial magnitude (~30 Tg N y^{-1}). Naqvi et al. [1992] proposed that this deficit may be made up as a result of water exchange with the Indian Ocean, mostly through the inflow of bottom water. This requires a bottom water transport of ~2 x 10^6 m^3 s^{-1}, which is roughly half of the estimated transport of bottom water into the Somali Basin through the Amirante Passage [Barton and Hill, 1989]. This clearly shows that the extent of nitrogen input required by the mass balance calculations is not only possible but most likely. Naqvi et al. [1992] also suggested that the high primary productivity of the Arabian Sea and the associated "nutrient trap" at mid-depth could result principally from a high rate of nutrient supply with the northward flowing deep waters.

The nitrogen budget can be utilized to determine the unknown term in the carbon budget, i.e. input with the bottom water (Table 1). As the bottom water contains approximately 65-70 times more carbon on a molar basis than nitrogen, the nitrogen input for maintaining a steady state would require a net carbon flux of ~1,750 Tg annually. As is the case for nitrogen, the excess carbon should also be carried to the upper layers by upwelling. An unknown fraction of this will be transported outside the Arabian Sea. The large input of carbon to the Arabian Sea as suggested by the mass balance computations would greatly increase pCO_2 in the upper layers, sustaining a high rate of CO_2 emission from the sea to the atmosphere.

There have not been many measurements of carbonate properties in surface waters of the northern Indian Ocean, although the region has long been regarded as a source of atmospheric CO_2 [Keeling, 1968; Keeling and Waterman, 1968: Batrakov et al., 1981; Broecker et al., 1986; Etcheto and Merlivat, 1988; Takahashi, 1989; Tans et al., 1990; Cochran, 1991]. The northwestern Indian Ocean is expected to be characterized by extremely large seasonal changes in the extent of carbon dioxide emission to the atmosphere caused by variations in both the ocean-air pCO_2 gradient [Tans et al., 1990; Cochran, 1991] and the CO_2 exchange coefficient [Etcheto et al., 1991]. Significantly, both the exchange coefficient and the pCO_2 gradient reach peak values at the same time - during the southwest (SW) monsoon - when widespread upwelling occurs off Somalia, Arabia and, to a smaller extent, off southwest India. Very high pCO_2 values (450 μatm) at the surface have been observed in the western Arabian Sea during the SW monsoon; such high concentrations must be the consequence of intense upwelling which probably overwhelms the opposing effect expected from the high rates of photosynthesis [Cochran, 1991]. Given the strong winds that persist during this period (the mean for July sometimes exceeds 17 m s^{-1}) [Hastenrath and Lamb, 1979], it would correspond to an atmospheric flux reaching up to 40 mmol m^{-2} d^{-1}, among the highest recorded for the world's ocean. Still the overall significance of the northwestern Indian Ocean as a source of CO_2 for the atmosphere is poorly quantified due to the sparse and uneven coverage of the region in space and time. The estimate included in Table 1 is based on the monthly mean speeds, averaged for the entire Arabian Sea from the climatology of Hastenrath and Lamb [1979], coupled with the assumed pCO_2 gradients of 50 μatm during the SW monsoon (June-August) and 25 μatm during the rest of the year. As expected, the bulk (~87%) of the CO_2 flux occurs during the three SW monsoon months. One would arrive at similar conclusions using the exchange coefficients of Etcheto et al. [1991] computed from the satellite-derived wind data; however, the latter values would be lower as they represent an anomalously weak SW monsoon year (1987). Although the estimate of CO_2 flux included in Table 1 compares favourably with that (60 Tg y^{-1}) of Takahashi [1989] for the Indian Ocean north of Lat. 10°N, we regard our estimate as conservative because of the nature of the relationship between the CO_2 exchange coefficient and wind speed [Liss and Merlivat, 1986]. Clearly, more observations are needed to accurately quantify the CO_2 flux to the atmosphere from the Arabian Sea and to understand its temporal variability on both the seasonal and interannual scales.

A high rate of CO_2 pumping from the deep to the surface layers is partly compensated by the transports of carbon to the deep sea with sinking particles and as suspended and dissolved organic carbon (DOC). A crude estimate of the vertical flux can be made as follows. Qasim [1977] estimated the total primary production in the Arabian Sea as 1,064 Tg C y^{-1}. If we assume that at least 20% of this represents new production [Somasundar et al., 1990], then the organic matter sinking beneath the surface layer will be ~200 Tg C y^{-1}. Further, given that the ratio between the flux of carbon leaving the surface as organic matter and as $CaCO_3$ is ~4 [Broecker and Peng, 1982], the inorganic carbon flux should be around 50 Tg y^{-1}. Thus we get a total of ~250 Tg

C as the amount of carbon exported out of the surface layer of the Arabian Sea annually. For comparison, Somasundar et al. [1990] estimated the flux at 1,000 m depth as 47.6 Tg y^{-1}.

ROLE OF DOC

In addition to the vertical sinking flux, significant inputs of organic carbon to the deep sea may also occur as DOC that could be transported from the surface layer by advection and diffusion. Kumar et al. [1990] found that the DOC levels in the Arabian Sea, determined following a high-temperature catalytic oxidation method, were higher by a factor of 3-4 than the values reported previously by Menzel [1964] from this region based on the more conventional wet oxidation method. The results of Kumar et al. [1990], reproduced in Figure 1 for a north-south section, reveal large horizontal gradients in DOC; the vertical gradients are small, however, particularly in the north. The observed DOC concentrations show good negative correlation with total carbon dioxide (TCO_2) (Figure 2). Although the general relationship between the DOC and apparent oxygen utilization (AOU) was found to be rather weak [Kumar et al., 1990], improved negative correlations could be seen when the data from various regions are treated separately [Figure 3]. Interestingly, the DOC was found to exhibit linear relationship with nitrate deficit (a measure of the extent of denitrification) showing minimal values (~60 µM) within the core of the denitrifying layer (DNL) at the northernmost stations [Kumar et al., 1990]. These observations suggest a non-conservative behaviour of DOC in the Arabian Sea in conformity with the results from some other oceanic areas [e.g., Sugimura and Suzuki, 1988; Suzuki et al., 1992]. The northward decrease in DOC is believed to be caused by its increased utilization during respiration by an increasingly abundant heterotrophic bacterial population [Kumar et al., 1990]. It may be pointed out, however, that there currently exists considerable uncertainty concerning the analysis of DOC in seawater following the high-temperature catalytic oxidation methods [Suzuki et al., 1992]. Although the operational parameters and the catalyst for oxidation in the method followed by Kumar et al. [1990] differed from those of Sugimura and Suzuki [1988], the DOC levels observed in the two studies are comparable.

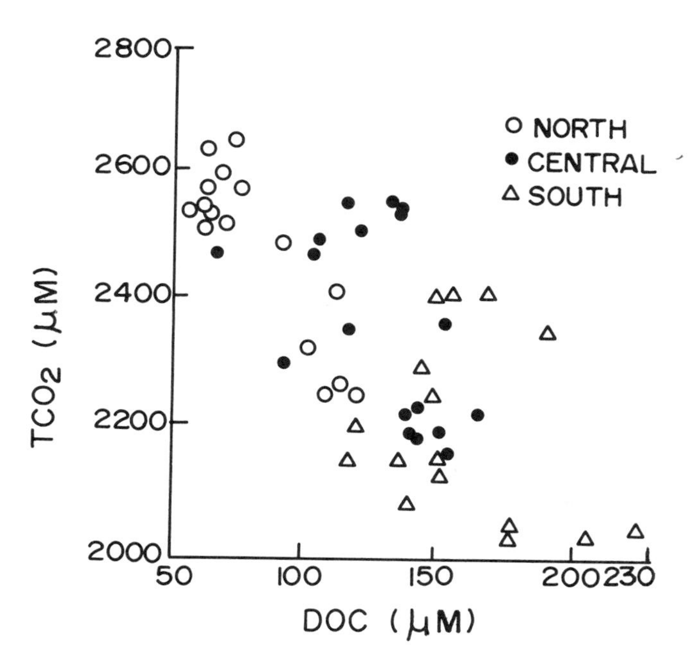

Fig. 2. Relationship of DOC with total carbon dioxide at three selected stations in the Arabian Sea (from Kumar et al., 1990).

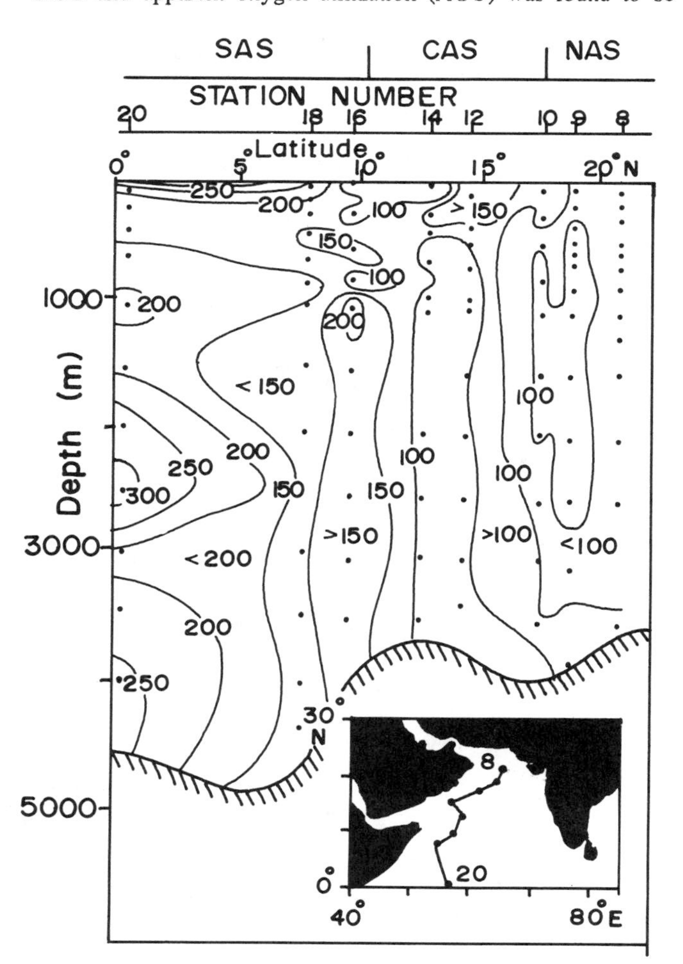

Fig. 1. Vertical section of DOC (µM) in the northwestern Indian Ocean. The area has been divided into the northern, central and southern Arabian Sea at the top of the Figure (from Kumar et al., 1990).

The possible role of DOC in fuelling subsurface respiration including denitrification is also supported by recent measurements of the activity of the respiratory electron transport system (ETS) by Naqvi and Shailaja [1992]. They found that the ETS-based denitrification rates do not follow the pattern expected from the data on primary productivity, being maximal beneath offshore area of low primary production (Figure 4). These results indicate a decoupling of subsurface denitrification from primary productivity of the overlying surface waters. This is consistent with the data on nitrogen system as the highest nitrite concentrations are also observed in areas of relatively low production [Naqvi, 1991a]. Based on computations involving global relationship between

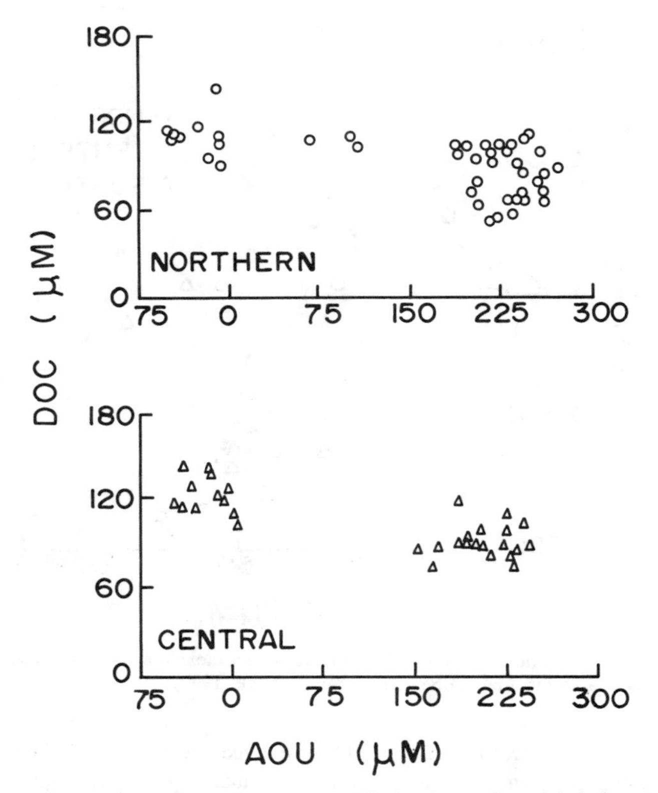

Fig. 3. Relationship of DOC with AOU for the northern (circles) and central (triangles) parts of the Arabian Sea (modified from Kumar et al., 1990).

primary productivity and the sinking organic flux [Betzer et al., 1984], Naqvi and Shailaja [1992] demonstrated that the vertical, particle-associated fluxes could be inadequate to sustain the calculated denitrification rates. These results have been interpreted to point to additional modes of supply of organic carbon, probably through quasi-horizontal processes, as suspended and/or dissolved organic matter [Naqvi et al., 1992; Naqvi and Shailaja, 1992].

Regeneration of CO_2 from Biogenic Debris

Total carbon dioxide (TCO_2) contents of subsurface waters in the Arabian Sea show steady increases northward at all the depths [Kumar et al., 1992]. This is probably a combined effect of an increased rate of CO_2 regeneration from both the soft tissues and skeletal materials due to a general northward increase in productivity, and the aging of subsurface waters as they move northward. This is evident in Figure 5 which gives the distribution of CO_2 added due to the regenerative processes. The regenerated CO_2 exceeds 500 μM at intermediate depths in the northern Arabian Sea, being higher by a factor of about 1.5 at any depth at 21°N than that produced at the equator. Total CO_2 also exhibits good linear correlations with AOU and phosphate (Figure 6), reflecting the dominant role of the oxidative processes in CO_2 regeneration. However, the slopes of the TCO_2- AOU and TCO_2-PO_4 regression lines vary systematically from north to south [Kumar et al., 1992]. This implies regionally variable contributions by regeneration from the organic matter and from $CaCO_3$. Kumar et al. [1992] have attempted to resolve the two following Kroopnick [1985]. Their results, reproduced in Figure 7, reveal that the extent of skeletal carbonate dissolution (ICO_2) also increases northward. For example, it was higher by a factor of 3 in the northern Arabian Sea as compared to that in the equatorial region. ICO_2 also decreases from east to west (Figure 7). The observed general northward increase in ICO_2 is in accordance with the productivity pattern, while the decrease from east to west is not. The ratio of CO_2 produced by the soft tissue decomposition to that by skeletal dissolution was more in the south and western regions than in the north and east (Table 2). An increased rate of $CaCO_3$ supply coupled with a higher rate of $CaCO_3$ dissolution due to a progressive northward increase in the degree of $CaCO_3$ undersaturation may explain the decrease in the ratio in the northern Arabian Sea. As a result of limited river runoff, the supply of $CaCO_3$ from river runoff would be small. Although the atmospheric input is unknown, it is also not expected to be very large. Thus, the $CaCO_3$ supply to the subsurface layers may be regulated mainly by the biological production in surface layers which, of course, is closely related to upwelling and vertical mixing caused by the monsoon winds.

It is well known that the northeastern Arabian Sea experiences a more severe oxygen depletion than the western Arabian Sea [Wyrtki, 1971]. One would, therefore, expect the deep waters in the northeastern region to be more corrosive to $CaCO_3$. This may in part be responsible for the observed higher OCO_2/ICO_2 in the west (Table 2). As proposed by Kumar et al. [1992], it is also possible that the skeletal material (such as cocolith tests) may be

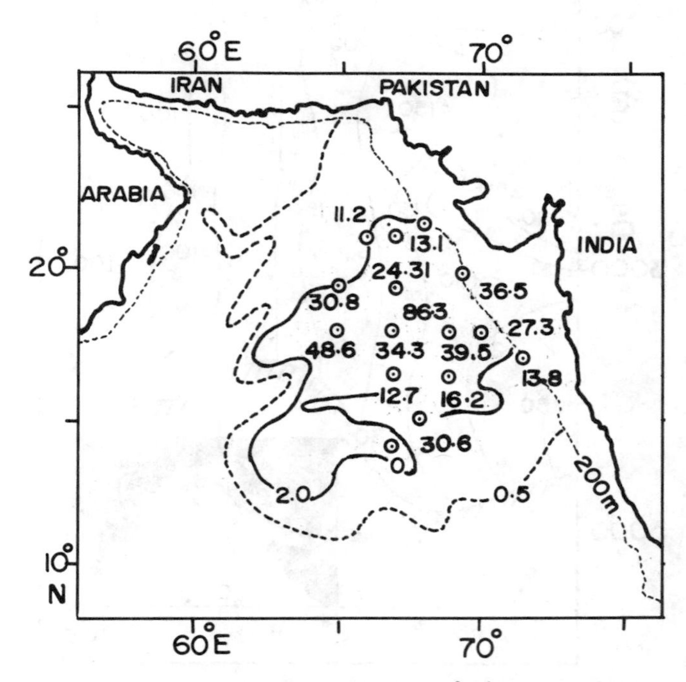

Fig. 4. ETS-based denitrification rates (gN m^{-2} y^{-1}) integrated over the denitrifying layer. The contours represent the distribution of nitrite (μM) at the secondary maximum (from Naqvi and Shailaja, 1992).

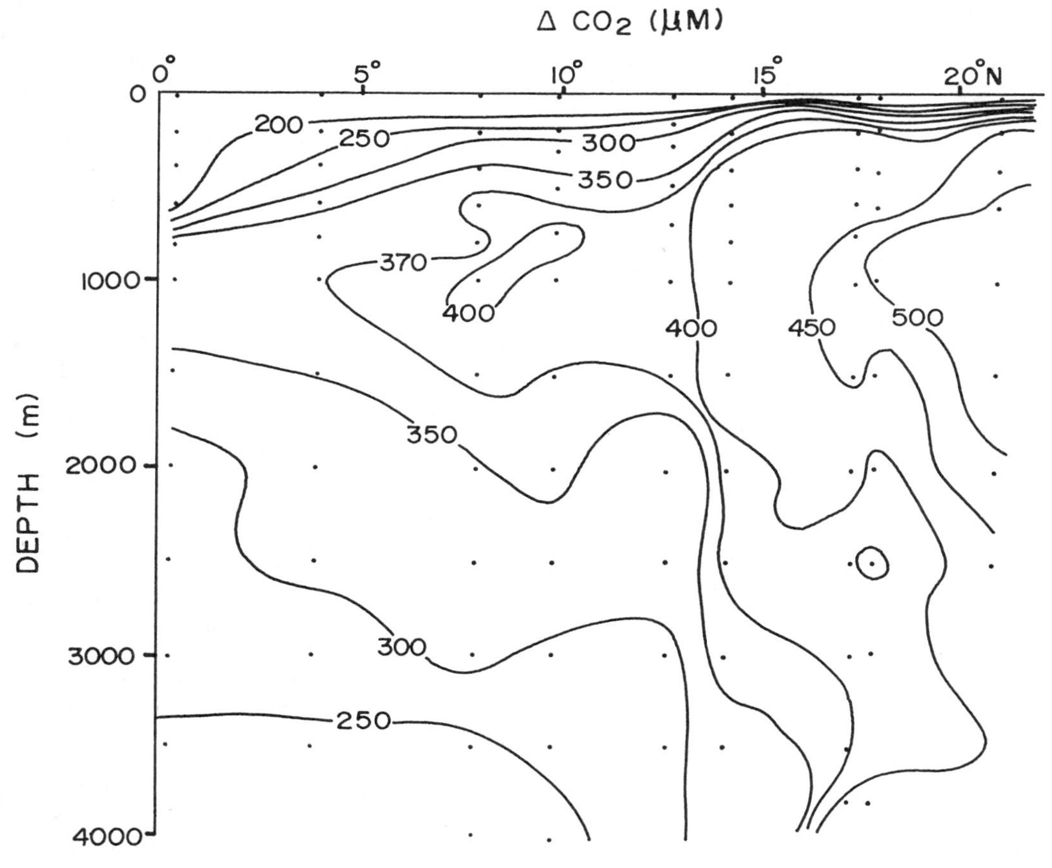

Fig. 5. Extent of the regenerated CO_2 along a north-south section (as in Figure 1) in the Arabian Sea (from Kumar et al., 1992).

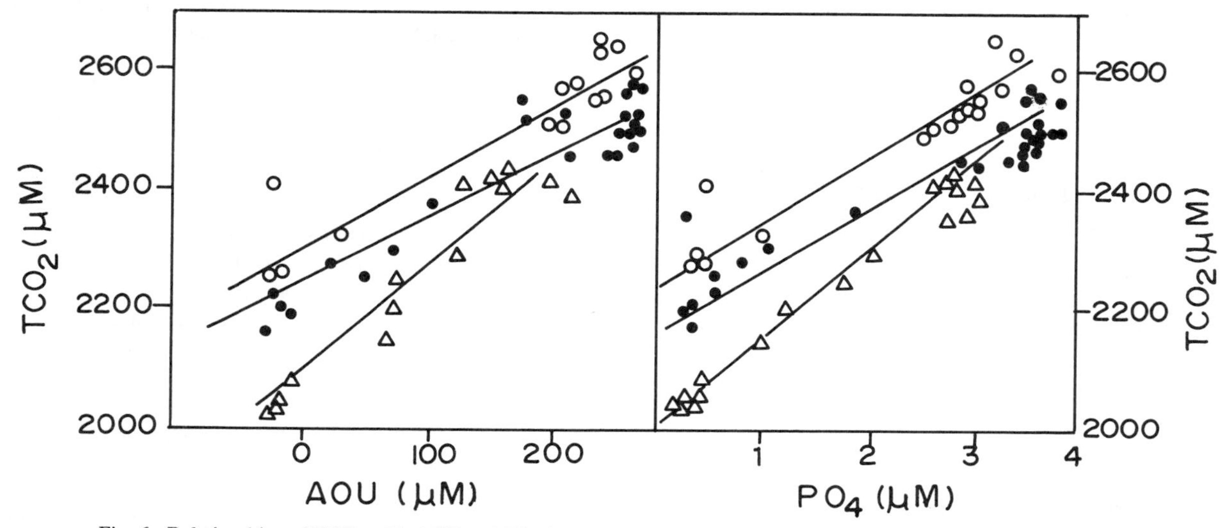

Fig. 6. Relationships of TCO_2 with AOU and PO_4 in the Arabian Sea. Open circles, closed circles and triangles represent observations from the northern, central and southern regions, respectively (from Kumar et al., 1992.)

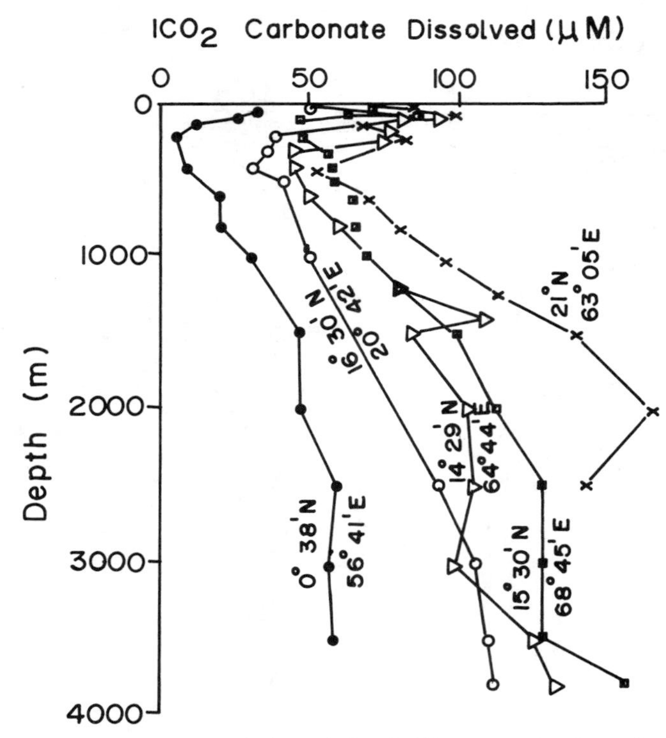

Fig. 7. Regional and depthwise variations in CO_2 regeneration due to the solution of skeletal material (from Kumar et al., 1992).

rapidly incorporated into the fecal pellets and consequently undergo deposition with a lower degree of dissolution in the western Arabian Sea, which experiences intense upwelling.

Nitrous Oxide Cycling

Measurements of nitrous oxide in the air and surface waters of the northwestern Indian Ocean were first made during the GEOSECS cruises by Weiss [1978]. He found the surface waters to be consistently supersaturated with N_2O north of about 10°S latitude (i.e. the Hydrochemical Front [Wyrtki, 1973]). While the average supersaturation for the world's oceans based on the analyses of 4,200 samples was ~4%, the supersaturation observed in the Arabian Sea was considerably higher (~25%). Recent observations by Law and Owens [1990] and Naqvi and Noronha [1991] support these results, and provide more comprehensive data sets on N_2O cycling in the northwestern Indian Ocean.

TABLE 2. Geographical variations in the ratio Org. CO_2/ Inorg. CO_2 (from Kumar et al., 1992)

Sta. No.	Location	800m	1500m	2500m
8 (North)	21°00'N 63°05'E	3.3	2.2	1.8
2 (East)	15°30'N 68°45'N	3.9	2.5	1.8
3 (Central)	14°29'N 64°44'E	4.0	2.6	2.0
12 (West)	14°21'N 57°40'E	7.5	3.4	2.4
20 (South)	00°38'N 56°41'E	9.1	6.0	2.7

Law and Owens [1990] found a high degree of N_2O saturation (167±46%) in surface waters throughout the northwestern Indian Ocean in the period following the SW monsoon (September-October, 1986). The N_2O fluxes across the air-sea interface estimated by these authors were consequently high (15.51±7.66 μmol m^{-2} d^{-1} based on a stagnant-film model and 8.64±4.32 μmol m^{-2} d^{-1} calculated from the CO_2 transfer velocities assuming an annual average wind speed of 5 m s^{-1}). The total atmospheric flux for the region bounded by the latitudes 15 and 25°N and longitudes 50 and 72°E was estimated as 0.22-0.39 Tg N_2O. Law and Owens [1990] considered it to be 5-18% of the total marine source from merely 0.43% of the total oceanic area.

Naqvi and Noronha [1991] also observed high degrees of saturation of N_2O in surface waters of the northeastern and central Arabian Sea during December 1988. The average saturation observed by them (186%) was remarkably similar to that reported by Law and Owens [1990] north of 15°N latitude. However, the atmospheric fluxes estimated by Naqvi and Noronha [1991] (4.46±2.60 μmol m^{-2} d^{-1}) were lower because these were based on lower wind speeds. Indeed, their estimate for the entire Arabian Sea (0.44 Tg N_2O) is comparable with the total flux north of 15°N computed by Law and Owens [1990]. The average wind speed chosen by Law and Owens [1990] for the computation of the atmospheric flux is close to the annual mean for the entire Arabian Sea (4.64 m s^{-1}) calculated from the data of Hastenrath and Lamb [1979]. If we combine this with the 167% saturation observed by Law and Owens [1990], the N_2O flux to the atmosphere from the Arabian Sea works out to be ~0.3 Tg N_2O y^{-1}. However, since the studies conducted so far have not sampled intense upwelling waters, where, as mentioned earlier, most vigorous outgassing is expected to occur, and because of the nature of the relationship between the transfer coefficient and wind speed, the above estimates of the N_2O flux from the Arabian Sea (0.30-0.44 Tg) may be conservative. Nevertheless, the available data clearly show that the northwestern Indian Ocean serves as a large source for atmospheric N_2O.

The vertical distribution of N_2O in the Arabian Sea is very similar to that observed in the other oxygen-depleted environments; the results obtained in the two studies are in excellent agreement with each other [Law and Owens, 1990; Naqvi and Noronha, 1991]. As expected, minimal values (<10 nM) occur at depths where a secondary nitrite maximum is also observed (Figure 8, reproduced from Naqvi and Noronha [1991]), presumably due to the consumption of N_2O during denitrification. The maxima in the N_2O profile (> 50 nM) are located at the peripheries of DNL. Outside the denitrifying zone, only one broad maximum in N_2O is observed. The increased production of N_2O at low oxygen concentrations (in waters other than those which are strongly reducing) has been well known, but has been interpreted differently by different authors to support its production through nitrification or denitrification (see Naqvi and Noronha [1991] and Codispoti et al. [1992] for discussions). Naqvi [1991b] and Naqvi and Noronha [1991] argued

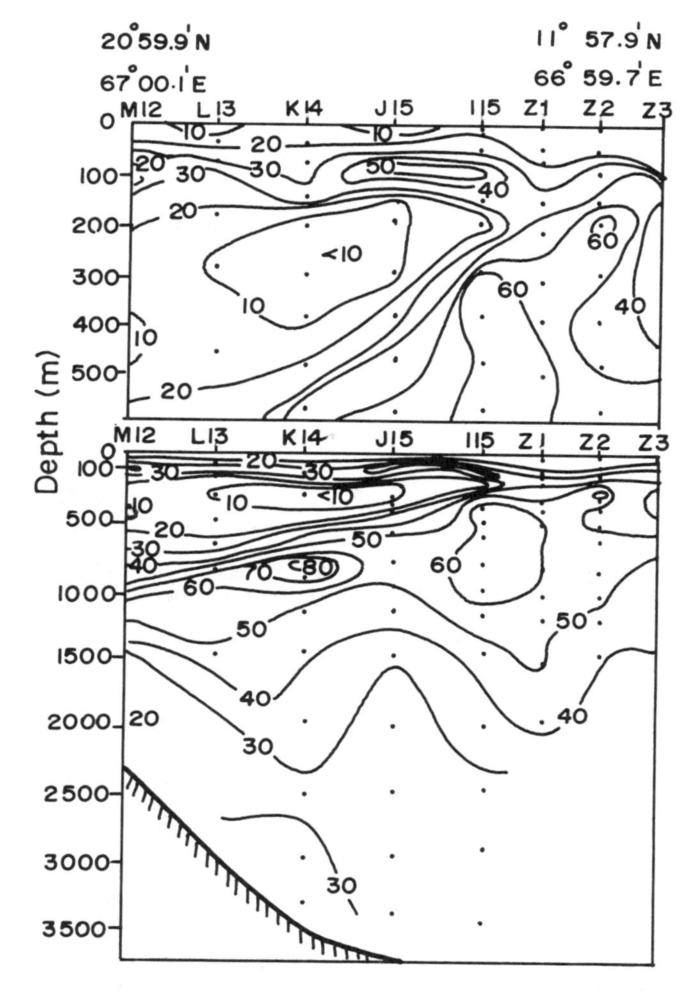

Fig. 8. Vertical section showing distribution of nitrous oxide (nM) along 67°E longitude (from Naqvi and Noronha, 1991).

that as both the classical nitrification and denitrification pathways are likely to lead to a depletion of ^{15}N in N_2O, these processes may not dominantly contribute to the oceanic N_2O production in view of the observed enrichment of the heavier isotopes in dissolved nitrous oxide within the oxygen minimum of the Pacific Ocean [Yoshida et al., 1989; Kim and Craig, 1990]. Instead, these authors proposed a possible coupling of nitrification and denitrification through nitric oxide (NO), i.e. production of NO through nitrification with hydroxylamine as an intermediate and its reduction to N_2O within the micro-environments, to be the mechanism responsible for N_2O accumulation at low oxygen concentrations. Although speculative, this pathway ($NH_4^+ \rightarrow NH_2OH \rightarrow NO \rightarrow N_2O$) can explain both the isotopic composition and the distribution of N_2O within the oxygen minimum zones.

The probable loss of N_2O through denitrification may be a quantitatively significant term in the nitrogen budget. The large horizontal and vertical gradients in N_2O concentrations at the boundaries of the reducing zones are expected to result in the transport and subsequent reduction of N_2O within the oxygen- deficient zones. Naqvi and Noronha [1991] estimated that ~1.4 Tg N_2O-N could be lost within the water column in this manner. An interesting feature of N_2O distribution in the Arabian Sea is that there occur large onshore-offshore gradients (decrease toward the continental margin) both off India [Naqvi and Noronha, 1991] and Oman [Law and Owens, 1990]. It would appear from these observations that the marginal sediments in contact with the low- oxygen (not anoxic or suboxic) waters in the Arabian Sea could also serve as a significant sink for N_2O. From the observed gradients, Naqvi and Noronha [1991] estimated a minimal loss of N_2O to the eastern margin (2000 km x 1 km) as 0.8 Tg N y^{-1}. It is possible, however, that the observed gradients could also result from the advection of lower-N_2O waters off the continental margin, but the similarity of distributions off Oman and India makes it rather unlikely. Thus, as summarized in Table 1, the net losses of N_2O from the Arabian Sea are probably >2 Tg N y^{-1}. It is evident from the N_2O distribution that the Arabian Sea as a whole is a net source for oceanic N_2O, requiring an annual gross production rate well over 2 Tg N. If we compare this with the recent estimates of the extent of oceanic N_2O production (~5 Tg N y^{-1}) [Oudot et al., 1990], it would appear that the N_2O turnover in the oceans may be more rapid than realized so far.

The relationship between the apparent N_2O production [$\Delta N_2O = (N_2O)_{obs} - (N_2O)_{eq}$], where the subscripts "obs" and "eq" refer to the N_2O concentrations measured and expected from equilibrium with atmospheric N_2O, respectively, and AOU in the Arabian Sea appears to be linear, although within two ranges [Law and Owens, 1990; Naqvi and Noronha, 1991] as in other oceanic areas [Yoshinari, 1976; Elkins et al., 1978; Cline et al., 1987; Oudot et al., 1990]. There are two aspects of the N_2O-AOU relationship which should be mentioned. First, the slope of the ΔN_2O-AOU regression lines in the deep Arabian Sea (depth >1 km in Naqvi and Noronha [1991] and AOU >197 µM in Law and Owens [1990]) are significantly higher than most other values found in the literature [Cohen and Gordon, 1978; Elkins et al., 1978; Oudot et al., 1990]. This implies a higher rate of N_2O production in the deep Arabian Sea. Secondly, when the data from the Arabian Sea are fitted to a regression function similar to that of Elkins et al. [1978], the relationship obtained [Naqvi and Noronha, 1991] shows an independent variation of ΔN_2O with temperature which is much weaker than that seen elsewhere [Elkins et al., 1978; Butler et al., 1989]. This is in conformity with the results of Law and Owens [1990]. It is possible that temperature serves as a surrogate tracer at ammonification. That is, N_2O production rate is largely a function of labile carbon, which may generally follow a depth dependence similar to temperature. The little variability of the organic carbon fluxes with depth in the deep Arabian Sea [Nair et al., 1989] could thus result in an apparently weaker temperature dependence of N_2O.

REFERENCES

Babenerd, B., and J. Krey, *Indian Ocean: Collected Data on Primary Production, Phytoplankton Pigments and Some Related Factors, 521* pp., Universitatsdruckerei Kiel, Kiel, 1974.

Barton, E.D., and A.E. Hill, Abyssal flow through the Amirante Trench (Western Indian Ocean), *Deep-Sea Res., 36,* 1121-1126, 1989.

Batrakov, G.F., A.A. Bezborodov, V.N. Yeremeyev, and A.D. Zemlyanoy, CO_2 exchange between Indian Ocean waters and the atmosphere, *Oceanology, 21*, 39-43, 1981.

Betzer, P.R., W.J. Showers, E.A. Laws, C.D. Winn, G.R. DiTullio, and P.M. Kroopnick, Primary productivity and particle fluxes on a transect of the equator at 153°W in the Pacific Ocean, *Deep-Sea Res., 31*, 1-11, 1984.

Broecker, W.S., and T.H. Peng, *Tracers in the Sea*, 690 pp., Eldigio Press, Palisades, 1982.

Broecker, W.S., J.R. Ledwell, T. Takahashi, R. Weiss, L. Merlivat, L. Memery, T.H. Peng, B. Jahne, and K.O. Munnich, Isotopic versus micrometeorologic ocean CO_2 fluxes: a serious conflict, *J. Geophys. Res., 91*, 10517-10527, 1986.

Butler, J.H., J.W. Elkins, and T.M. Thompson, Tropospheric and dissolved N_2O of the West Pacific and East Indian Oceans during the El Nino Southern Oscillation event of 1987, *J. Geophys. Res., 94*, 14865-14877, 1989.

Cline, J.D., D.P. Wisegarver, and K. Kelly-Hansen, Nitrous oxide and vertical mixing in the equatorial Pacific during the 1982-1983 El Nino, *Deep-Sea Res., 34*, 857-873, 1987.

Cochran, J.K., Sources and sinks for geochemical tracers in the northern Indian Ocean, in *U.S. JGOFS: Arabian Sea Process Study*, U.S. JGOFS Planning Report No. 13, pp. 66-74, Woods Hole Oceanographic Institution, Woods Hole, 1991.

Codispoti, L.A., and J.P. Christensen, Nitrification, denitrification, and nitrous oxide cycling in the eastern tropical South Pacific Ocean, *Mar. Chem., 16*, 277-300, 1985.

Codispoti, L.A., G.E. Friederich, C.M. Sakamoto, J. Elkins, T.T. Packard, and T. Yoshinari, Nitrous oxide cycling in upwelling regions underlain by low oxygen waters, in *Oceanography of the Indian Ocean*, edited by B.N. Desai, Oxford & IBH, New Delhi, in press, 1992.

Cohen, Y., and L.I. Gordon, Nitrous oxide in the oxygen minimum of the eastern tropical North Pacific: evidence for its consumption during denitrification and possible mechanisms for its production, *Deep-Sea Res., 25*, 509-524, 1978.

Elkins, J.W., S.C. Wofsy, M.B. McElroy, C.E. Kolb, and W.A. Kaplan, Aquatic sources and sinks for nitrous oxide, *Nature, 175*, 602-606, 1978.

Etcheto, J., and L. Merlivat, Satellite determination of the carbon dioxide exchange coefficient at the ocean-atmosphere interface: a first step, *J. Geophys. Res., 93*, 15669-15678, 1988.

Etcheto, J., J. Boutin, and L. Merlivat, Seasonal variation of the CO_2 exchange coefficient over the global ocean using satellite wind speed measurements, *Tellus, 43*, 247-255, 1991.

Hastenrath, S., and P.J. Lamb, *Climatic Atlas of the Indian Ocean, Part I: Surface Climate and Atmospheric Circulation*, 97 pp., Wisconsin University Press, Madison, 1979.

Keeling, C.D., Carbon dioxide in surface ocean waters, 4, Global distribution, *J. Geophys. Res., 73*, 4543-4553, 1968.

Keeling, C.D., and L.S. Waterman, Carbon dioxide in surface ocean waters, 3, Measurements on Lusiad Expedition 1962-1963, *J. Geophys. Res., 73*, 4529-4541, 1968.

Kim, K.-R., and H. Craig, Two isotope characterization of N_2O in the Pacific Ocean and constraints on its origin in deep water, *Nature, 347*, 58-61, 1990.

Kroopnick, P.M., The distribution of ^{13}C of ΣCO_2 in the world oceans, *Deep-Sea Res., 32*, 57-84, 1985.

Kumar, M.D., A. Rajendran, K. Somasundar, B. Haake, A. Jenisch, Z. Shuo, V. Ittekkot, and B.N. Desai, Dynamics of dissolved organic carbon in the northwestern Indian Ocean, *Mar. Chem., 31*, 299-316, 1990.

Kumar, M.D., A. Rajendran, K. Somasundar, V. Ittekkot, and B.N. Desai, Processes controlling carbon components in the Arabian Sea, in *Oceanography of the Indian Ocean*, edited by B.N. Desai, Oxford & IBH, New Delhi, in press, 1992.

Law, C.S., and N.J.P. Owens, Significant flux of atmospheric nitrous oxide from the northwest Indian Ocean, *Nature, 346*, 826-828, 1990.

Liss, P., and L. Merlivat, Air-sea gas exchange rates: introduction and synthesis, in *The Role of Air-Sea Exchange in Geochemical Cycling*, edited by P. Buat-Menart, pp. 113-128, D. Reidel Pub. Co., Dordrecht, 1986.

Menzel, D.W., The distribution of dissolved organic carbon in the western Indian Ocean, *Deep-Sea Res., 11*, 757-766, 1964.

Nair, R.R., V. Ittekkot, S.J. Manganini, V. Ramaswamy, B. Haake, E.T. Degens, B.N. Desai, and S. Honjo, Increased particle flux to the deep ocean related to monsoons, *Nature, 338*, 749-751, 1989.

Naqvi, S.W.A., Some aspects of the oxygen deficient conditions and denitrification in the Arabian Sea, *J. Mar. Res., 45*, 1049-1072, 1987.

Naqvi, W.A., Geographical extent of denitrification in the Arabian Sea in relation to some physical processes, *Oceanol. Acta, 14*, 281-290, 1991a.

Naqvi, S.W.A, N_2O production in the ocean, *Nature, 349*, 373-374, 1991b.

Naqvi, S.W.A., and R.J. Noronha, Nitrous oxide in the Arabian Sea, *Deep-Sea Res., 38*, 871-890, 1991.

Naqvi, S.W.A., and M.S. Shailaja, Activity of the respiratory electron transport system and respiration rates within the oxygen minimum layer of the Arabian Sea, *Deep-Sea Res., 39*, in press, 1992.

Naqvi, S.W.A., R.J. Noronha, K. Somasundar, and R. Sen Gupta, Seasonal changes in the denitrification regime of the Arabian Sea, *Deep-Sea Res., 37*, 593-611, 1990.

Naqvi, S.W.A., R.J. Noronha, M.S. Shailaja, K. Somasundar, and R. Sen Gupta, Some aspects of the nitrogen cycling in the Arabian Sea, in *Oceanography of the Indian Ocean*, edited by B.N. Desai, Oxford & IBH, New Delhi, in press, 1992.

Olson, D.B, G.L. Hitchcock, R.A. Fine, and B.A. Warren, Maintenance of the low-oxygen layer in the central Arabian Sea, *Deep-Sea Res., 39*, in press, 1992.

Oudot, C., C. Andrie, and Y. Montel, Nitrous oxide production in the tropical Atlantic Ocean, *Deep-Sea Res., 37*, 183-202, 1990.

Qasim, S.Z., Biological productivity of the Indian Ocean, *Indian J. Mar. Sci., 6*, 122-137, 1977.

Qasim, S.Z., Oceanography of the northern Arabian Sea, *Deep-Sea Res., 29*, 1041-1068, 1982.

Sen Gupta, R., and S.W.A. Naqvi, Chemical oceanography of the Indian Ocean, north of the equator, *Deep-Sea Res., 31*, 671-706, 1984.

Somasundar, K., A. Rajendran, M.D. Kumar, and R. Sen Gupta, Carbon and nitrogen budgets of the Arabian Sea, *Mar. Chem., 30*, 363-377, 1990.

Sugimura, Y., and Y. Suzuki, A high temperature catalytic oxidation method of non-volatile dissolved organic carbon in seawater by direct injection of liquid sample, *Mar. Chem., 24*, 105-131, 1988.

Suzuki, Y., E. Tanoue, and H. Ito, A high temperature catalytic oxidation method for the determination of dissolved organic carbon in seawater: analysis and improvement, *Deep-Sea Res., 39*, 185-198, 1992.

Takahashi, T., The carbon dioxide puzzle, *Oceanus, 32*, 22-29, 1989.

Tans, P.P., I.Y. Fung, and T. Takahashi, Observational constraints on the global atmospheric CO_2 budget, *Science, 247*, 1431-1438, 1990.

Weiss, R.F., Nitrous oxide in surface water and marine atmosphere of the North Atlantic and Indian Oceans, *Eos Trans. AGU, 59*, 1101, 1978.

Wyrtki, K., *Oceanographic Atlas of the International Indian Ocean Expedition*, 531 pp., National Science Foundation, Washington D.C., 1971.

Wyrtki, K., Physical oceanography of the Indian Ocean, in *The Biology of the Indian Ocean*, edited by B. Zeitzschel, pp. 18-36, Springer-Verlag, Berlin, 1973.

Yoshida, N., H. Morimoto, M. Hirano, I. Koike, S. Matsuo, E. Wada, T. Saino, and A. Hattori, Nitrification rates and ^{15}N abundances of N_2O and NO_3 in the western North Pacific, *Nature, 342*, 895-897, 1989.

Yoshinari, T., Nitrous oxide in the sea, *Mar. Chem., 4*, 189-202, 1976.

S.W.A. Naqvi, R. Sen Gupta and M. Dileep Kumar, National Institute of Oceanography, Dona Paula, Goa - 403 004, India.

Re-Evaluation of the Global Energy Balance

ATSUMU OHMURA and HANS GILGEN

Swiss Federal Institute of Technology Zurich (E.T.H.), Winterthurerstrasse 190, CH-8057 Zurich, Switzerland

Energy balance on the earth's surface is reevaluated based mainly on the instrumentally measured fluxes. The basic data source for the present work is the Global Energy Balance Archive (GEBA), WCP-Water Project A7. The present article explains the content of the GEBA and presents the first results of the reevaluation, which are centred around radiation. Regional and global distributions of global radiation, surface albedo and absorbed shortwave radiation are investigated and their latitudinal distributions and global means are calculated. The annual mean of global radiation for the earth is about 50 % of the extraterrestrial solar radiation. The Northern Hemisphere receives slightly more global radiation than the Southern Hemisphere. The globally averaged global radiation reaches a maximum and minimum in December and June, respectively. The monthly albedo of the earth's surface is evaluated for the land surfaces taking the snow cover into account, for the open water surface taking solar altitude and cloud amount into account and for the sea ice covered surface taking ice concentration and seasonal albedo change of the ice floe into account. The global mean earth's surface albedo is 0.16. The new value for the global mean absorbed shortwave radiation is 142 Wm^{-2} or 42 % of the extraterrestrial solar radiation. The annual mean absorbed shortwave radiation is the same for both hemispheres. The global mean absorbed shortwave radiation reaches the maximum and minimum in December and June, respectively.
The problems of underestimation of longwave measurement by the conventional installation method for pyrradiometers and pyrgeometers are discussed and the degree of the underestimation of the longwave incoming radiation is estimated at 10 to 15 Wm^{-2}. This puts the annual mean net radiation for the earth's surface at 102 Wm^{-2} or 30 % of the extraatmospheric solar radiation.

INTRODUCTION

The earth's energy balance plays a fundamental role in the formation of the circulations and the thermal conditions of the atmosphere and the ocean, hence climate. Within the earth's climate system, the earth's surface plays an especially important role, because this is the place where the most active radiative absorption and energy transformation take place. Therefore, accurate knowledge of the energy exchange on the earth's surface is essential for unterstanding the formation of the present climate as well as the processes of the climatic changes of the past and the future.

Despite progress in energy balance climatology during the last forty years, there are several reasons why the reevaluation of the global energy balance should be made. It is presently possible to improve the geographical distribution of the energy balance components and the possibility to investigate their secular variation. The following grounds for the reevaluation are detailed:

Interactions Between Global Climate Subsystems, The Legacy of Hann
Geophysical Monograph 75, IUGG Volume 15

1. Recent satellite-based observation of the planetary albedo indicates lower values [Barkstrom et al., 1990] than those used for previous global evaluations [Budyko et al., 1963]. This development must yield more absorption within the atmosphere or by the earth's surface or both.
2. Previously published atlases require numerical improvements and regional supplementations. The basic problem of these evaluations is their heavy dependence on empirical formulae which are affected by regionalities. In addition, polar regions have not been adequately dealt with.
3. The latent heat of melt, which has been ignored in previous works, should be taken into consideration, as this is the major heat sink on the melting surfaces, such as seasonal snow cover, sea ice and glaciers.
4. There are important secular variations in energy fluxes. Such secular variations were first found in global radiation [Ohmura and Lang, 1988] and are presently known to exist also for such non-radiative components as latent heat flux. This phenomenon may be important in understanding the processes of climatic changes.
5. Recently there have been great improvements in the quality and quantity of the instrumental measurements of energy fluxes. This trend has been intensified since the IGY.

30d42'N, 111d05'E	Yichang, CN	50 m ams

Surface Characteristics: grass

Period(s) of measurement: 1958 - 1960

Measured Components: Q, q, Gl, a, R

Results: Kcalcm-2mon-1 and Kcalcm-2y-1

	Jan	Feb	Mar	Apr	May	Jun	Jul	Aug	Sep	Oct	Nov	Dec	Year
Q	1.63	3.33	2.76	3.30	5.11	6.38	7.74	5.87	6.67	2.38	2.31	2.85	49.73
q	3.36	4.17	5.33	6.29	6.49	6.31	6.72	5.99	4.20	4.99	3.11	3.15	60.71
Gl	4.99	7.50	8.09	5.59	11.50	12.69	14.46	11.86	10.87	7.37	5.42	6.00	110.44
a	0.21	0.21	0.20	0.19	0.19	0.20	0.20	0.20	0.21	0.21	0.21	0.21	0.20
R	0.52	2.44	3.28	4.86	6.29	7.47	9.78	7.31	5.19	2.11	0.38	1.29	50.88

Methods: Q: At-50 type pyrrheliometer; q: AC-3X3 type pyranometer;
R: Yanishevskii net radiometer; a: by inverting the q sensor.

The data are hourly spot (instantaneous) readings with a currentmeter
(Ld+Lu as calculated with the Berlyand equation are also presented)

Source: Pan, S.-W., 1962: Radiation-climatic characteristics of the mi d-upper reaches of the Yangtse. Acta Mete orologica Sinica, Vol. 32, No. 3, 119-213

Fig. 1. GEBA data sheet for non-permanent stations. Data evaluated for non-permanent stations are organized as monthly means over the period of measurement. An example in the figure shows monthly mean direct solar, diffuse sky, global radiation, albedo and net radiation measured for three years at Yichang, China.

6. The satellite radiometry is shown to be useful in retrieving the radiative fluxes at the earth's surface. This method produces better results for surfaces with a rather homogenous or constant albedo and emissivity, such as oceans for which instrumental measurements are scarce.

All these new conditions contribute to the understanding of the energy balance of the earth. For the reevaluation, the authors developed a worldwide data base for the instrumentally measured energy fluxes for the earth's surface, in a project of the World Climate Programme-Water in 1986.The data base possesses 150'000 station month data for about 1'600 sites. In the following sections, new aspects of the global energy balance will be presented mainly based on the WCP-Water Project A7, Global Energy Balance Archive (GEBA). The present work concentrates on radiative components. The non-radiative fluxes will be dealt with in later publications.

GLOBAL ENERGY BALANCE ARCHIVE (GEBA)

The objective of the Global Energy Balance Archive (GEBA) is to organize the instrumentally measured energy fluxes; it does not include empirically estimated values. Under the instrumentally measured fluxes all radiometers are included except the bimetalic actinographs and the Bellani spherical pyranometers. For the turbulent fluxes measurements, the gradient-flux type approximations with more than two-level measurements are accepted, but one level bulk approximation is excluded from the GEBA.

The main sources of data are periodicals, monographs, data reports and unpublished data. Therefore, the quality of the data is variable and a rigorous quality control has been applied.

The fluxes archived are direct radiation, diffuse sky-radiation, global radiation, shortwave reflected radiation or albedo, longwave incoming radiation, longwave outgoing radiation, longwave net radiation, net radiation, sensible and latent heat fluxes, latent heat of melt and subsurface heat flux. The minimum duration of the measurements considered is one month.

Energy flux data were first extracted from the above mentioned sources. This procedure necessarily involved a certain standardization and formalization, for which purpose a standard sheet for monthly fluxes were used (Fig. 1). Measurements with more than ten years duration are classified as the observation at permanent stations and are organized as a time-series of monthly means for each year (Fig. 2). These fluxes are accompanied with information about the site, i.e., coordinates, altitude, geographical region and country, observational period, surface characteristics, observational methods and the instruments, and the source of the data. As this information is stored in the computer, the entire copy of the original printed sources is also archived in our library.

60d24′ N 5d19′E 45 m   Bergen   NO   Ld, ly / day

	Jan	Feb	Mar	Apr	May	Jun	Jul	Aug	Sep	Oct	Nov	Dec	Year
1965				618	632	724	691	717	721	706	586	591	666
1966	580		630	574	661	730	699	719	723	689	627	619	656
1967	618	639	661	634	682	704	740	752	722	692	691	636	682
1968	615	594	647	630	660	725	716	701	732	686	607	618	661
1969	655	550	576	654	692	730	768	771	750	734	608	595	673
1970	589	576	608	596	658	674	708	713	684	672	619	620	644
1971	637	616			699	710	736	723	697	661	605	659	676
1972	538	567	601	619	668	708	715	692	644	660	607	627	639
1973	630	614	659	615	702	726	772	730	717	644	613	620	671
1974	651	647	575	609	656	709	726	738	715	612	628	648	660
1975	642	586	575	618	671	694	746	758	714	668	649	648	665
1976	583	601	546	614	673	725	745	715	667	664	644	533	637
1977	575	551	654	614	644	708	731	731	679	699	655	648	661
1978	615	552	616	595	655	742	763	753	692	690	643	533	654
1979	552	548	608	648	682	730	729	729	702	644	626	592	651
1980	576	587	585	643	682	764	775	748	724	611	597	628	661
1981	606	572	611	604	702	731	749	729	732	668	647	526	657
1982	595	608	627	639	674	678	739	741	695	642	625	598	656
1983	617	559	598	622	691	707	731	722	696	678	644	610	658
1984	543	613	550	631	674	712	745	737	692	703	652	641	662
1985	576	593	606	626	674	708	744	748	686	707	569	606	654
1986	546	515	647	594	707	693	723	689	682	700	674	621	651
1887	551	599	589	656	653	712	717	721	695	687	646	643	656
1988	638	613	581	614	686	702	751	742	735	684		647	668
1989	678	634	645	601	665	684	689	716	687	656	612	592	655

Source: The Radiation Observatory , Geophysical Institute, University of Bergen, 1990: Radiation Observations in Bergen, Norway, 1989. University of Bergen, Geophysical Insti tute, Bergen, 1990.

Fig. 2. GEBA data sheet for permanent stations. An example of longwave incoming radiation for Bergen, Norway is presented in the figure.

The distribution of the GEBA stations is uneven (Fig. 3). The ocean surfaces are extremely poorly investigated, except for the Western Pacific and the Northeastern Atlantic. The land surface are covered, in general, densely enough so that global radiation can be mapped with the measured data alone. The landbased sites are located in 124 countries grouped in the seven regions which are identical to those of the WRDC/Leningrad. A serious data shortage exists in parts of Latin American and Arab countries.

The importance of the individual stations differ widely. Stations are presented in Table 1, with respect to energy balance components and the duration of the measurements. The most widely observed component is global radiation. Historically viewed, global radiation is also the oldest regularly observed component. Global radiation is currently observed for more than 10 years consecutively at about 600 stations. There are about 50 stations worldwide with more than 30 years measurement of global radiation. The next well observed component is net radiation with 177 sites with more than one year of observations and 75 locations with more than 10 years. There is a serious lack of information for albedo. Although the albedo is currently measured at a number of locations, it is mostly for the short grass surface at meteorological stations. The lack of albedo over natural vegetations, especially shrubs and forests, is a serious drawback for radiation climatology. The seasonal variation of

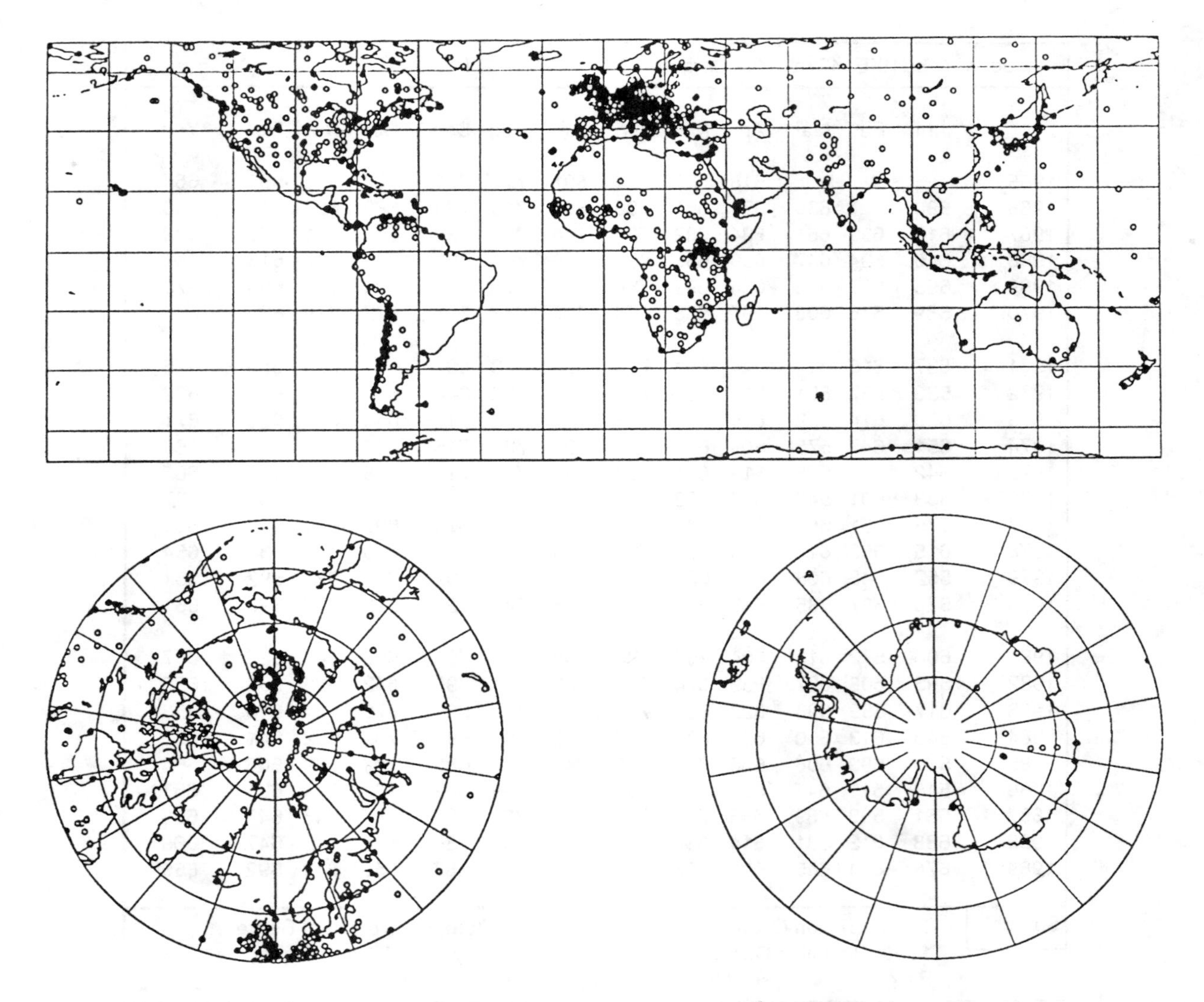

Fig. 3. Distribution of the stations archived in the GEBA as of August 1st 1991. There are about 1'600 sites.

the albedo for these surfaces can be very different from that for short grass. A wide range of the albedo change happens with the onset of the snow cover. The albedo on short grass may reach 0.8 after a short snow fall, while over the neighbouring evergreen coniferous forest, it may stay below 0.25 throughout the winter. The lack of such elementary information is one of the most serious hurdles to overcome before completing the computation of absorbed global radiation.

Table 1 also shows that the turbulent fluxes are rarely measured for a long period. There are seven sites so far archived, where all energy balance components are measured at least for one year. All these sites are located in the Northern Hemisphere.

A relatively new addition to the GEBA is a set of energy balance measurements over the ocean surfaces, usually accomplished as a part of oceanographic investigations.

The present state of the archive shows that the global radiation can be charted for continental surfaces without help of other means such as the model calculation with satellite data. For other components, however, the GEBA data can be used to check the reliability of model and bulk computations for selected locations. Detailed descriptions of the structure of the GEBA are presented in Ohmura et al. (1989) and Ohmura and Gilgen (1991).

GLOBAL RADIATION

Global radiation is by far the most frequently measured component of all the energy balance fluxes. Presently there

TABLE 1. The number of stations in the GEBA by energy balance component and duration of the measurement (≥ 360 means more than or equal to 360 months record; 52/53 means 52 permanent stations and 53 permanent as well as non-permanent stations).

	component	≥ 360	≥ 240	≥ 120	≥ 12	≥ 1
Gl	global radiation	52/53	308/318	595/612	1141/1258	1200/1553
Q	direct solar rad.		2/2	5/6	29/39	29/43
q	diffuse sky rad.	2/2	12/12	33/72	59/72	59/76
a	albedo			0/2	1/15	4/180
r	reflected short-wave r.	1/1	7/7	9/10	17/36	17/59
L↓, Ld	inc. long-wave r.	1/1	2/2	2/4	7/20	9/44
L↑, Lu	outg. long-wave r.	1/1	1/1	1/3	4/15	4/34
L + L	long-wave net r.		1/1	1/2	1/12	1/34
R	radiation balance	1/1	39/39	72/75	139/177	145/249
H	sensible heat flux			0/2	0/8	0/57
lE	latent heat flux			1/3	7/26	8/79
G	subsurface heat flux			0/2	0/15	0/50
M	latent heat of melt			0/1	0/4	0/28
UV	ultraviolet radiation				1/4	1/5
Gl + r	absorbed short-w. r.			2/3	8/15	8/32

are about 2000 stations, equipped with thermo-elecric pyranometers. The estimated accuracy of most pyranometers is better than +/- 15 Wm^{-2}, if the instruments are properly maintained. In this sense, global radiation is the most accurately known energy flux.The global coverage, however, falls short for ocean surfaces.

There is only one site with more than one year's continuous record of global radiation, observed aboard ship, namely Ocean Weather Station I. Semi-regular observation on shipboard is just starting (H. Otobe, personal communication) in the Pacific. The satellite-based estimation of the surface global radiation is still in methodological development. Previously published investigations with computed surface energy fluxes on the oceans are, therefore, compared to the observed global radiation in the GEBA. Two outstanding works on the ocean energy fluxes were published by Oberhuber (1988) and Bottomley et al. (1990). Both works are based on simple empirical computational schemes with standard meteorological and oceanographical data, with relatively accurate results. The computed values in these works for 75 maritime sites are compared against the GEBA data for monthly and annual means. An example of the annual mean values is presented in Table 2. Mean values for the 75 sites indicate a general overestimation in these empirical values of 11 Wm^{-2} or 6 %, well within the present limit of accuracy for pyranometers in field conditions. Both works show, however, a marked regional difference in quality. Bottomley et al.(1990) shows better agreement with the observed values for the Atlantic Ocean and the Tropical Pacific regions, while Oberhuber's (1988) computations appear better for the Mediterranean, the Northwestern Pacific and the Indian Oceans. Regional works such as Hastenrath and Lamb (1978) and Iwasaka and Hanawa (1990) offer considerably improved values for the Eastern Tropical Pacific and Northern Pacific, respectively.

New distributions of global radiation for various regions of the world are attempted for monthly and annual means. These distributions are based on about 900 land-based measurements with more then 5 years of records, supplemented with the above mentioned ocean surface fluxes for the regions in best agreement with the GEBA. An example of the new map of annual mean global radiation for Middle Europe is presented in Fig. 4.

These large-scale, high resolution maps make the bases for the new global distribution maps for various energy components. Examples of global radiation for the entire earth are presented in Figs. 5, 6 and 7 for monthly and annual means.

The new distribution maps show significant differences from previous works for the regions of Africa, Southeast Asia, U.S.A., and especially for ocean and polar regions. Global annual mean global radiation based on the present results is 169 Wm^{-2} (50 % of the extraatmospheric solar radiation) and slightly smaller than previous results [e.g. Budyko, 1978]. With respect to the hemispheric annual means, the Northern Hemisphere with 172 Wm^{-2} receives slightly more global radiation than the Southern Hemisphere at 167 Wm^{-2}. The difference of 5 Wm^{-2} is considered to be caused by differences in cloud conditions and surface albedo between both hemispheres. Viewed globally, the earth receives more global radiation in December (176 Wm^{-2}) than in June (163 Wm^{-2}). The December-June difference of 12 Wm^{-2} in global mean global radiation is equivalent to a fluctuation of +/- 3.4%, which is exactly the same as the annual fluctuation of the global extraatmospheric solar radiation, resulting from the present eccentricity of the earth's orbit of 0.0167. A final

TABLE 2. Comparison of the annual mean global radiation for oceans among the computed (MO / MIT: Bottomley et al., 1990 and MPIM: Oberhuber, 1988) and the measured values in the GEBA).

	Wm^{-2}			
	MO / MIT	MPIM	GEBA	Obs. Years
Atlantic Ocean				
Ocean Weather Station I	80	70	84	64-67
Dunstaffnage	95	100	94	70-87
De Kooy	120	125	112	78-87
St. Pierre	120	110	122	72-79
Sable Island	130	120	133	69-87
Noirmoutier	140	125	137	85-87
Genova	180	150	138	64-87
Montpellier	180	150	158	75-87
Pisa	180	150	152	64-87
Tunis	180	180	185	64-67
Tampa	200	175	202	52-74
Mindelo	200	175	238	57-74
Dakar	200	200	227	64-87
Lamentin	205	205	196	74-79
Isla Orchila	205	205	192	64-80
Pacific Ocean				
Wakkanai	120	125	121	74-80
Nemuro	115	130	137	74-80
Miyako	135	135	137	74-80
Fukuoka	160	140	135	59-87
Hachijojima	160	135	132	74-80
Kagoshima	165	145	148	74-80
Naha	175	150	149	74-80
Ishigakijima	175	160	177	74-80
Chichijima	190	160	174	74-80
Minamitorishima	200	180	206	74-80
Hong Kong	175	160	162	69-78
Johnston	200	180	235	75-76
Invercargill	110	125	131	65-87
Nandi	170	175	195	65-87
Suva / Laucala Bay	170	175	183	85-87
Whenuapai	145	150	162	65-69
Aukland	145	150	165	69-87
Ohakea	140	150	157	65-87
Taita	135	145	144	65-68
Wellington	135	140	147	77-79
Christchurch	120	130	145	65-87
Pearl Harbor	185	180	230	50-52
Honolulu	185	180	238	50-80
Wake Island	205	180	244	51-76
Guam	195	190	199	78-80
Rabaul	170	175	191	64-69
Port Moresly	170	175	226	65-73
Townsville	185	190	223	65-69
Kuomac	175	180	218	80-87
Gladstone	180	180	255	64-66
Sidney	160	160	168	82-83
Aspendale	140	140	167	66-81
Indian Ocean				
Khormaksar	240	225	291	64-67
Djibouti	235	230	227	73-79
Pinang	170	200	196	75-87
Gan Island	185	190	223	67-75
Karachi	220	215	207	66-87

TABLE 2. Continued

	Wm^{-2}			
	MO / MIT	MPIM	GEBA	Obs. Years
Bhaunagar	215	215	219	67-87
Vishakhapatnam	190	190	213	64-87
Goa	200	200	214	64-87
Madras	200	190	216	64-87
Trivandrum	200	190	212	64-87
Zanzibar Island	220	215	191	64-87
Dar es Salaam	220	215	187	64-74
Beira	200	205	219	57-86
Gillot	190	190	210	73-87
Inhambane	195	200	207	64-87
Maputo	190	185	202	55-87
Durban	180	175	173	52-76
Port Elisabeth	160	160	191	57-76
Marion Island	140	125	133	54-75
Kuala Lumpur	165	170	180	73-87
Singapore	165	175	172	65-84
Djakarta	180	175	155	70-73
Dili	200	200	209	64-74
Denpasar	190	190	196	70-73
Perth	180	190	201	65-69

confirmation of the annual fluctuation of global radiation must be made after detailed investigations of the seasonal attenuation of solar radiation. It is interesting to note that the present radiometric network revealed percentwise the same magnitude of the annual fluctuation of global radiation on the earth's surface as at the upper boundary of the atmosphere.

ALBEDO OF THE EARTH'S SURFACE

Little is known about the albedo of the earth's surface on a global scale. The GEBA has presently archived albedo for about 200 sites. These measurements are made mostly over the short grass at meteorological stations and are often far from representative of the surrounding surfaces. Other problems were mentioned in the introduction. Selecting the best currently available sources the earth's surface albedo is reevaluated for the following three surfaces, land, ice-free and ice-covered ocean surfaces.

The land surface albedo are evaluated mainly based on the digitized land-use information by Matthews (1985) and the monthly mean snow cover thickness by K. Masuda (personal communication), seasonal albedo by Davies (1965), McFadden and Rogotzkie (1967), Kondrat'ev (1973), and Roth (1985) in addition to the GEBA and the extensive airborn albedometry conducted during the last seven years by the authors' group.

The albedo of the open water surface represents about 65 % of the entire earth's surface. The ice-free ocean surface albedo is calculated taking into account the altitude of the sun and total cloud amount, as expressed in the following formula:

$$A = A_G\ (1 - N) + A_D\ N \qquad (1)$$

where A is the monthly mean albedo of water surface, A_G is the Grishchenko's monthly fair weather albedo for open water as presented by Cogley (1979), A_D is the open water albedo for diffuse sky-radiation assumed as 0.10, and N is the monthly mean total cloud amount [Warren et al.1986,1988].

The albedo for the sea ice covered portion of oceans is poorly known. The albedo values reported for sea ice are usually obtained over the real ice surfaces, avoiding leads and paddles. The areally averaged albedo for the sea ice must be measured from low-flying aircraft at least on a monthly basis. Lacking this kind of information, the authors estimated the regional distribution of the albedo over the sea ice area, based on a number of albedo measurements over the drier portion of the sea ice available from the GEBA and the ice concentration as compiled by NAVY / NOAA-JIC for the period of 16 years from 1973 to 1988 for Antarctic and 17 years from 1972 to 1988 for the Arctic. The 2 lat. x 2 long. degrees areal mean albedo is calculated as a weighted mean of the albedo of the unpaddled ice surface and the albedo of the water surface as discussed above, with respect to the monthly ice concentration.

Examples of the global albedo of the earth's surface for June and December are presented in Figs. 8 and 9.

Annual Mean Global Radiation for Middle Europe

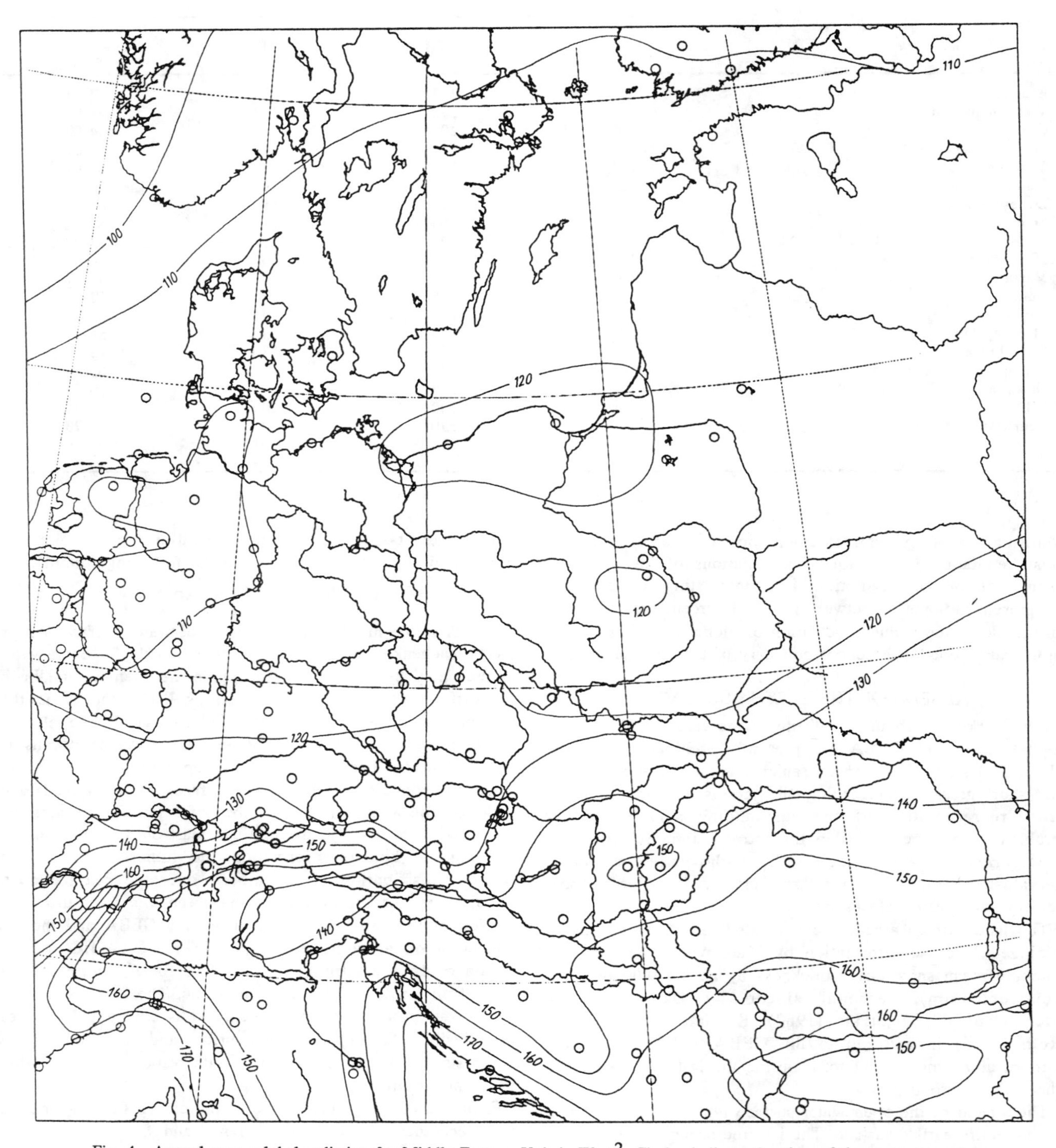

Fig. 4. Annual mean global radiation for Middle Europe. Unit in Wm^{-2}. Circles indicate the sites of the measurement.

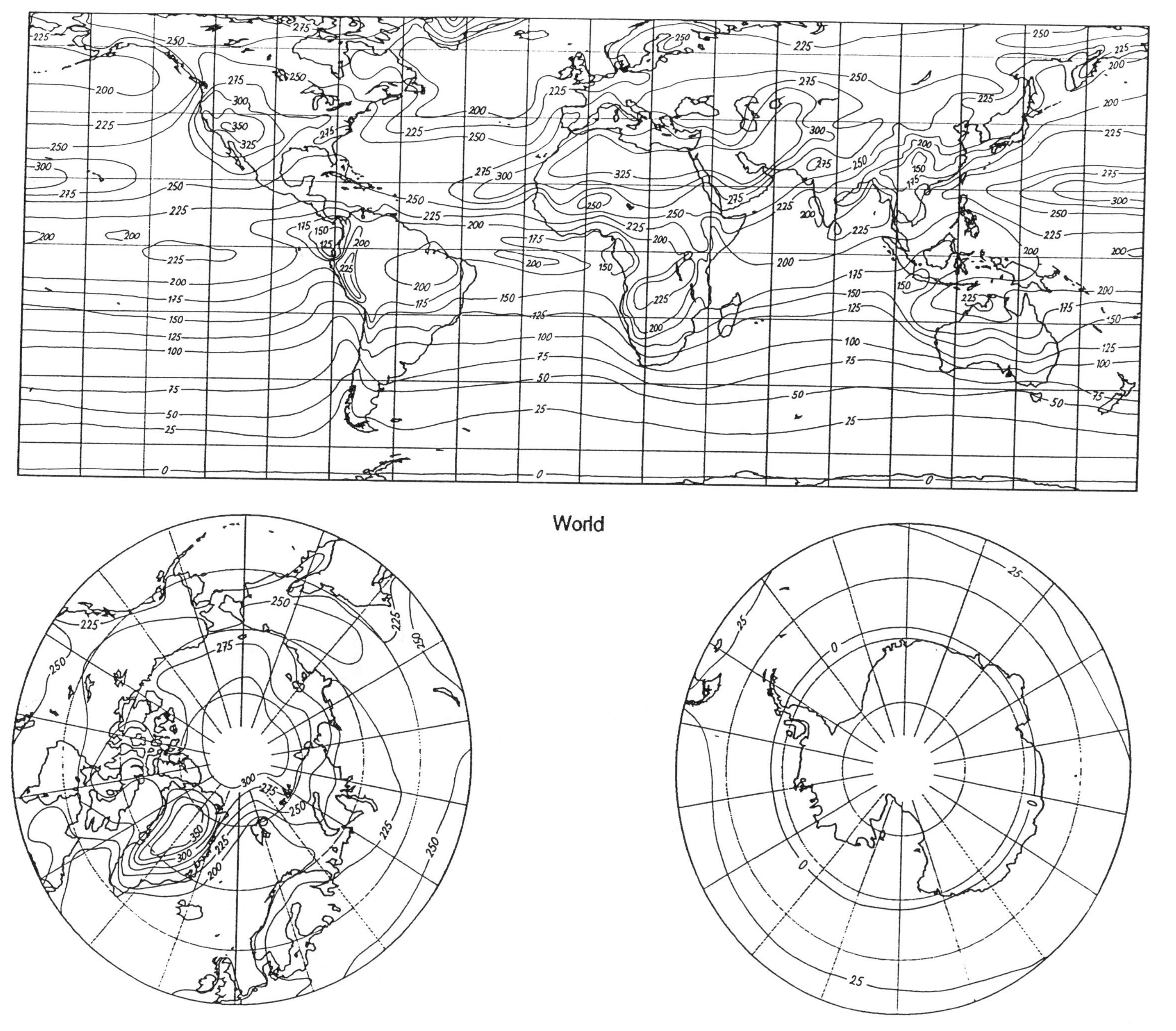

Fig. 5. Monthly mean global radiation for the globe: June. Unit in Wm^{-2}.

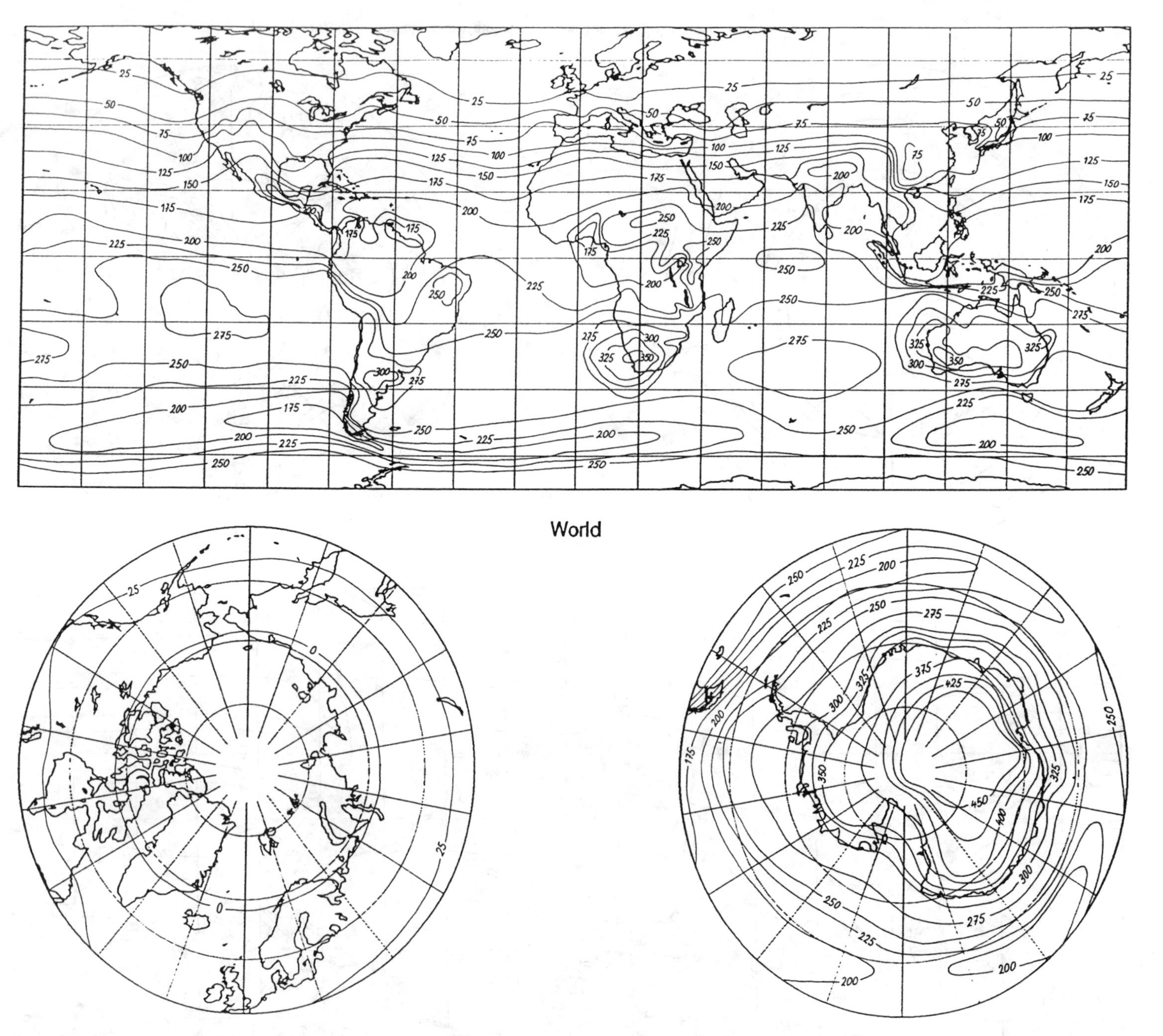

Fig. 6. Monthly mean global radiation for the globe: December. Unit in Wm^{-2}.

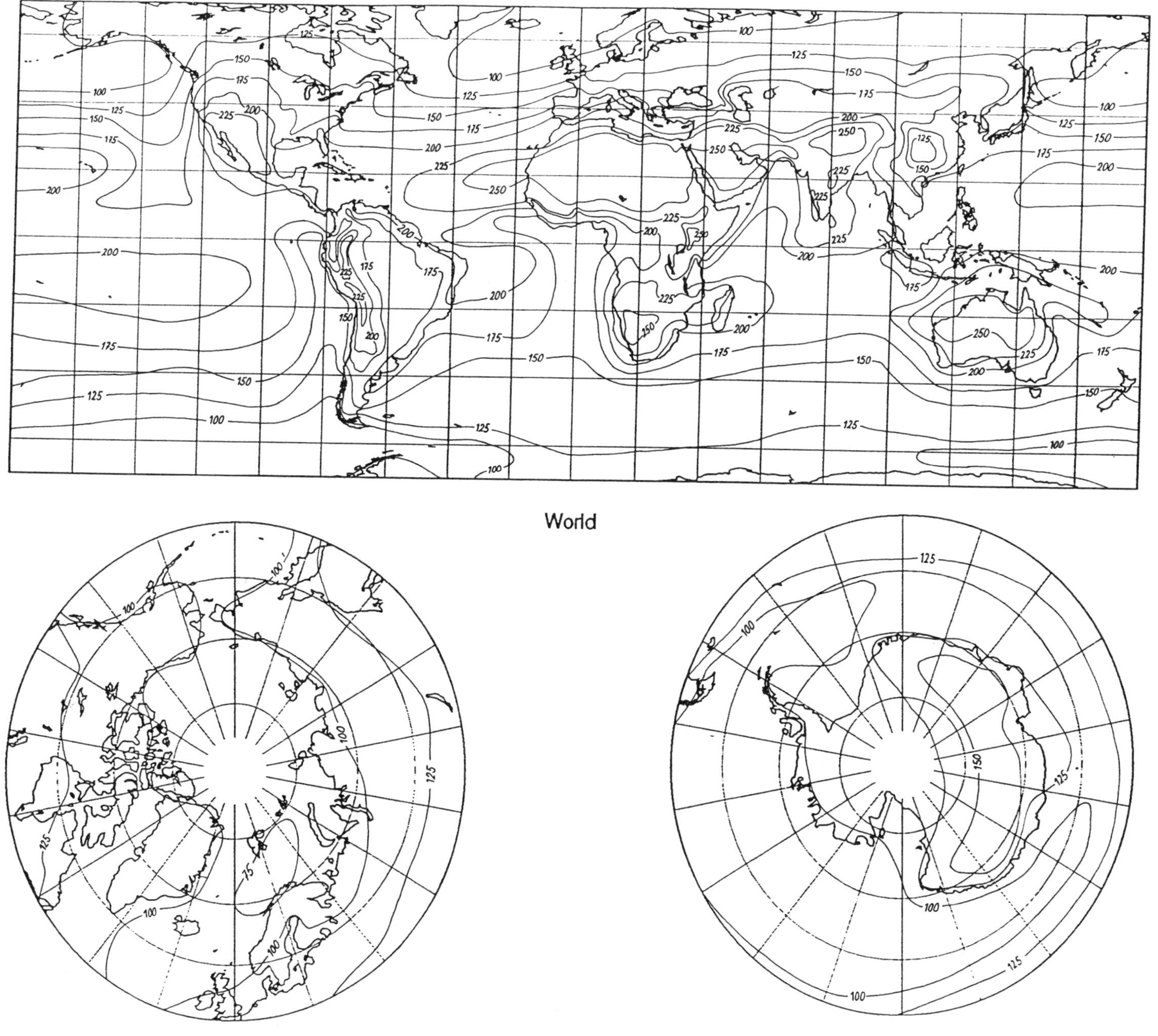

Fig. 7. Annual mean global radiation for the globe. Unit in Wm^{-2}.

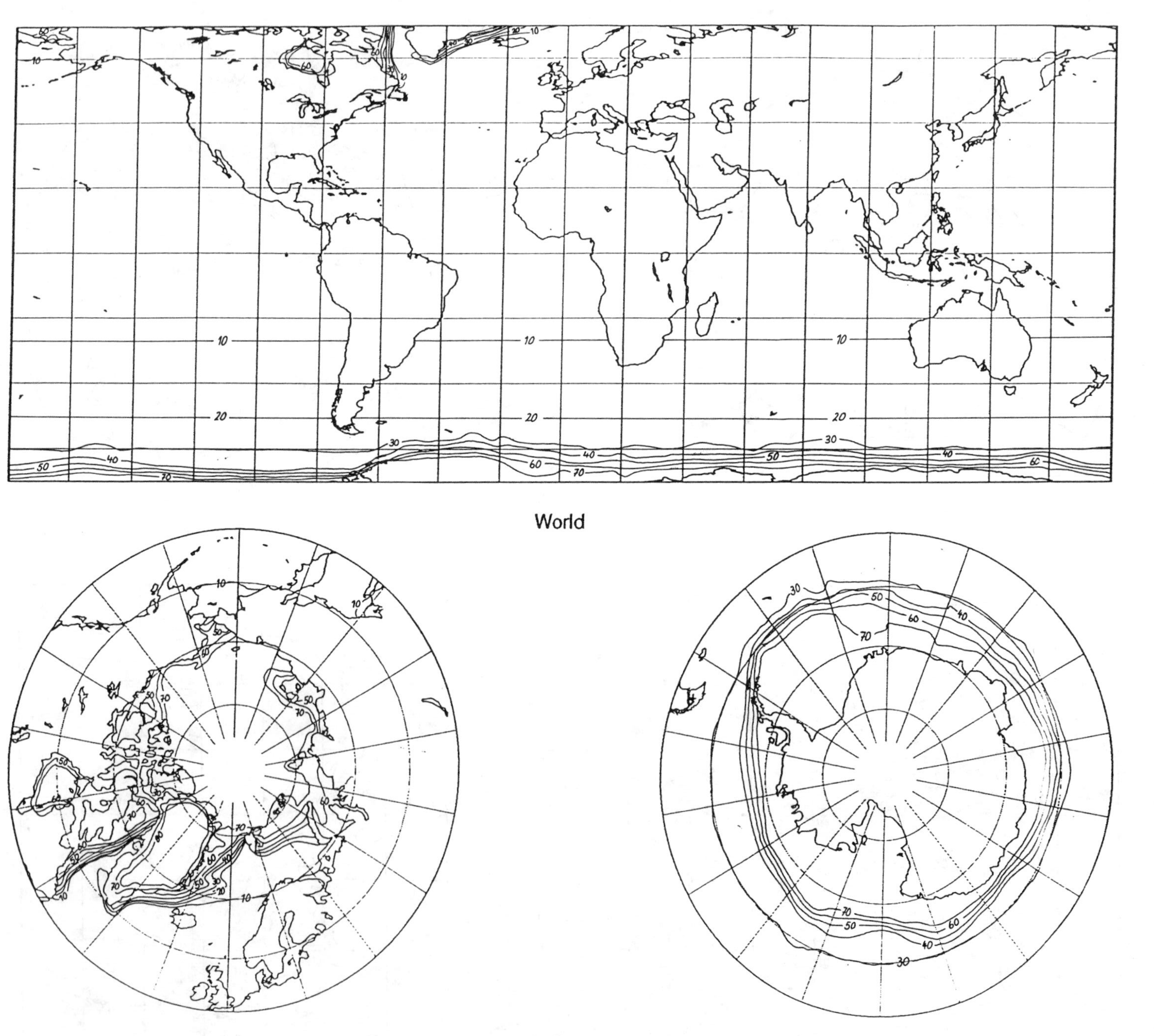

Fig. 8. Monthly mean albedo for polar regions: June.

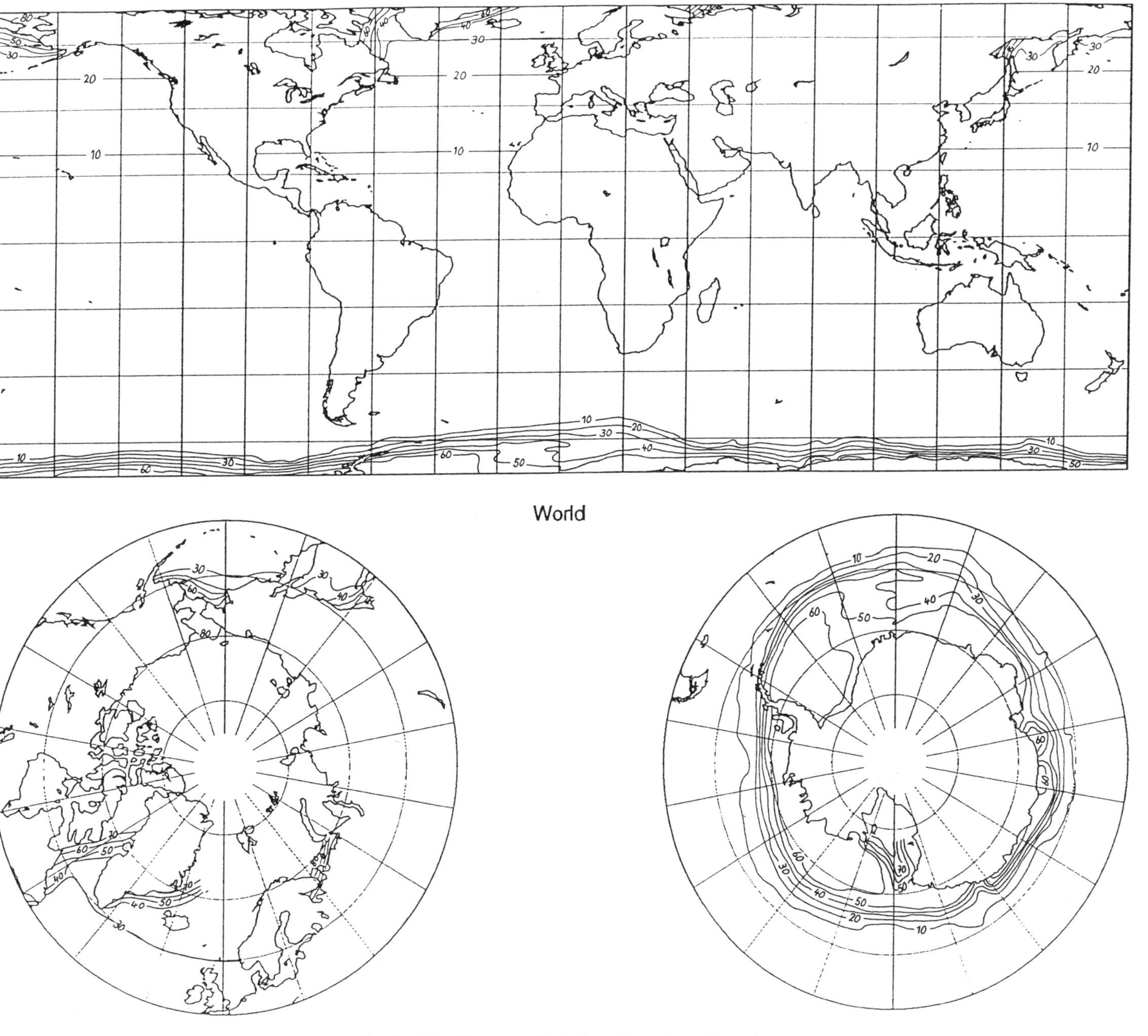

Fig. 9. Monthly mean albedo for polar regions: December.

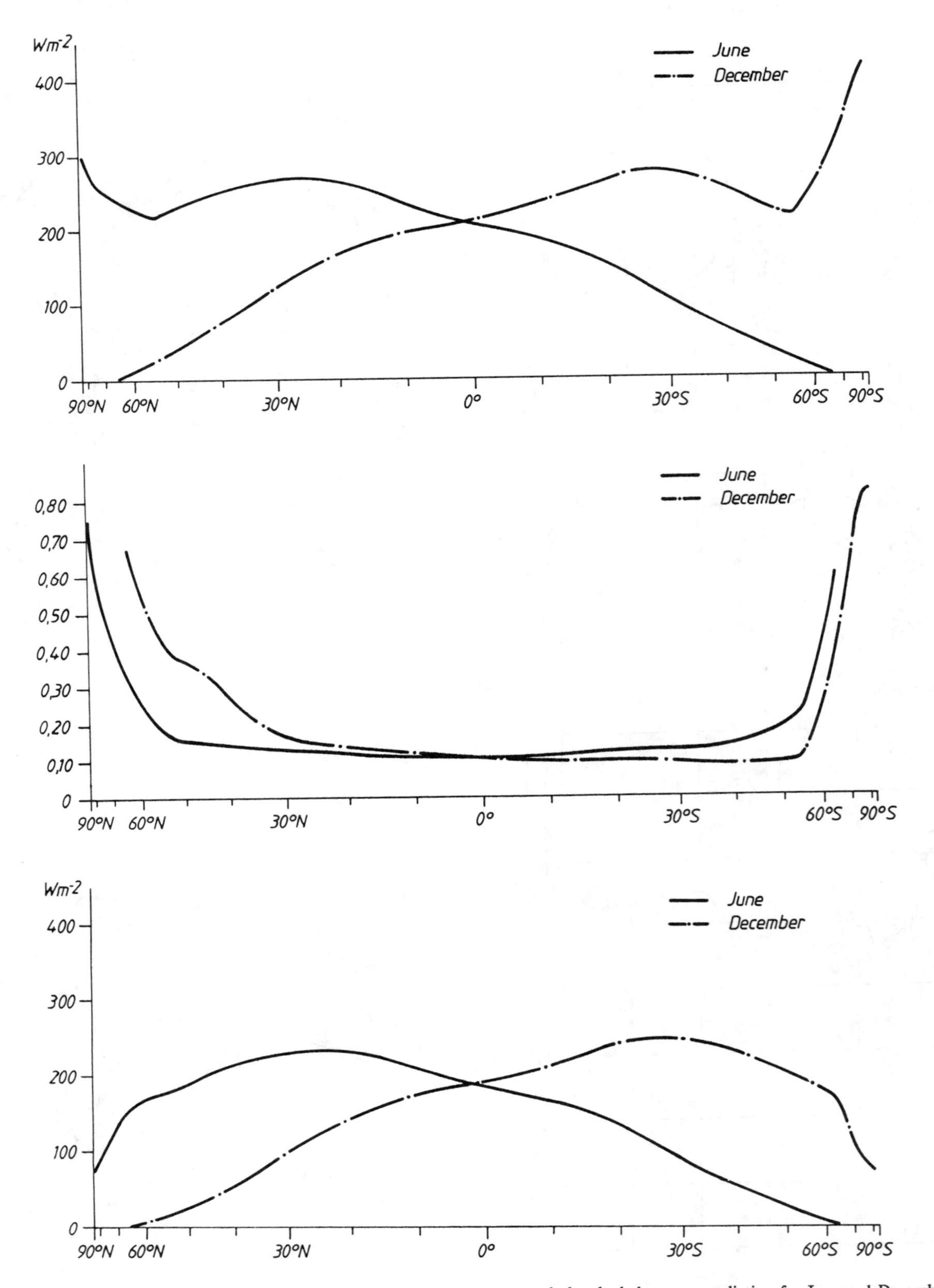

Fig. 10. Latitudinal distribution of monthly global radiation, albedo and absorbed shortwave radiation for June and December.

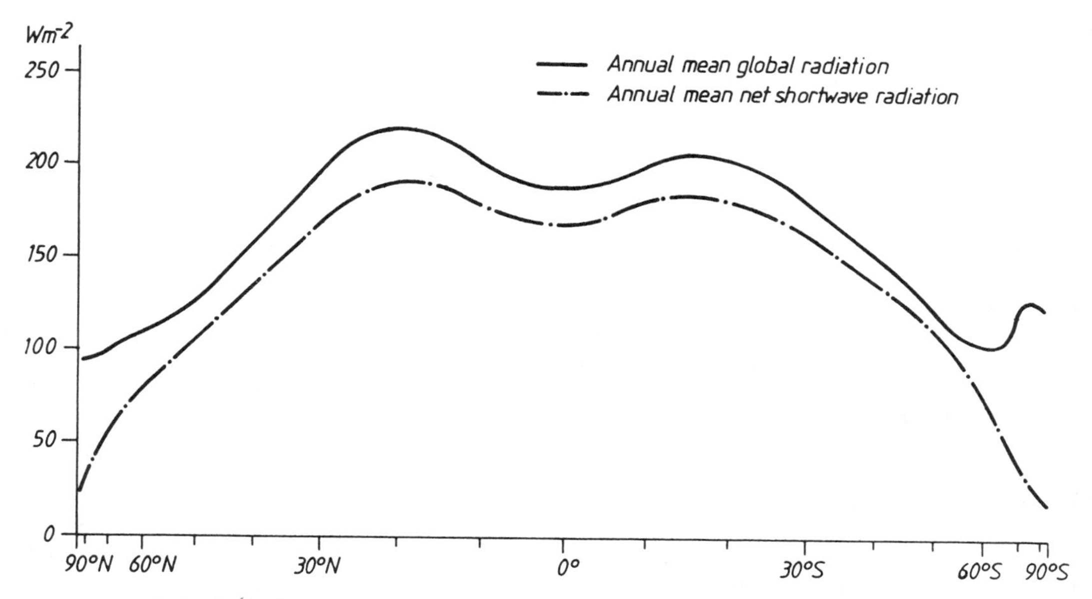

Fig. 11. Latitudinal distribution of annual global radiation and absorbed shortwave radiation.

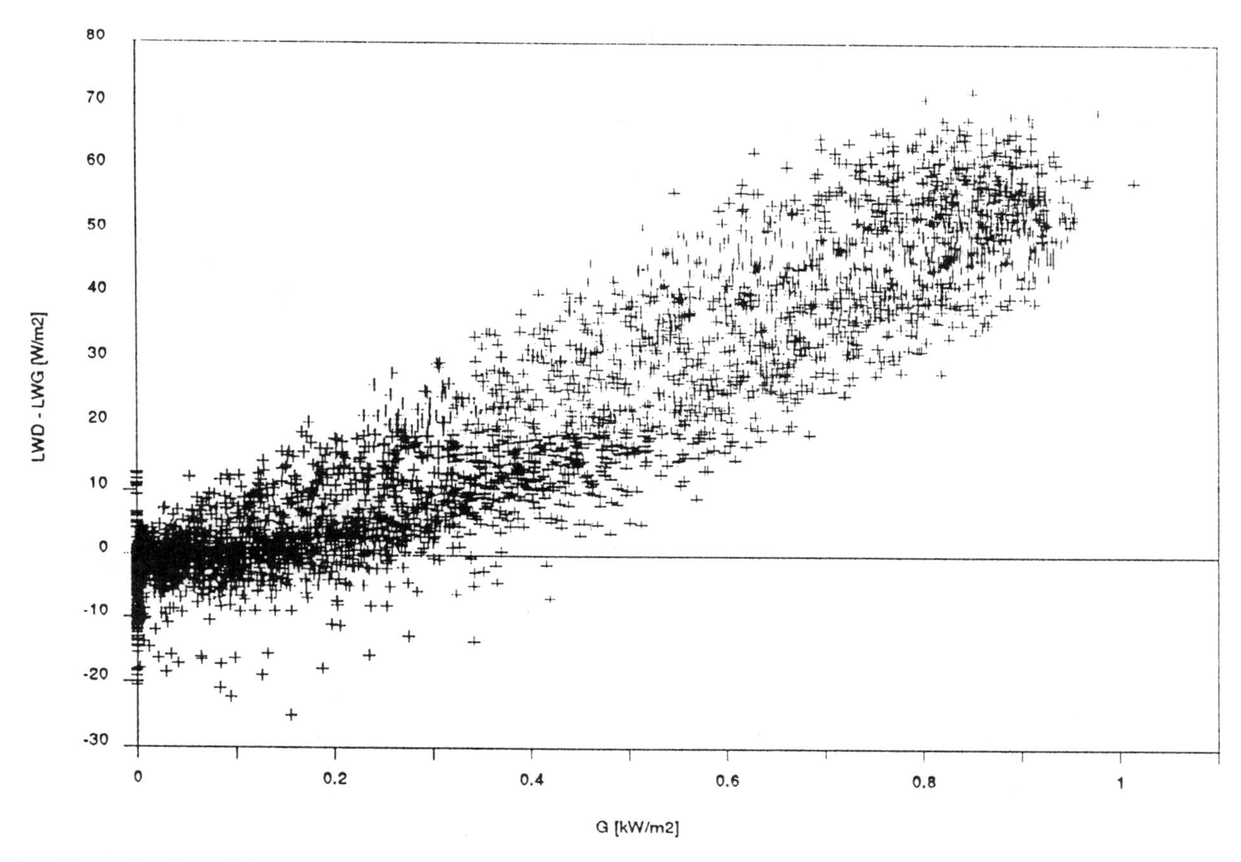

Fig. 12. Overestimation of longwave incoming radiation by a conventionally installed pyrradiometer, as a function of global radiation. Each point represents the difference in hourly mean fluxes measured by the shaded and ventilated pyrradiometer (LWD) and the pyrradiometer without a shading disc and ventilation (LWG) on the dome (units in Wm^{-2}).

LATITUDINAL DISTRIBUTION OF SHORTWAVE RADIATION

Based on the above presented global radiation and surface albedo, the monthly and annual zonal means of the absorbed shortwave radiation are calculated. The zonal means are computed based on the 10° lat. x 10° long. area-averages. The latitudinal mean global radiation, albedo and absorbed shortwave radiation are presented for June and December in Fig. 10. The same quantities for the annual mean are presented in Fig. 11.

Compared with previous works such as Budyko et al. (1978), the present results show smaller global and absorbed shortwave radiation for the equatorial and tropical regions and larger absorbed shortwave radiation for the higher latitudes in the Northern Hemispere. The smaller global radiation for the lower latitudes and especially near the equator in the present work results from the use of more observed global radiation and data quality control. The larger absorbed radiation, north of 55° N is the result of the new albedo distribution of the present work. The global annual mean absorbed shortwave radiation is estimated at 142 Wm^{-2}, or 42 % of the extraatmospheric solar radiation, 17 Wm^{-2} smaller than the previous estimates. The annual course of absorbed shortwave radiation fluctuate with amplitude of 4 Wm^{-2}, reaching a maximum in December and a minimum in June. The hemispheric annual mean absorbed radiation turns out to be the same magnitude for both hemispheres. The smaller global radiation for the Southern Hemisphere is compensated for by the larger surface albedo of the Northern Hemisphere. The present work also indicates that the global mean surface albedo of the earth is 0.16 +/- 0.01, instead of 0.11 or 0.14 [e.g., Budyko, 1963,1978].

LONGWAVE RADIATION AND RADIATION BALANCE

The accuracy of the longwave measurements is presently disputed. The estimated accuracy of the pyrgeometers and pyrradiometers which are currently used at permanent stations varies between 10 and 30 Wm^{-2}, depending on how the instruments are calibrated, installed and maintained. The sites for longwave incoming radiation archived in the GEBA number only 50, because measurements which are obviously

TABLE 3. Comparison of the hourly mean longwave incoming radiation under the cloudless sky among two measured and two computed fluxes. From the left to right, the measured flux by the pyrradiometer without a shading disc and ventilation, the measured flux by the pyrradiometer with a shading disc and ventilation, the computed flux by LOWTRAN 7, and the computed flux by graphic integration on the 2nd revised Elsasser radiation chart. The last column is added for educational purposes.

		Wm^{-2}			
daytime	date	M(SMI)	M(ETH)	LOWTRAN7	ELS.
Midday (12h)	22. Feb. 88	201.6	235.8	235.7	244
	09. Mar. 88	175.4	215.2	212.3	222
	22. Apr. 88	245.3	292.9	289.8	300
	6. May 88	269.5	319.4	305.6	318
	22. May 88	222.2	280.7	276.1	286
	24. June 88	284.5	319.9	312.6	323
	25. Jul. 88	287.0	329.0	321.7	332
	6. Aug. 88	288.9	333.3	318.5	329
	26. Sep. 88	285.6	321.1	320.7	331
	3. Nov. 88	253.5	256.3	243.3	252
	24. Nov. 88	206.0	217.7	222.8	232
	16. Dec 88	210.3	215.3	216.1	226
	Mean	244.1	278.1	272.9	283
Midnight (0h)	22. Feb. 88	234.2	229.4	226.4	232
	09. Mar. 88	229.4	225.1	215.4	225
	22. Apr. 88	277.9	274.0	270.8	280
	6. May 88	299.4	296.7	292.9	299
	22. May 88	268.6	264.4	253.4	263
	24. June 88	304.5	305.1	294.7	305
	6. Aug. 88	303.8	300.3	298.5	304
	26. Sep. 88	302.1	300.7	309.1	309
	3. Nov. 88	280.9	280.2	260.6	268
	24. Nov. 88	255.4	256.6	222.0	231
	16. Dec 88	246.0	246.1	222.1	236
	Mean	272.9	270.8	260.5	268

so inaccurate are not archived. Most of these 50 sites deploy the instruments as they are and the detailed care necessary for longwave measurements is not taken. Numerical estimations of the errors when exposing the instrument to the direct solar radiation have been made [Ohmura and Schroff, 1980]. In Fig.12 the difference in the hourly longwave radiation measured by the conventional installation and by our own installation is presented. Our installation is made so that firstly the sensor polyethylene dome is shaded from direct solar radiation by a shading disc, and secondly the external surface of the dome is ventilated. With such additional case, it is possible to keep the sensor temperature lower than the surrounding atmospheric temperature, whereby convective heat loss from the sensor is prevented, and the additional emission from the dome can be kept at a minimum. The figure shows that the underestimation by the conventional installation increases as global radiation increases.The results of these two installations are further compared against the theoretical computation of longwave incoming radiation under cloud-free conditions using the LOWTRAN 7 and the data from the radiosonde and the ozon-sonde at Payerne, in Switzerland (Table 3). The comparison clearly shows that the discrepancy between the two methods is insignificant at midnight but the difference at midday is very large. The fact that the LOWTRAN 7 computations come within 5 Wm^{-2} of the measurement by the shaded and ventilated pyrradiometer indicates that the conventional use of a pyrradiometer can underestimate longwave atmospheric radiation by as much as 70 Wm^{-2} as hourly means and often by 10 to 15 Wm^{-2} as monthly means. The same problem exists also for pyrgeometers and net radiometers which are widely in use. These new findings suggest that the present global distribution of the longwave incoming radiation or longwave net radiation is an underestimation of the same order of magnitude as mentioned above. The longwave atmospheric radiation data are presently being reevaluated by this group through close examination of pyrradiomtric data and also by computation of the LOWTRAN 7 with global radiosonde data which have been compiled by the ECMWF. Preliminary results indicate that the global mean annual longwave net radiation is close to -40 Wm^{-2}, instead of -50 Wm^{-2} which is presently accepted. The most likely global net radiation for the earth's surface then becomes 102 Wm^{-2} or 30 % of the extraterrestrial solar radiation.

CONCLUSIONS

The reevaluation of the global distribution of radiative fluxes for the earth's surface is made based as much as possible on the radiometric measurements with the following results. Global mean annual shortwave incoming radiation is 169 Wm^{-2}. The Northern Hemisphere receives 5 Wm^{-2} more than the Southern Hemisphere. The global mean shortwave radiation is larger in December than in June by 12 Wm^{-2}. Annual mean absorbed shortwave radiation is 142 Wm^{-2}, which makes the mean earth's surface albedo 0.16. The global average longwave net radiation is estimated at -40 Wm^{-2} which makes the mean net radiation of the earth's surface 102 Wm^{-2} or 30 % of the primary solar radiation at the top of the atmosphere. The differences between the present results and previous works are mainly owing to detailed quality control of the measured fluxes and the close examination of instruments undertaken in WCP-Water Project A7.

Acknowledgements. The present work is financed by the Swiss National Science Foundation, through Grants No. 2.307-0.86, 20-25271.88 and 20-29826.90 for the re-evaluation of the global energy balance. The authors are indebted to Dr. A. Junod, Director of the Swiss Meteorological Institute and the radiation team at Payerne Aerological Station, in particular Dr. A. Heimo and Mr. P. Wasserfallen for extremely productive joint works on radiation. The authors were very fortunate to receive help from a number of individuals in the course of the preparation of the article: Mr. Pierluigi Calanca for calculating albedo for ice-covered sea, Dr. Guido Müller for evaluating the results of the comparison of measurements of longwave radiation at Payerne Aerological Station, Mr. Stefan Casanova and Urs Sutter for typing the manuscript and preparing the figures.

The authors are indebted to the following individuals for providing magnetic tapes: Dr. D. Carson (MO / MIT computed heat fluxes for ocean), Dr. J.M. Oberhuber (MPIM computed heat fluxes for ocean), Dr. W. Rossow (NASA / Goddard surface vegetation and land-use), Prof. R.G. Barry and Dr. S. Kondo (NAVY / NOAA-JIC monthly sea ice concentration) and Prof. J. London (Global cloud amount and types).

REFERENCES

Barkstrom, B., Harrison, E.F., and Lee, R.B. III, 1990: Earth Radiation Budget Experiment. EOS, Vol. 71, No. 9, 297 - 305.

Bottomley, M., Folland, C.K., Hsiung, J., Newell, R.E., and Parker, D.E., 1990: Global Ocean Surface Temperature Atlas. Meteorol. Office and MIT, 20 P.

Budyko, M.I. (Ed.), 1963: Atlas teplovogo balansa zemnogo shara (Atlas of the heat balance of the earth). Moscow, Akademiya Nauk SSSR.

Budyko, M.I. et al., 1978: Heat Balance of the Earth. Leningrad, Gidrometeoizdat, as guoted in Budyko, M.I., 1980: The Earth's Climate: Past and Future, Intern. Geophys. Ser., Vol. 29, N.Y. and London, Academie Press.

Cogley, J.G., 1979: The albedo of water as a function of laltitude. Mon. Wea. Rev., Vol. 107, No. 6, 775 - 781.

Davies, J.A., 1965: Albedo investigations in Labrador - Ungava. Arch. Met. Geophys. Biokl., Ser. B, Bd. 13, 137 - 151.

Hastenrath, S., and Lamb, P.J., 1978: Heat Budget Atlas of the Tropical Atlantic and Eastern Pacific Oceans. Madison, Uni. Wisconsion Press. 13 P.

Iwasaka, N., and Hanawa, K., 1990: Climatologies of marine meteorological variables and surface fluxes in the North

Pacific computed from COADS. Tohoku Geophys. Jour., Vol. 33, No. 3 &4, 185 - 239.

Kondrat'ev, K.Ya. (Ed.) 1973: Radiation Characteristics of the Atmosphere and the Earth's surface. New Delhi & N.Y., Amerind Publ. Co., 580 P.

Matthews, E., 1985: Atlas of Archived Vegetation, Land - Use and Seasonal Albedo Data Sets. NASA Tech. Mem. 86199, N.Y. Goddard Space Flight Center, 13 P.

McFadden, J.D., and Ragotzkie, R.A., 1967: Climatological significance of albedo in Central Canada. Jour. Geophys. Res., Vol. 72, No. 4, 1135 - 1143.

Oberhuber, J.M., 1988: The Budgets of Heat, Buoyancy and Turbulent Kinetic Energy at the Surface of the Global Ocean. Report No. 15, Hamburg, Max-Plank-Institut für Meteorologie, 20 P.

Ohmura, A., and Schroff, K., 1980: Physical characteristics of the Davos-type pyrradiometer for short- and long-wave radiation. Arch. Met. Geophys. Biokl., Ser. B, Bd. 33, 57 - 76.

Ohmura, A., and Lang, H., 1988: Secular variation of global radiation in Europe. in Lenoble, J., and Geleyn, J.-F. (Eds.): IRS'88: Current Problems in Atmospheric Radiation. Hampton, Deepak Publ., 298 - 301.

Ohmura, A., Gilgen, H., and Wild, M., 1989: Global Energy Balance Archive (GEBA), World Climate Program - Water Project A7, Rep. 1: Introduction. Zurich, Verlag der Fachvereine, 62 P.

Ohmura, A., and Gilgen, H., 1991: Global Energy Balance Archive (GEBA), World Climate Program - Water Project A7, Rep. 2: The GEBA Database, Interactive Application, Retrieving Data. Zurich, Verlag der Fachvereine, 60 P.

Roth, M., 1985: Albedomessung über dem Schweizerischen Mittelland. Diplomarbeit, ETH Zürich, 114 P.

Warren, S.G., Hahn, C. J., London, J., Chervin, R.M., and Jenne, R.L., 1986: Global Distribution of Total Cloud Cover and Cloud Type Amounts over Land. DOE / ER / 60085-H1, NCAR / TN-273+STR, 29 P.

Warren, S.G., Hahn, C. J., London, J., Chervin, R.M., and Jenne, R.L., 1988: Global Distribution of Total Cloud Cover and Cloud Type Amounts over the Ocean. DOE / ER-0406, NCAR / TN-317+STR, 42 P.

WMO, 1991: Radiation and Climate, Workshop an Implementation of the Baseline Surface Radiation Network, Washington, DC, U.S.A., 3 - 5 December 1990. WCRP-54, WMO / TD-NO.406 Geneva, WMO, 8 P.

A. Ohmura and H. Gilgen, Swiss Federal Institute of Technology Zurich (E.T.H.), Winterhurerstrasse 190, CH-8057 Zurich, Switzerland.

Interannual Variations in the Stratosphere of the Northern Hemisphere: A Description of Some Probable Influences

H. VAN LOON

The National Center for Atmospheric Research, Boulder, CO

K. LABITZKE

Meteorologisches Institut, Freie Universität, Berlin, Germany

Abstract. The longest continuous set of daily analyses of stratospheric constant pressure levels covers 34 years, but the levels are all below 25 km. These analyses are for the Northern Hemisphere and have no equivalent on the Southern Hemisphere. Data from single stations go back another five to eight years. The attempts here to link qualitatively some of the interannual variability in the stratosphere to forcings from outside the stratosphere therefore deal with samples that are not necessarily representative of long periods.

In addition to the random interannual variability which is inherent in the atmosphere-ocean system, some of the interannual changes in the stratospheric circulation are associated with the following:

1. The Quasi-Biennial Oscillation in the stratospheric winds above the equator. This oscillation is forced from the troposphere.
2. The Southern Oscillation, which is defined as a seesaw in sea level pressure between the Indian and Pacific Oceans but has widespread effects over the globe.
3. Major volcanic eruptions, of which there were three during the period analyzed.
4. A 10–12 year oscillation which is present in the data of the last 40 years, during which time it was in phase with the 11-year solar cycle.

We shall describe each of the four, but emphasize the 10–12 year oscillation.

Introduction

Daily historical analyses of the temperature and geopotential height at standard pressure levels in the stratosphere of the Northern Hemisphere began with the International Geophysical Year 1957/1958. The longest continuous sequences, at present 34 years long, have been made in the Stratospheric Research Group at the Freie Universität in Berlin and are for levels below 25 km. Shorter series for higher levels exist, analyzed by the same group, and in later years daily operational analyses have become available from various meteorological centers.

Interactions Between Global Climate Subsystems, The Legacy of Hann
Geophysical Monograph 75, IUGG Volume 15

There is no equivalent for the Southern Hemisphere of the long series from the Stratospheric Research Group because of the meager network of radiosonde stations over the southern oceans. This paper therefore deals only with the Northern Hemisphere, but even here the comparatively short series, containing only small samples of the low frequency variations, cannot furnish stable statistical relationships.

With these drawbacks in mind, we shall describe four contributors to the interannual variability of the stratosphere: The Quasi-Biennial Oscillation (QBO), the Southern Oscillation (SO), volcanic eruptions, and the Ten–to–Twelve Year Oscillation (TTO). The last one will receive the main share of attention, for it was discovered only recently and is thus still relatively unknown, and also unexplained. It should be noted that such identifiable influences on the variability of the stratosphere are superposed on random variability which is intrinsic to the atmosphere-ocean system.

Figure 1 typifies the latitudinal distribution of interannual variability in the stratosphere. It shows the standard deviations of monthly mean 30-mb temperature along the Greenwich meridian. The annual march of the variability at the lowest latitudes has a semi-annual component, but the farther north one goes the more marked the annual cycle becomes, for there is little change with latitude of the low variability in summer, whereas the numbers rise with increasing latitude during the rest of the year, especially in January and February.

The Quasi-Biennial Oscillation

The QBO was discovered 30 years ago [Reed et al., 1961; Veryard and Ebdon, 1961], and can be followed back to the early 1950s. The oscillation appears in the stratosphere above the equator as persistent east or west winds during alternate, very variable periods (Fig. 2), whose average at present is 27.7 months for the levels between 10 and 70 mb [Naujokat, 1986]. The QBO propagates downward, with the westerlies descending faster than the easterlies; its amplitude is 40 to 50 ms^{-1}, largest at 15–20 mb, and it is as variable as the period. The temperature has a corresponding oscillation.

The development of the current theory of the QBO, by Holton and Lindzen, is described by the latter in a paper which also contains an extensive bibliography [Lindzen, 1987]. The theory depends on propagation from the troposphere of equatorial Kelvin and Rossby-gravity waves for eastward and westward acceleration, respectively.

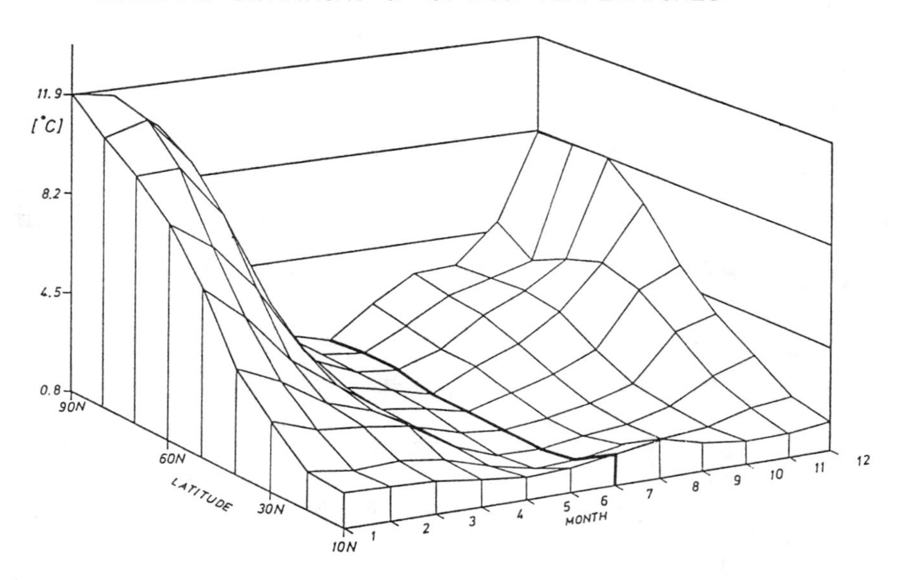

Fig. 1. Standard deviations (C) of monthly mean 30-mb temperature along the Greenwich meridian.

The waves interact with the mean flow such that their momentum is deposited at the level where their phase speed equals the zonal wind speed; it is not certain, however, that this mechanism can account fully for the observed wind speeds.

The influence of the QBO on the interannual variability in the stratosphere is not limited to low latitudes but is observed as far north as the pole [e.g., van Loon and Labitzke, 1987]. The mark of the oscillation on the circulation in winter is seen in Fig. 3a, which shows the difference in 50-mb height between winters with westerlies and winters with easterlies in the lower (40–50 mb) equatorial stratosphere. To minimize the influence of the Southern Oscillation, only winters that were not in an extreme of the Southern Oscillation were used. The map has the sign of the west phase (the average of eight winters in the east phase was subtracted from that of six winters in the west phase). It shows that the polar vortex tends to be deeper and the heights between 40°N and 60°N to be higher in the west years, and the polar night westerly jet stream therefore tends to be stronger in the west than in the east years of the QBO. The heights in the west years are on an average lower in the subtropics and higher near the equator than in the east years.

The difference at middle and high latitudes between the east and west years is largest in early winter, when the vortex is cold and intense in the west years, but warmer and weaker in the east years. The difference is smaller in January–February because of the major midwinter warmings, which occur in both phases of the QBO [van Loon and Labitzke, 1987, their Fig. 9].

The Southern Oscillation. Volcanic eruptions

The Southern Oscillation has been known since the turn of the century [see Rasmusson and Carpenter, 1982, for a historical review]. At its core it consists of a seesaw of atmospheric mass between the South Pacific Ocean and the Indian Ocean–Australia, but its influence is worldwide e.g., [Walker and Bliss, 1932; Berlage, 1957; van Loon and Madden, 1981]. Its extremes are associated with marked anomalies of sea surface temperature along the coast of Peru ("El Niño" in the warm extreme) and along the equator in the Pacific. In the following the extremes of the SO are called Warm or Cold Events for positive or negative deviations of sea surface temperature on the equator.

The mean difference between the 50-mb height anomalies associated with the extremes of the Southern Oscillation in winter (Warm minus Cold Events in Fig. 3b) shows that the polar vortex tends to be weaker in Warm than in Cold Events. The pattern in many respects resembles the QBO anomalies in Fig. 3a, but the latter are more zonally symmetric. In contrast to the QBO, the difference between the Warm and Cold Events in the SO is considerably stronger at the end than at the beginning of the winter, because the polar vortex tends to break down in January or February during Warm Events, but to stay cold and intense in Cold Events [van Loon and Labitzke, 1987, their Figs. 2, 6, and 8].

The troposphere at low latitudes is warmer than normal in Warm Events [Horel and Wallace, 1981], in large measure because of the increased convection associated with the abnormally high sea surface temperature. At the same time, the temperature in the lower tropical stratosphere is, on an average, below normal owing to the cooling associated with the rising dry air at tropopause levels over the convective systems and to radiative cooling from the cloud tops. There are, however, three exceptions in the records: Figure 4 is a time series of the mean 30-mb temperature in December at 10°N, from 1963 to 1991. During this period there were eight Warm Events, which are marked in Fig. 4 by WE in the year when they began. Five of the eight events have temperatures well below normal, whereas three (1963, 1982, and 1991) were about two standard deviations above the 18-year mean. In those three years tropical volcanoes erupted: Mt. Agung in Indonesia in 1963, El Chichón in Mexico in 1982, and Mt. Pinatubo in the Philippines in 1991. The effect of the volcanic aerosols, as described by Newell, [1970] and Labitzke et al. [1983], was to raise the temperature in the tropical stratosphere (Fig. 4).

A possible influence of the eruptions on the polar cyclonic vortex is shown in Figs. 5a,b. Instead of a weak vortex with above normal temperature in the center as in the other Warm Events, the vortex in

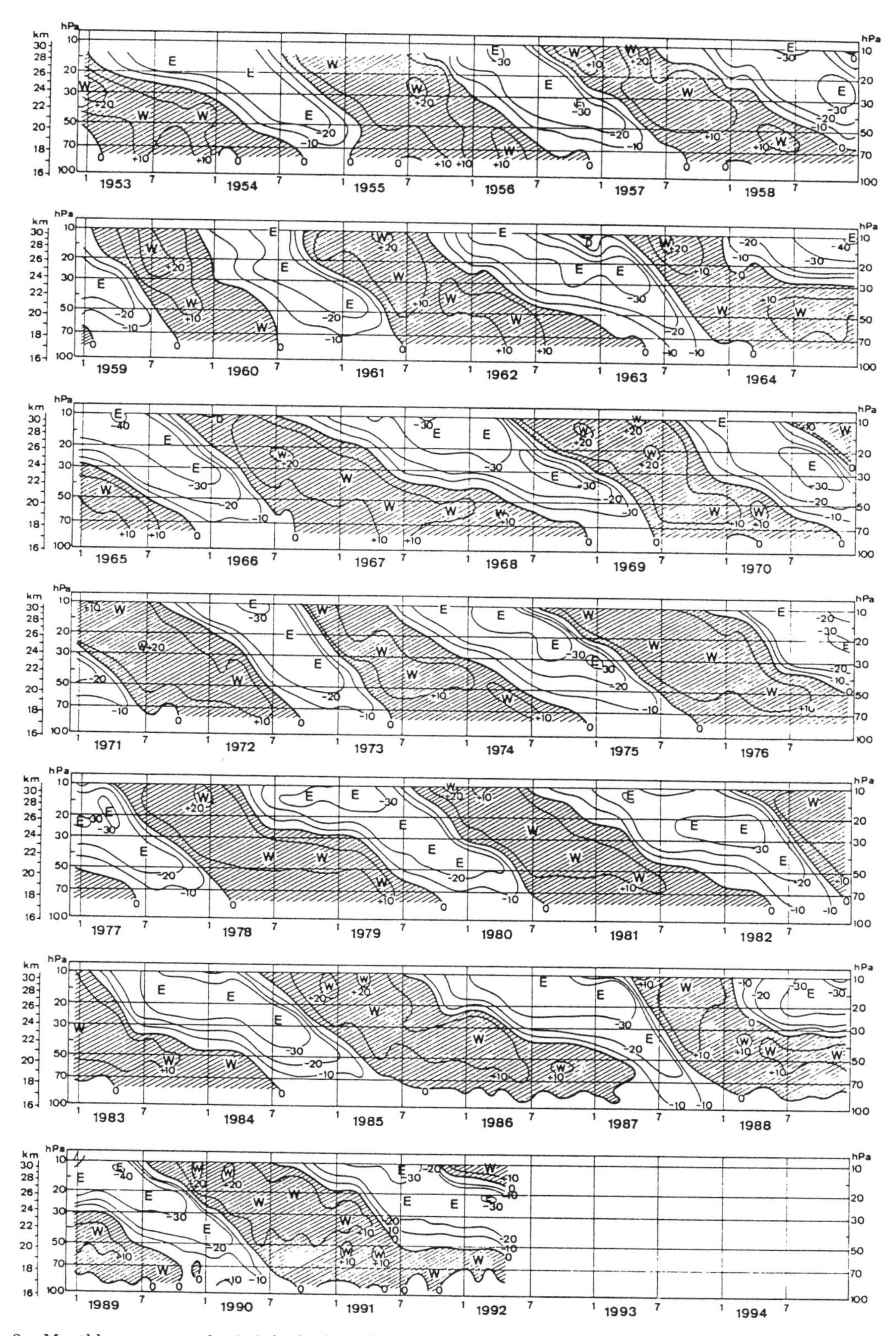

Fig. 2. Monthly mean zonal wind (m/sec) in the stratosphere above the equator. Updated from Naujokat [1986].

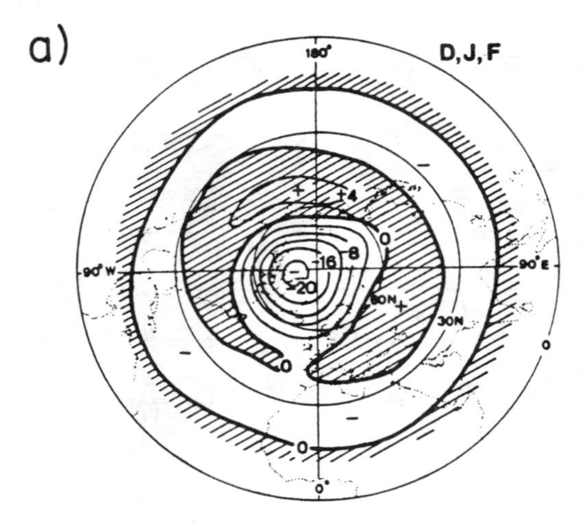

b) D,J,F,

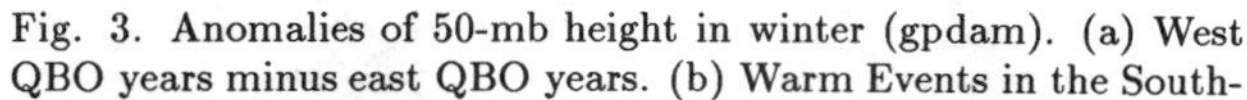

Fig. 3. Anomalies of 50-mb height in winter (gpdam). (a) West QBO years minus east QBO years. (b) Warm Events in the Southern Oscillation minus Cold Events. From van Loon and Labitzke [1987].

the winters of 1963/1964 and 1982/1983 was cold and intense [Labitzke and van Loon, 1989]. One may speculate that the coldness of the vortex was due to the presence of volcanic aerosols in the arctic whose effect, in this instance, overcame the influence of the Warm Event.

The 10–12 Year Oscillation

a. Summer. Whereas the knowledge about the three items discussed above (the QBO, the SO, and volcanic eruptions) is fairly well established, the Ten–to–Twelve Year Oscillation (TTO) was not known till recently. The evidence for its existence is comparatively meager, for the data at hand allow at most four periods to be analyzed, and only on the Northern Hemisphere. Its share of the interannual variability in the stratosphere in summer can be gauged from Fig. 6 which shows time series of the area-weighted averages of 30-mb heights and temperatures between 10°N and the North Pole. The three-year running means outline the TTO well in both heights and temperatures and indicate that its amplitude is comparable to changes at higher frequencies. Note that the warming associated with

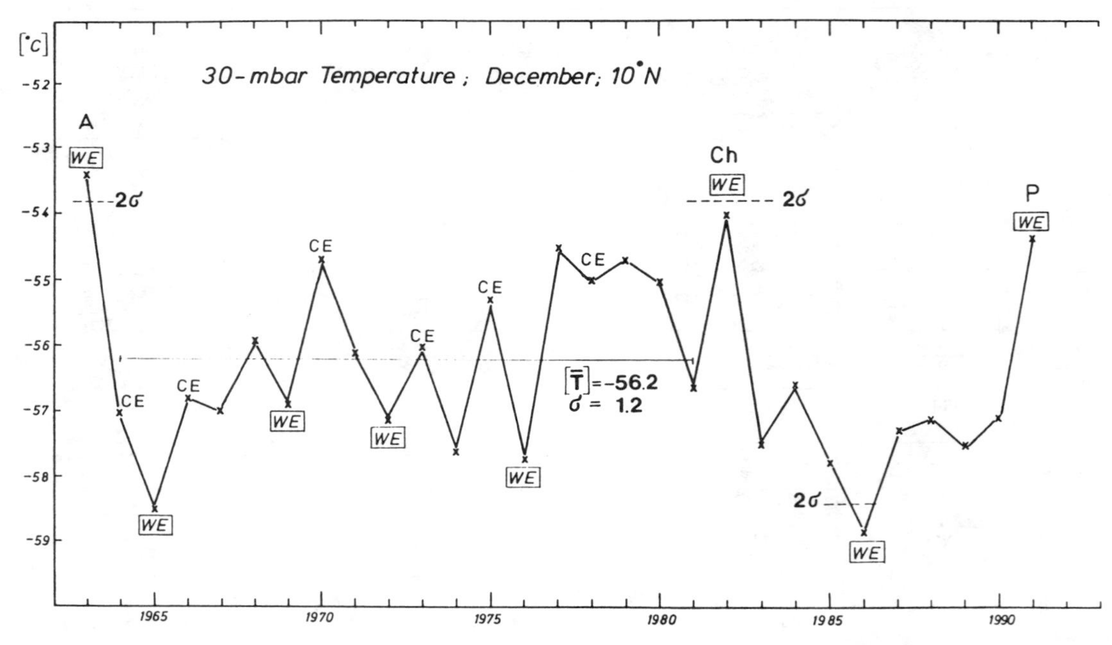

Fig. 4. Time series of 30-mb temperature (C) at 10°N in December. "A" stands for Mt. Agung, "Ch" for El Chichón, and "WE" for Warm Event in the Southern Oscillation. The mean temperature is the 18-year mean for 1964–1981.

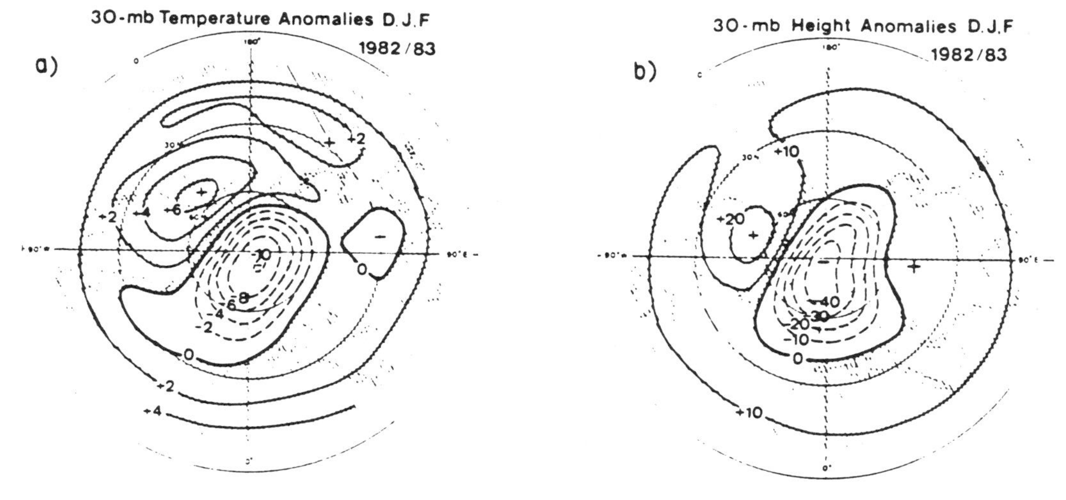

Fig. 5. (a) Anomalies of 30-mb temperature (C) in the winter of 1982/1983. (b) As (a) but for 30-mb height (gpdam). From Labitzke and van Loon [1989].

Mt. Agung's eruption depressed the TTO because it occurred in a minimum of the oscillation, whereas it was enhanced by the warming that took place after El Chichón's and Pinatubo's eruptions, which were near TTO maxima.

The dashed curves in Fig. 6 are time series of the 10.7 cm solar flux and are used to represent the 11-year solar cycle. The TTO has been in phase with the solar cycle as long as stratospheric data have been available, and it is convenient to outline it in terms of its correlation with the cycle. This is done in Fig. 7 where the 30-mb height at grid points north of 10°N has been correlated with the solar flux. The shape of the correlation pattern is the same as in Fig. 7 in all months of the year, but the correlations are largest from April to December and statistically most significant in summer [van Loon and Labitzke, 1990]. The map shown here is field significant at better than the 1%

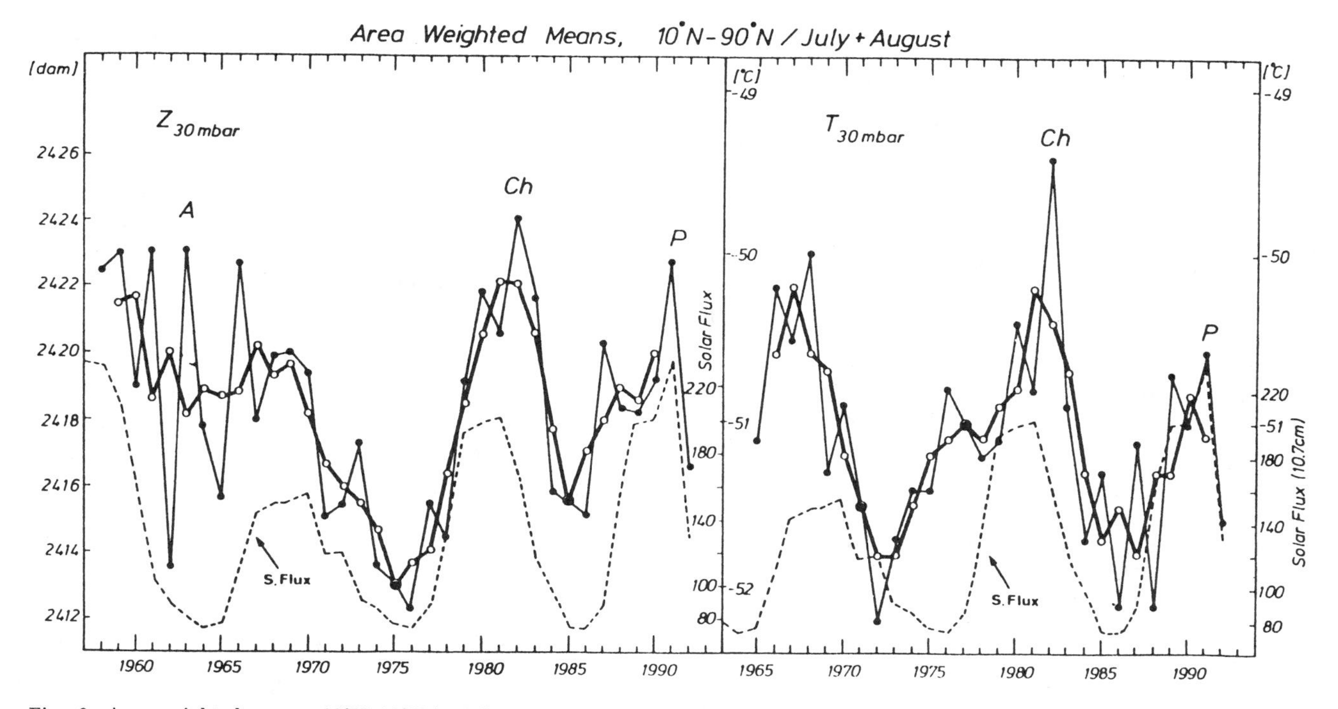

Fig. 6. Area weighted means, 10°N–90°N in July–August, of 30-mb height (gpdam, left) and 30-mb temperature (C, right). "A" stands for Mt. Agung and "Ch" for El Chichón. The dashed curve is a time section of the 10.7 cm solar flux. From Labitzke and van Loon [1991], updated.

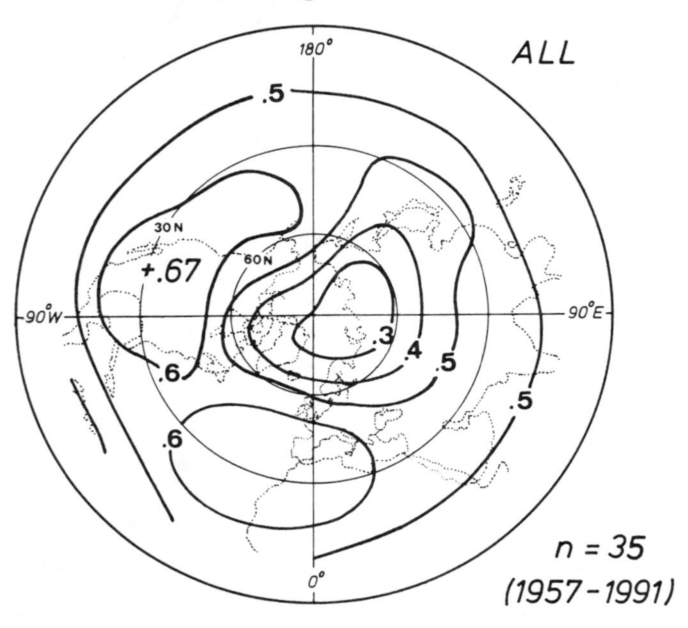

Fig. 7. Lines of equal correlation between 30-mb height in July–August and the 10.7cm solar flux. From van Loon and Labitzke [1990], updated.

level when tested by a Monte Carlo field significance test devised by Livezey and Chen [1983].

For such an oscillation to exist in the heights of the stratospheric constant pressure levels, the temperature in the layers below must oscillate with the same period. Figure 8 is an example of this relationship; it contains time series of 100- and 30-mb heights and 200-mb temperature at Charleston (33°N, 80°W), which lies in the area of highest correlation in Fig. 7. The temperature at 200 mb in Fig. 8 is available as early as 1951, and the TTO is conspicuous in this temperature during the 40 years since then. One can thus infer that the TTO existed as far back as 1951 in the heights between 100 mb and 30 mb over the area for which this station is representative.

The vertical structure of the TTO is illustrated in Figs. 9 and 10. Figures 9a and b are vertical meridional sections of the temperature and height differences between the extremes in the solar cycle, and also between the extremes in the TTO since the TTO is in phase with the solar cycle. Two-year means of the extremes have been used to suppress the variations at higher frequencies. The sections in Fig. 9 run across North America from the tropics to the arctic and show the basis for the 30-mb correlation pattern in this region of Fig. 7. The layer between 500 mb and 150 mb, which the 200-mb temperature in Fig. 8 represents, is warmer in solar maxima than in minima at middle and low latitudes. The difference between the solar extremes is zero or negative at 100 mb—the average level of the tropopause in these latitudes—and the temperatures in the stratosphere are higher in the solar maxima. The oscillation does not appear in the geopotential heights of the lower troposphere (Fig. 9b), but becomes increasingly evident with rising elevation in response to the accumulated effect of the oscillation in the temperature [Labitzke and van Loon, 1991].

The stations south of 45°N in Fig. 9 are representative of an appreciable part of the earth's circumference, as indicated by the mean vertical profiles in Fig. 10: These profiles show the average temperature difference between extremes in the solar cycle at stations as far apart as Truk and Curaçao, a distance of 140 deg. long. At all these stations the peaks of the oscillation are warmer than the valleys in

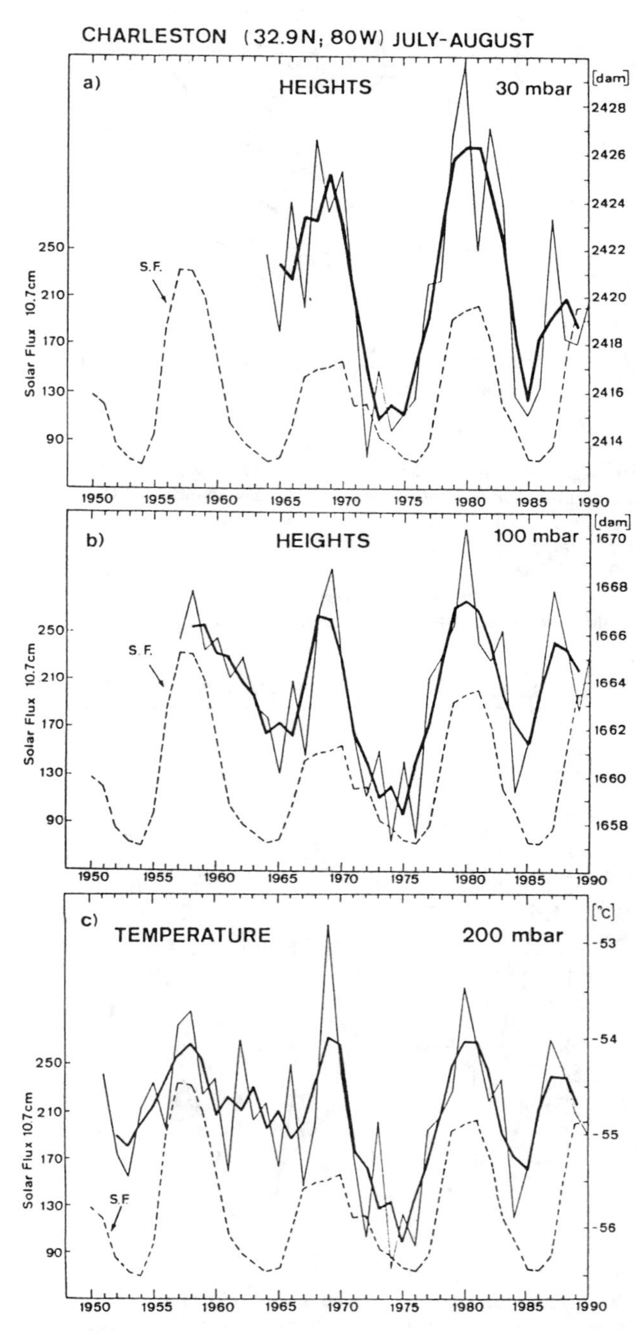

Fig. 8. Time series of 30- and 100-mb height (gpdam) and 200-mb temperature (C) at Charleston in July–August. The heavy lines are three-year running means, and the dashed lines are the 10.7 cm solar flux. From van Loon and Labitzke [1990].

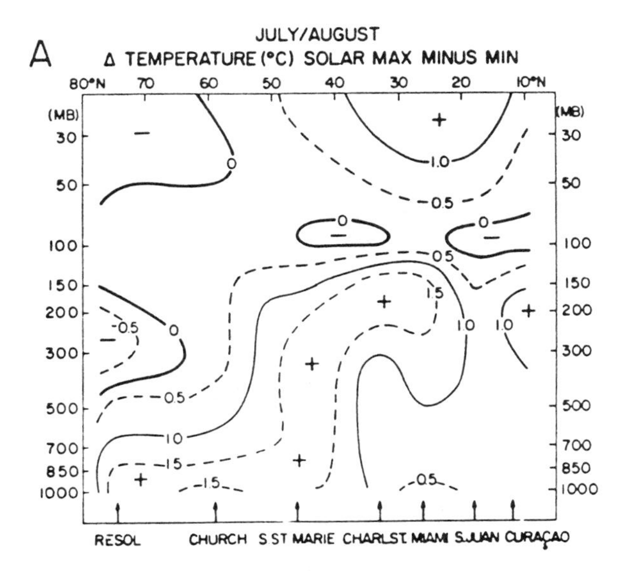

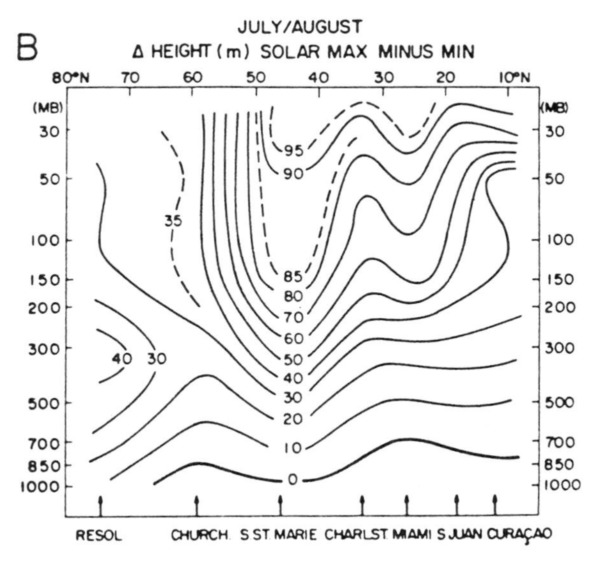

Fig. 9. Vertical meridional sections across North America for July–August of the difference between maxima and minima in the 11-year solar cycle of (a) Temperature (C), and (b) Height (gpdam). From Labitzke and van Loon [1991].

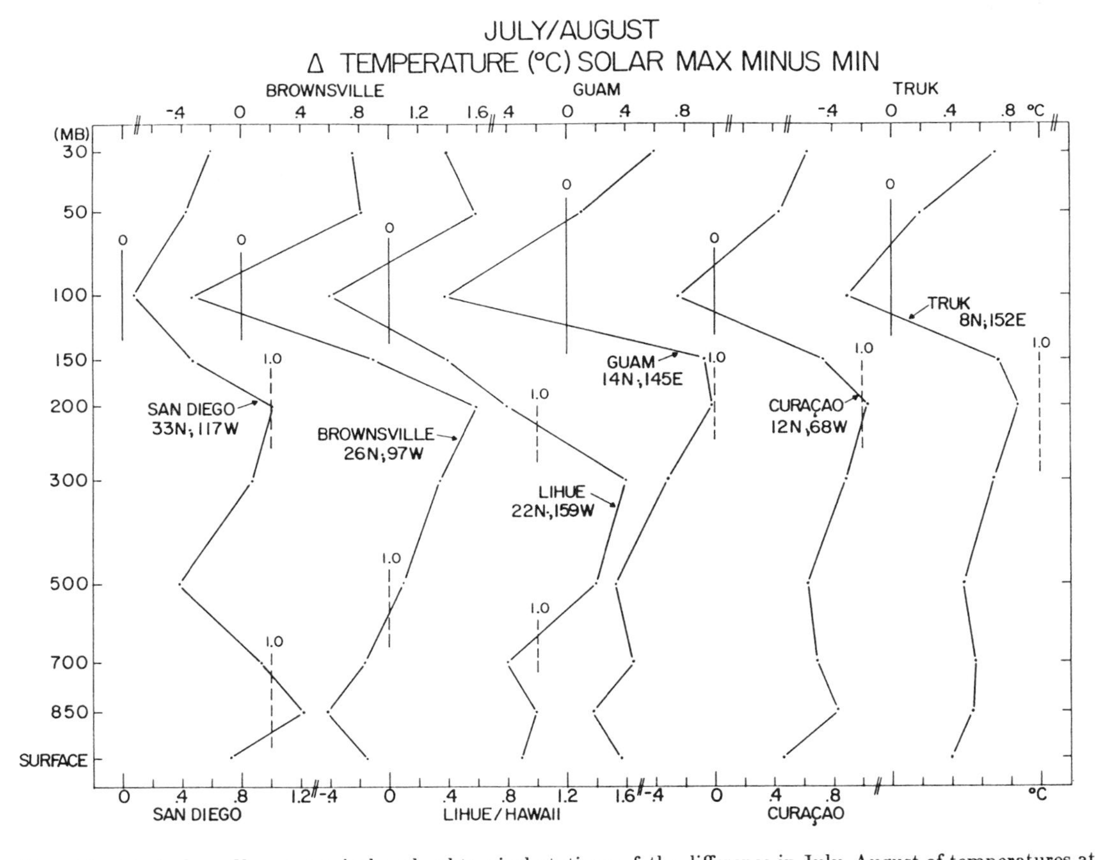

Fig. 10. Vertical profiles at tropical and subtropical stations of the difference in July–August of temperatures at standard pressure levels. From Labitzke and van Loon [1991].

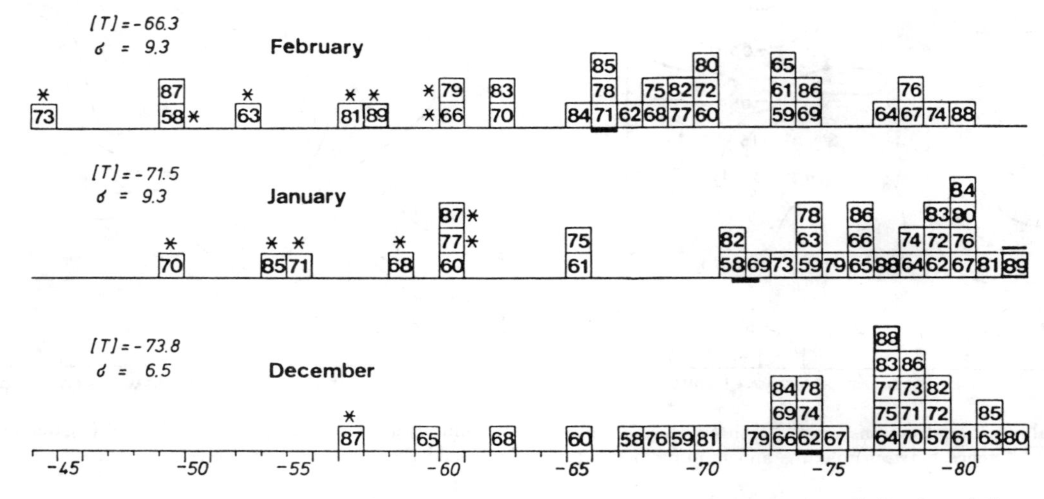

Fig. 11. Monthly mean 30-mb temperatures at the North Pole in winter. The mean is underlined, and the asterisks denote major midwinter warmings.

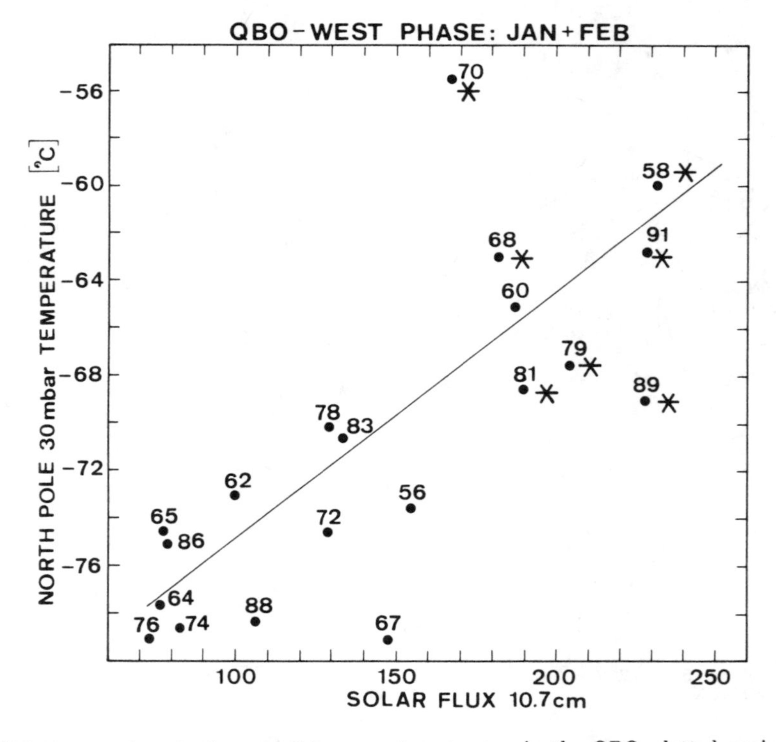

Fig. 12. North Pole temperature in January–February of west years in the QBO plotted against the 10.7 cm solar flux. The asterisks denote major midwinter warmings. From van Loon and Labitzke [1990], updated.

the upper half of the troposphere, the difference is zero or negative at 100, but positive again in the stratosphere. With at most four periods available, three being a more usual number, the samples are too small to lend themselves to statistical testing, but they are physically consistent over large parts of the hemisphere, as witnessed by their homogeneous vertical structure.

b. Winter. As noted above, the interannual variability in the lower arctic stratosphere is high in winter (Fig. 1), in contrast to the antarctic where the vortex in midwinter varies little from one year to another. The frequency distribution of the north polar temperature in January and February (Fig. 11) shows a wide scatter, skewed toward the warm side. This distribution comes about because of the frequent major midwinter warmings—the years marked with asterisks—during which the higher latitudes become warmer than the surrounding areas, and the polar cyclone breaks down and is replaced by an anticyclone. As mentioned in the section about the QBO, in winters when the equatorial winds in the lower stratosphere blow from the west, the polar vortex on an average is cold and intense. There is, however, a comparatively frequent exception to this rule [Labitzke, 1987]. In those QBO west years which fall in a peak of the 11-year solar cycle, major midwinter warmings do occur in January and February, contrary to theoretical expectations: Holton and Tan [1980] demonstrated that, considering the dynamics, the north polar vortex in winter should be cold and intense in the west phase of the QBO, and warmer and weaker in the east phase. This expectation is met in solar minima but not in solar maxima. There were seven major midwinter warmings with associated breakdowns of the polar westerly vortex in the 20 west years since 1958 (Fig. 12), and they all occurred in peaks of the 11-year solar cycle. The correlation of 0.8 between the arctic temperature and the solar cycle is thus due to the contrast between the 12 low temperatures in the solar minima and the seven years with major warmings in solar maxima. In the east years of the QBO the major midwinter warmings occurred mainly in solar minima. The stratospheric temperatures in the north polar regions are therefore not correlated with the solar cycle if one uses the complete time series, Fig. 13a; but if the west and east years are correlated separately (Figs. 13b and c), the west years are positively and the east years are negatively correlated with the cycle.

A Monte Carlo test in Labitzke and van Loon [1988] showed that splitting a series like the one in Fig. 13a, which is uncorrelated with the solar cycle, into two series according to the phase of the QBO, and thereby obtaining large correlations of opposite sign as in Figs. 13b and c, would be possible in less than 4 out of 1000 instances. It has been suggested that these correlations are due to aliassing caused by dividing the data according to the phase of the QBO [e.g., Salby and Shea, 1991; Teitelbaum and Bauer, 1990], but the question is if the problem should be considered a purely statistical one, even if some of the high correlation were due to aliassing. The map in Fig. 14a shows lines of equal correlation between the 30-mb height in January–February and the 11-year solar cycle (10.7 cm flux). This map is not statistically significant in a field significance test, but it has the same pattern as the corresponding maps in all other months (cf., Fig. 7). When the stratospheric data used in Fig. 14a are divided into the east and west years of the QBO and correlated with the solar cycle, one gets the correlations in Figs. 14b and c. The polar regions are positively correlated in the west years and negatively in the east years, as in the case of the temperature in Fig. 13. Outside the polar regions there are no noteworthy correlations in the west years, Fig. 14a, but in the east years there are strong positive correlations and the pattern is the same as that obtained with ungrouped data in the warmer months of the year, as for instance in Fig. 7. (If the years in Fig. 7 are divided into the east and west years of the QBO, the correlation pattern is the same in both as in the undivided series). The fact that there is no unique difference between the sign of the correlations in low latitudes in the east and west years in Figs. 14a and b speaks against aliassing's being the reason why the arctic correlations are of opposite sign.

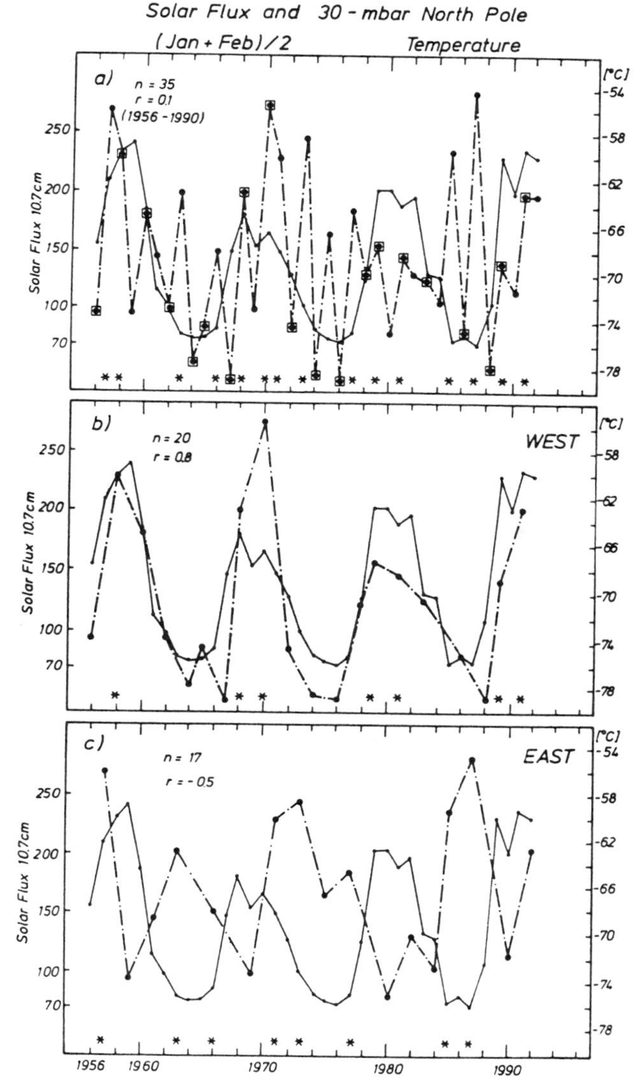

Fig. 13. Time series in January–February of the 10.7 cm solar flux (dashed lines) and (a) 30-mb temperature (C) at the North Pole, the asterisks denote years with major midwinter warmings and the squares are years of westerly QBO. (b) The west years in the QBO. (c) The east years. From Labitzke and van Loon [1988], updated.

The west years in winter are thus the only exception to the basic pattern of correlation with the solar cycle. The polar cyclone in the west years breaks down and warms in maxima of the solar cycle instead of acting according to theory (and to several recent model simulations) by staying cold and intense. In the east years, the vortex is cold and intense at solar maximum and not warm and weak as the theory demands. We suggest that these deviations from theory happen because the two phases of the QBO represent two different

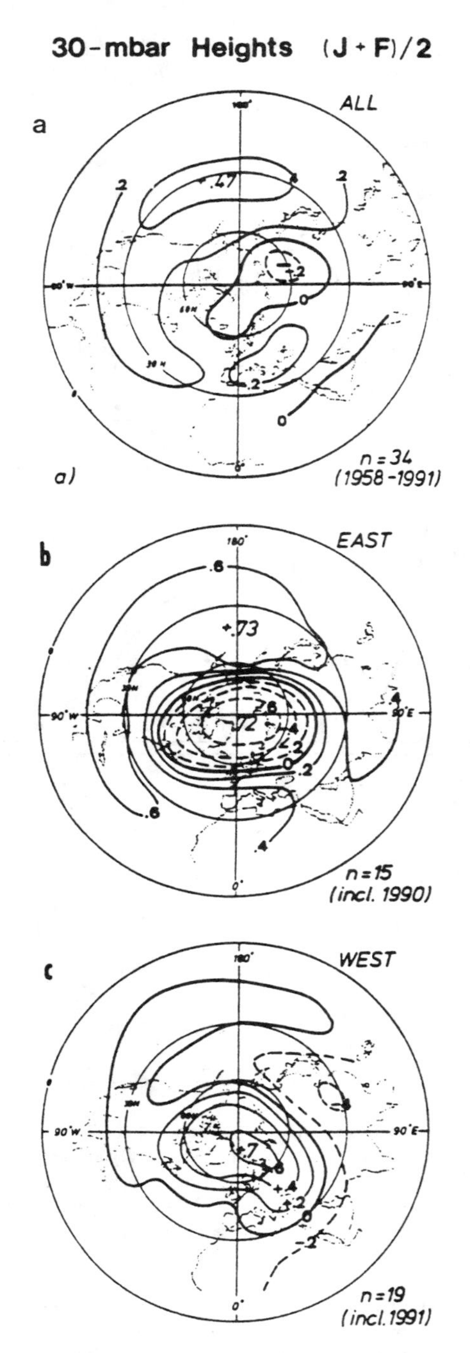

Fig. 14. Lines of equal correlation between the 10.7 cm solar flux and 30-mb height in January–February. (b) As (a) but for the QBO west years. (c) As (a) but for the east years. From Labitzke and van Loon [1988], updated.

states of the stratosphere in winter, each reacting differently to the maxima of the solar cycle. In other words, at peaks in the 11-year solar cycle the stratospheric vortex in QBO west years behaves as theory says east years should do, and east years act as west years should do. In solar minima both east and west years follow theory and modelling [Labitzke and van Loon, 1988].

Because of the high positive correlations in middle and low latitudes with the solar cycle in the east phase in winter, the time series of the area-averaged ungrouped data in January–February (Fig. 15) resemble those of the other seasons, with a TTO visible in the three-year running means. The oscillation becomes increasingly evident in the curves of ungrouped data after the winter months, while the stratospheric westerlies abate and become replaced by easterlies.

Labitzke and Chanin [1988] have outlined the vertical distribution of the TTO in the north polar area in winter by means of 20 years of observations from the rocket station on Heiss Island at 81°N, 58°E (Fig. 16). When they grouped the data according to the phase of the QBO they found the same sign of correlation in the lower stratosphere as in Fig. 13b,c. The correlations change sign near 40 km; and above 55 km they are positive in both east and west years. As a result there was no correlation with the solar cycle below 55 km when the complete time series was used, because the sign was opposite in west and east years.

The existence of the TTO makes it difficult to determine with certainty what the long term trends in the lower stratosphere are like. The shape of the time series in Figs. 6, 8, and 15 indicates that one can get trends which are unrepresentative of longer periods even with series as long as three to four decades if such a series begins in or near a valley of the TTO and ends at or near a peak, and vice versa.

The 10–to–12 Year Oscillation has not yet been explained, nor has the fact that it is, so far, in phase with the 11-year solar cycle.

Conclusion

In addition to what may be termed "Climate Noise", defined as random interannual variability inherent in the atmosphere-ocean system, the interannual variability in the stratosphere on the Northern Hemisphere is associated with at least the following four, not mutually exclusive features:

- The Quasi-Biennial Oscillation, which appears in the stratospheric winds above the equator as downward propagating, persistent east or west winds during alternate periods. The period is variable, but averaged over the last 40 years comes to 27.7 months. The amplitude is equally variable; it is largest between 25 and 30 km where its mean is between 40 and 50 ms^{-1}. In winter the influence of the QBO on the cyclonic vortex in the stratosphere is such that when the winds in the lower equatorial stratosphere (40–50 mb) are easterly, the vortex tends to be comparatively weak and warm (except in east years which fall in maxima of the 11-year solar cycle), whereas it is cold and intense in the west years—with the exception of west years that fall in maxima of the solar cycle. The theory of the QBO states that it is forced by equatorial, eastward moving Kelvin waves and westward moving Rossby-gravity waves, propagating from the troposphere and depositing their momentum at the level where their phase speed equals the zonal wind speed.
- The 10–to–12 Year Oscillation has been present in the data of the upper troposphere and low to middle stratosphere of the Northern Hemisphere in the 40 years with data available to describe it; and during that period it was locked in phase with the 11-year solar cycle. It is most marked in middle and low latitudes on the western, ocean-dominated side of the hemisphere. In winter it is weakest, but if the data in winter are grouped according to the phase of the Quasi-Biennial Oscillation, the pattern of the TTO in the east years is the same as in the undivided series in all other months, whereas it is reversed in the arctic in the west years. There is not yet an explanation of the TTO or of its being in phase with the 11-year solar cycle.

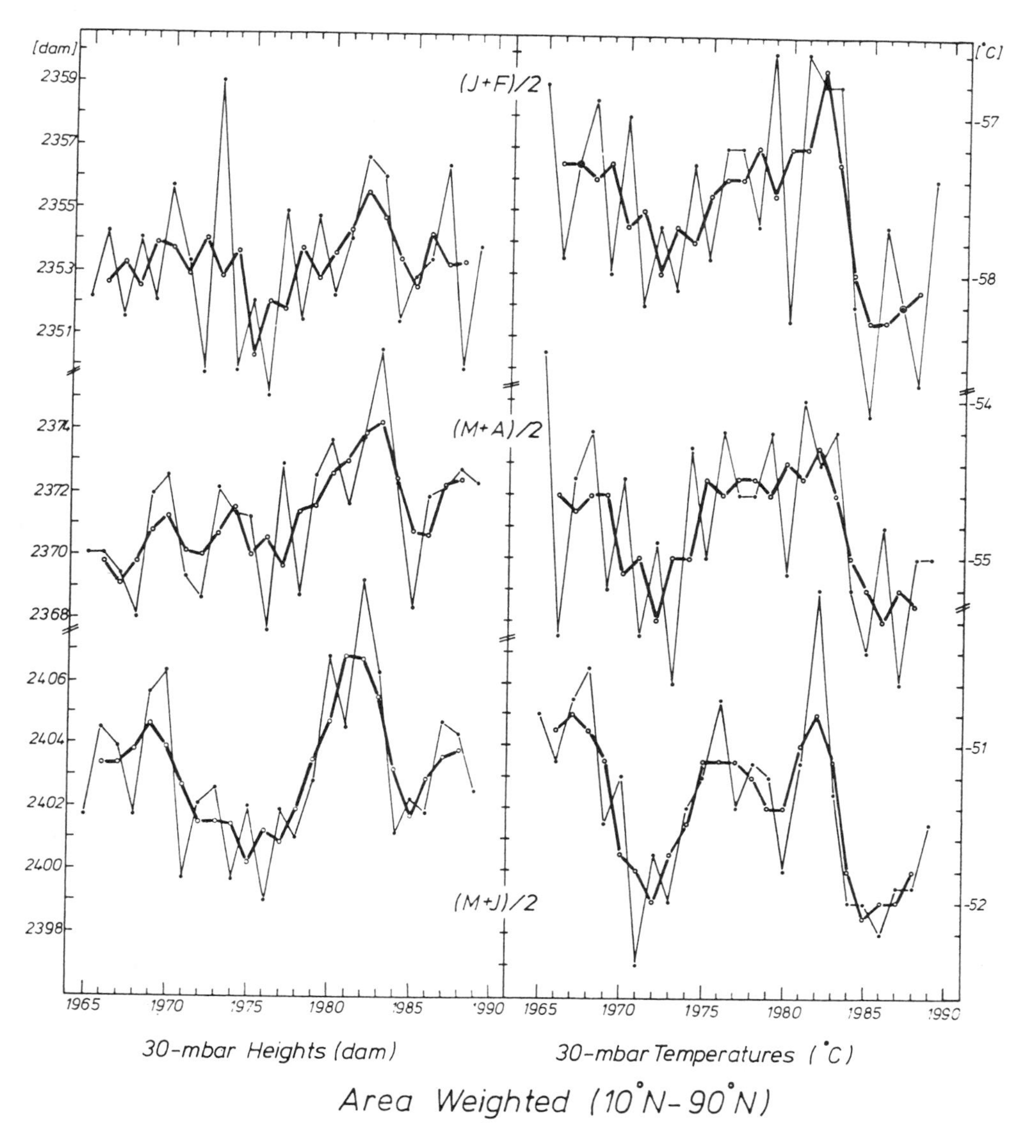

Fig. 15. Time series of two-month average 30-mb heights (gpdam, left) and temperatures (C, right), area weighted between 10°N and 90°N. The heavy lines are three-year running means.

• The Southern Oscillation is a swaying of atmospheric mass between the Pacific Ocean, especially the South Pacific, and the Australia-Indian Ocean region. It is associated with positive and negative anomalies in the surface waters along the equator in the Pacific Ocean, and its extremes affect the atmosphere worldwide. In the winter stratosphere the strength of the polar vortex on the Northern Hemisphere appears to be associated with the Southern Oscillation in such a way that when the Pacific equatorial water is colder than normal, the vortex tends to be cold and strong, but it is on an average weak and warm when the equatorial water temperature is well above normal. In the lower tropical stratosphere the warm extreme of the SO is associated with cooling; there are three exceptions to this in the comparatively short sample at hand: they are the warm extremes of the SO in 1963, 1982, and 1991, which brings us to the last item:

• Volcanic Eruptions. In each of these years a tropical volcano erupted and as a result the tropical temperature in the stratosphere rose and stayed above normal after the eruptions, even though these years were warm extremes in the Southern Oscillation. In addition, instead of warming and breaking down in 1963 and 1982 as in other warm extremes of the Southern Oscillation, the polar vortex stayed cold and intense through the winters of 1963/1964 and 1982/1983, though the Warm Event in 1982/1983 was the strongest recorded in the 20th century. It has been suggested that the failure of the stratospheric vortex to respond to the Warm Event in the Southern Oscillation in the winter following the volcanic eruption should also be ascribed to contamination of the stratosphere by the volcanoes.

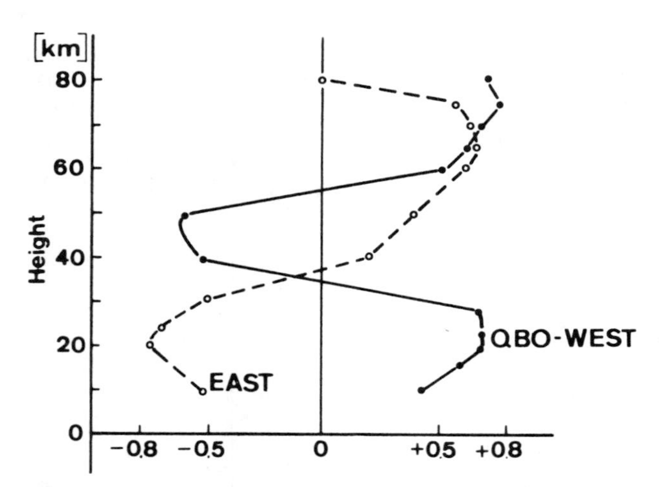

Fig. 16. Correlations between temperature in January–February and the 10.7 cm solar flux. Data above 30 km from the rocketsonde station Heiss Island (81°N, 58°E), and below 30 km from a nearby grid point in the Stratospheric Research Group analyses. From Labitzke and Chanin [1988].

References

Berlage, H. P., Fluctuations of the general atmospheric circulation of more than one year, their nature and prognostic value, *Mededelingen en Verhandelingen, 69*, K.N.M.I, 's-Gravenhage, 152 pp, 1957.

Holton, J. R., and H.-C. Tan, The influence of the equatorial Quasi-Biennial Oscillation on the global circulation at 50 mb, *J. Atmos., 37*, 2200–2208, 1980.

Horel, J. D., and J. M. Wallace, Planetary scale atmospheric phenomena associated with the Southern Oscillation, *Mon. Wea. Rev., 109*, 813–829, 1981.

Labitzke, K., Sunspots, the QBO, and the stratospheric temperature in the north polar region, *Geophys. Res. Letts., 14*, 535–537, 1987.

Labitzke, K., B. Naujokat, and M. P. McCormick, Temperature effects on the stratosphere of the April 4, 1982 eruption of El Chichón, Mexico, *Geophys. Res. Letts., 10*, 877–880, 1983.

Labitzke, K., and M.-L. Chanin, Changes in the middle atmosphere in winter related to the 11-year solar cycle, *Ann. Geophys., 6*, 643–644, 1988.

Labitzke, K., and H. van Loon, Associations between the 11-year solar cycle, the QBO and the atmosphere. Part I: the troposphere and stratosphere in the northern hemisphere in winter, *J. Atmos. Terr. Phys., 50*, 197–206, 1988.

Labitzke, K., and H. van Loon, The Southern Oscillation. Part IX: The influence of volcanic eruptions on the Southern Oscillation in the stratosphere, *J. Clim., 2*, 1223–1226, 1989.

Labitzke, K., and H. van Loon, Association between the 11-year solar cycle and the atmosphere. Part V: Summer, *J. Clim., 5*, 240–251, 1992.

Lindzen, R. S., On the development of the theory of the QBO, *Bull. Am. Met. Soc., 68*, 329–357, 1987.

Livezey, R. E., and W. Y. Chen, Statistical significance and its determination by Monte Carlo techniques, *Mon. Wea. Rev., 111*, 46–58, 1983.

Naujokat, B., An update of the observed Quasi-Biennial Oscillation of the stratospheric winds over the tropics, *J. Atmos. Sci., 43*, 1873–1877, 1986.

Newell, R. E., Stratospheric temperature change from the Mt. Agung volcanic eruption of 1963, *J. Atmos. Sci., 27*, 977–978, 1970.

Rasmusson, E. M., and T. H. Carpenter, Variations in tropical sea surface temperature and surface wind fields associated with the Southern Oscillation/El Niño, *Mon. Wea. Rev., 110*, 354–384, 1982.

Reed, R. J., W. J. Campbell, L. A. Rasmussen, and D. G. Rogers, Evidence of a downward-propagating annual wind reversal in the equatorial stratosphere, *J. Geophys. Res., 66*, 813–818, 1961.

Salby, M. L., and D. J. Shea, Correlations between solar activity and the atmosphere: An unphysical explanation, *J. Geophys. Res., 96, D12*, 22579–22595, 1991.

Teitelbaum, H., and P. Bauer, Stratospheric temperature eleven year variations: Solar cycle influence or stroboscopic effect, *Ann. Geophys., 8*, 239–242, 1990.

van Loon, H., and R. A. Madden, The Southern Oscillation. Part I: Global associations with pressure and temperature in northern winter, *Mon. Wea. Rev., 109*, 1150–1162, 1981.

van Loon, H., and K. Labitzke, The Southern Oscillation. Part V: The anomalies in the lower stratosphere of the Northern Hemisphere in winter and a comparison with the Quasi-Biennial Oscillation, *Mon. Wea. Rev., 115*, 357–369, 1987.

van Loon, H., and K. Labitzke, Association between the 11-year solar cycle and the atmosphere, Part IV: The stratosphere, not grouped by the phase of the QBO, *J. Clim., 3*, 827–837, 1990.

Veryard, R. G., and R. A. Ebdon, Fluctuations in tropical stratospheric winds, *Meteor. Mag., 90*, 125–143, 1961.

Walker, G. T., and E. W. Bliss, World Weather V, *Mem. Roy. Meteor. Soc., 4*, 53–84, 1932.

H. van Loon, The National Center for Atmospheric Research, Boulder, Colorado.

K. Labitzke, Meteorologisches Institut, Freie Universität, Berlin, Germany.

Changes of Total Solar Irradiance

CLAUS FRÖHLICH

Physikalisch-Meteorologisches Observatorium Davos, World Radiation Center, Davos Dorf, Switzerland

Measurements of the total solar irradiance during the last 13 years from satellites show variations over time scales from minutes to years and decades. The most important variance is in the range from days to several months and is related to the photospheric features of solar activity: decreasing the irradiance during the appearance of sunspots, and increasing it by faculae and the bright magnetic network. Long-term modulation by the 11-year activity cycle is observed conclusively with the irradiance being higher during solar maximum. These variations can be interpreted — at least qualitatively — as manifestations of activity related features on the photosphere. For the long-term variation the simultaneous changes of the frequencies of solar oscillations suggest a global origin of the variations. Indeed, it seems that the enhancement of the irradiance during maximum is a true luminosity increase. The relation to longterm climate changes is discussed and the possibilities of a relation between the coldest temperatures during the little ice-age and the Maunder minimum of solar activity explored.

INTRODUCTION

The irradiance from the Sun at the mean Sun-Earth distance, integrated over all wavelengths, hence total irradiance, is traditionally called "solar constant" S although it has been shown to vary on all time scales from minutes to decades (see e.g. Fröhlich et al. 1991). 'All wavelengths' means essentially the energetically important range from 200 nm to 5 μm containing 99.9% of S. Variations of S may influence the Earth's climate, but only if the changes persist over extended periods of time because of the large thermal capacity of the oceans which damps strongly higher frequency forcing. As the Earth's surface is four times its cross-section and ~30% of the sunlight is reflected back to space without being absorbed, the mean solar heating of the Earth is ~240Wm^{-2} with S=1367Wm^{-2}. A change of S by 0.1% is therefore a climate forcing of ~0.24Wm^{-2}. The forcing by a change of S is energetically very similar to a change in the magnitude of the greenhouse effect and their effect can be directly compared: a 2% increase of S and a doubling of CO_2 correspond both to a forcing of about 4-4.8 Wm^{-2} and would yield a change of the mean Earth' temperature of 1.5-5.5°C depending on the global circulation model used (see e.g. Hansen and Lacis, 1990). The climate sensitivity is thus 0.3-1.4 °C/Wm^{-2} for a change in S or of 0.8-3.3 °C/% with a mean of about 2 °C/%.

The measurements from satellites discussed in this paper have been performed by the sensors of the Earth Radiation Budget Experiment (ERB) of the NIMBUS-7 satellite in a near-polar orbit since November 16, 1978 (Hickey et al. 1988, Hoyt et al. 1992) and by ACRIM on the Solar Maximum Mission Satellite in a 27° inclined orbit since February 14, 1980 (Willson 1984). In these experiments the solar irradiance is measured by electrically calibrated cavity radiometers. The time series of the experiments are plotted in Fig.1 and illustrate the solar irradiance variability on all measured time scales. The differences in values among the two experiments (ERB was reduced by 4.25Wm$_{-2}$ before plotting Fig.1) are due to their absolute calibrations which are accurate to 'only' about 0.2% and do not reflect the precision and stability of the instruments which are obviously much better as illustrated by the excellent tracking in Fig.1. The new analysis of the NIMBUS data by Hoyt et al. (1992) have reduced the noise in the series and the agreement with ACRIM has been substantially improved, although some differences exist which are not readily explained.

VARIATIONS ON TIME SCALES UP TO MONTHS

The solar variance is best illustrated by the power spectral density as shown for the SMM/ACRIM data in Fig.2 for 1980 and '85. It covers the range from about 50 nHz (230 days period) to 86 μHz, the Nyquist frequency corresponding to the orbital period of SMM of about 90 minutes. At low frequencies the strength of solar activity changes the spectral density by up to a factor of ten (1980 at maximum compared to 1985 close to minimum) whereas the density at higher frequencies (>12 μHz, <1 day period) seems independent of solar activity. Spectra from simulated time series (Andersen 1991) taking observed contrast and time variations of granuation, meso- and supergranulation into account compare well with observations in this frequency range. This means that to first order the influence of solar activity is in this range negligible.

The time scales of days to months associated with solar magnetic active regions reveal extremely interesting variations of the total

Interactions Between Global Climate Subsystems, The Legacy of Hann
Geophysical Monograph 75, IUGG Volume 15

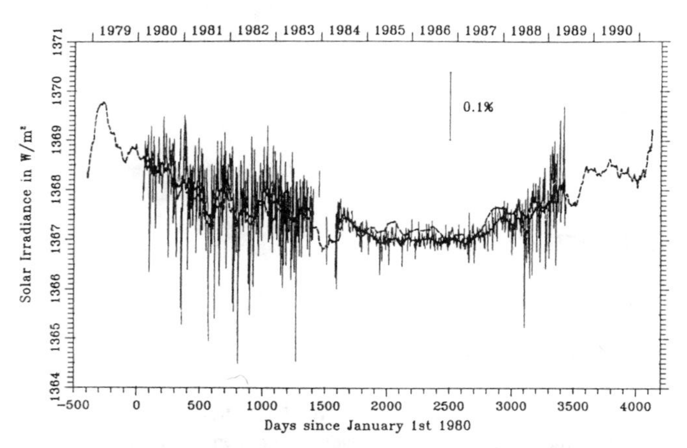

Fig.1 ACRIM and ERB total solar irradiance. The heavy full line is the 81–day running mean for ACRIM, the dashed full line the one for ERB (reduced by 4.25 Wm^{-2} to correct for the bias).

solar irradiance. The dominant feature in the time series for this range of periods is the 'dip', a negative excursion of a few days' length and a depth ranging up to a few tenths of a percent of the irradiance. These dips result from large sunspot groups rotating across the visible disk. Between the dips the total irradiance fluctuates on a wide range of time scales, as demonstrated by the broad-band character of the power spectrum. Prominent dips appeared during the first few months of data from the ACRIM instrument on board SMM. Willson et al. (1981) described them in terms of the Photometric Sunspot Index, the *PSI* function, similar to the models of sunspot darkness noted earlier by Foukal and Vernazza (1979) and Hoyt and Eddy (1982). Hudson and Willson (1982) define this photometric index as the sum of the projected areas of the sunspots multiplied by α, a factor taking into account the darkness of the sunspot relative to the photosphere, in the simplest possible way. This index, calculated from the synoptic data for sunspot areas and with textbook calculation of the value α (i.e., no adjustment of parameters whatsoever), removed about half of the variance of the total ACRIM time series in the active year 1980. This success was quite surprising in view of the grossly simplifying assumptions involved in the construction of *PSI*, but the correspondance between *PSI* and irradiance is quite obvious as illustrated in Fig. 3. The attempts to improve on the naive fit of the *PSI* have generally failed, owing perhaps to the additional noise in the correlation introduced by random and systematic errors in the synoptic data (Schatten et al. 1985).

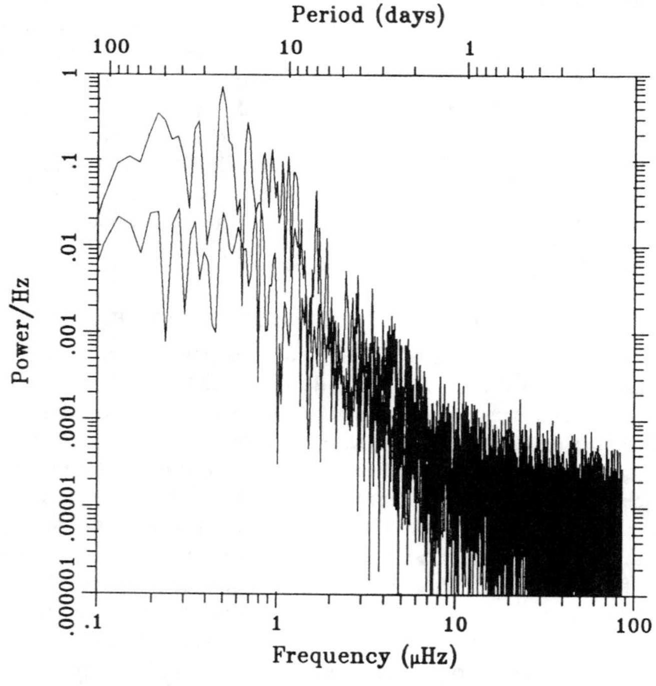

Fig.2 Powerspectrum of ACRIM for 1980 (upper curve) and '85 (lower curve). 11.6 μHz correspond to a period of 1 day.

Willson (1982) called attention to the effects of faculae in modulating the irradiance and having the potential to compensate part of the sunspot deficit. A detailed inspection of Fig.3 demonstrates the influence of faculae seen close to the limb on the photographs and as 'overshoots' on both sides of the dip in irradiance. More direct photometric approaches have yielded some insight into the relationship between the irradiance excesses due to faculae of an active region and the sunspot deficits (Chapman 1987, Foukal 1990, Lean 1991). The area and contrast of faculae are difficult to observe and are not as well known as those of the spots. Thus a facular photometric index is difficult to establish. Proxy data have to be used. The equivalent width of the He 1083 nm line as measured for the full disk, the He I index, is known to be representative of the magnetic elements, including both active-region and network facular elements (Harvey 1984). Foukal and Lean (1988) have identified this index as suitible proxy for the influence of faculae on total irradiance. An other, though less suitable parameter is the 10.7cm flux (F10.7). Both proxies are sensitive to magnetic fields in the photosphere and are often used reciprocally although some differences are observed: the F10.7 e.g. senses also the magnetic field of sunspots, the influence of which on irradiance is already covered by *PSI*.

Multi-variate spectral analysis is a powerful tool for the investigation of multiple influences on a time series, as for example the quasi-independent spot and facular contributions discussed above. Fröhlich and Pap (1989) have applied this technique to the ACRIM data using sunspot areas and the He I index. The results for 1980 and 1984/85 are shown in Fig.4. In that analysis the projected area of two types of sunspots were distiguished, namely active (young) and passive (old) spots. For the present discussion only the sum of both is taken into account which corresponds roughly to *PSI* with the difference that the former does not take the limb darkening effect into account. The results show that around 90% of the variance of the ACRIM data can be explained by the effects of sunspots and magnetic elements. The analysis reveals also the presence of power spectral peaks not explained by spots, faculae or magnetic network, as e.g. near 9 days in 1980, which may be due to heretofore unknown large scale effects.

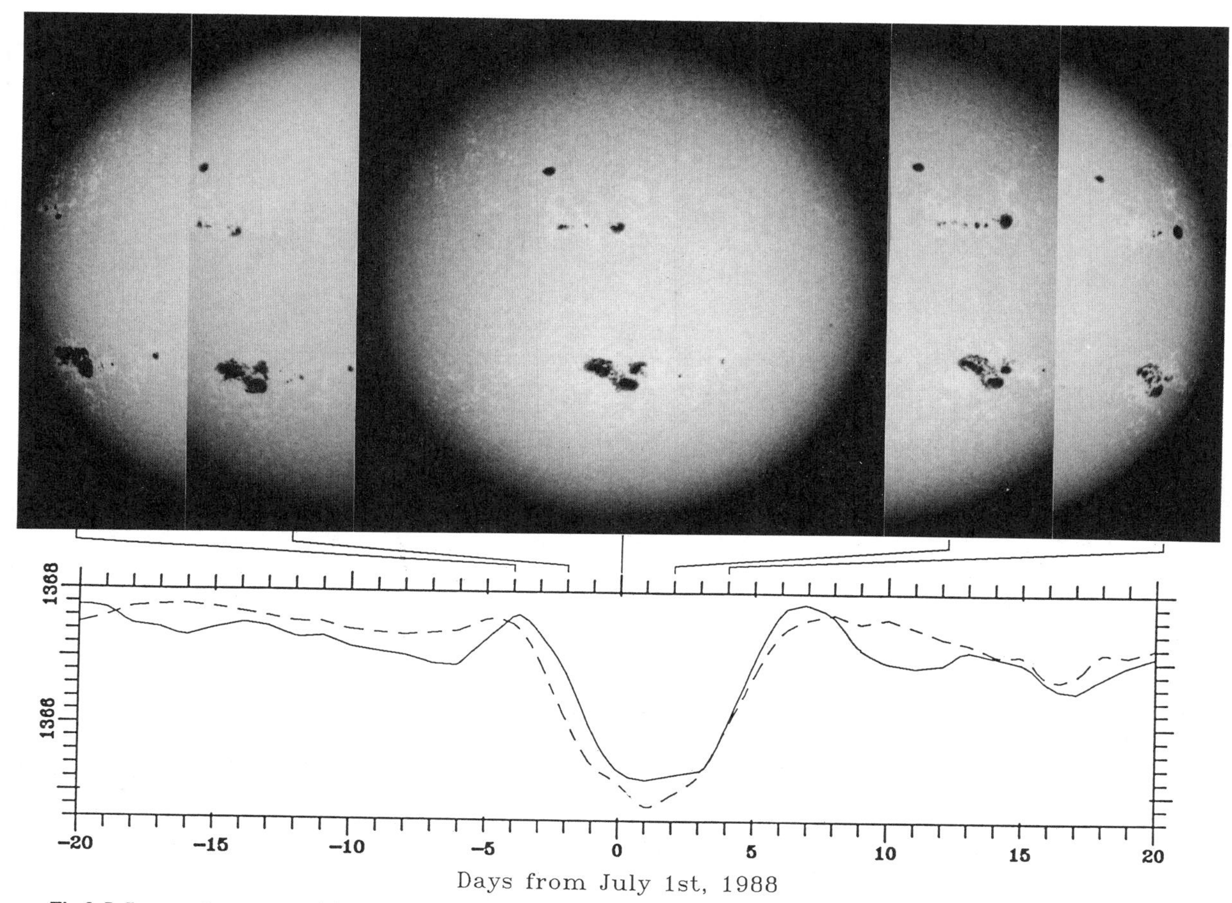

Fig.3 Influence of sunspots and faculae on the irradiance (solid line) and comparison with *PSI* (dashed line) for July 1988. Photographs by San Fernando Observatory, California State University, Northridge, U.S.A.

VARIATIONS ON TIME SCALES OF MONTHS TO YEARS

After smoothing the time series of the ERB and ACRIM radiometry (heavy solid and dashed lines in Fig.1) both data sets show variations on time scales of 4-9 months (Foukal and Lean, 1988), as well as a modulation with the solar cycle of the order of 0.1%. Several ideas and models have been put forward to explain these variations. One approach is to account for the variations purely in terms of the effects that magnetic flux tubes seem to have on the radiation and convection in the relatively shallow photospheric layers that emit most of the Sun's luminosity.

A relatively straightforward approach to both the 4-9 month and 11-year variations has been proposed by Foukal and Lean (1988) and Willson and Hudson (1988). In these studies, it was shown that the residual irradiance variations remaining after a correction for sunspot blocking is made correlate well with the time series of properly scaled He I index or F10.7. This is not surprising since facular area variations were previously shown to account also for shorter term variations in these residuals. Thus, one may conclude that the 6-9 month variations are caused by the tendency of major complexes of activity to persist for about this number of solar rotations. This time scale in persistence of solar activity episodes has been documented before in studies of the He I index time series (Harvey 1984) and may be related to the persistence of certain active longitudes on the solar surface, noted already by Dodson and Hedeman (1970). No well-accepted explanation of this 6-9 month activity time scale (or of the active longitudes) exists as yet, although the recent ideas of Wolff (1984) in terms of Rossby waves and surface pattern of g-mode oscillations and their possible interference are interesting, and deserve more attention. Wolff and Hickey (1987) have also proposed that such oscillations may modulate heat flow to the photosphere and thus contribute to irradiance variation directly.

These variations are also found by the multivariate analysis of the 9 year time series of ACRIM with *PSI* and He I as shown in Fig.5 (Pap and Fröhlich, 1992). The partial coherence ρ^2 of He I has a large peak at a period of 183 days (63 nHz) of close to 80%, whereas the corresponding ρ^2 of *PSI* has a dip reaching down to zero at the same period. This indicates that the 6-8 month modulation is indeed due to the bright network and faculae and not due to spots. There are similar peaks (high in He I, low in *PSI*) at 72 and 64 day periods the origin of which is not known. At low frequencies (<45 nHz, periods

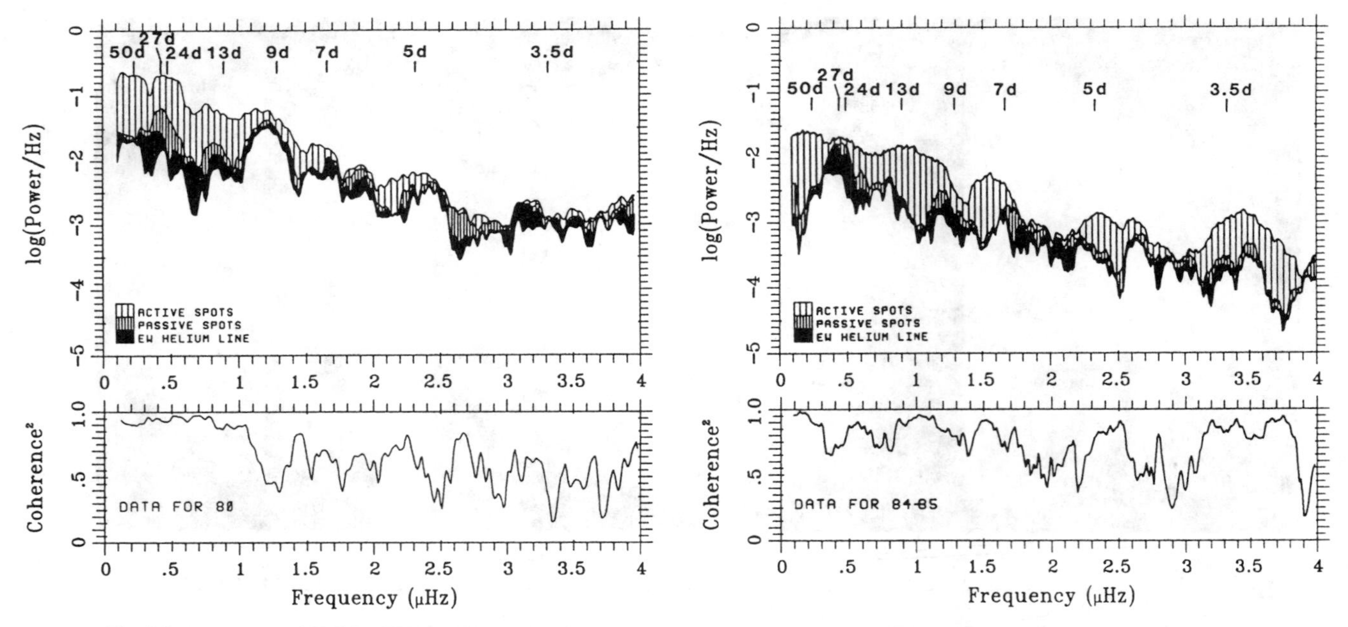

Fig.4 Power spectra of SMM/ACRIM 1980 with the shaded areas indicating the part of the irradiance explained by active and passive spot areas and the He I index.

longer than 260 days) He I explains about 50% whereas *PSI* is around 20% which seems to be already part of the solar cycle modulation. This latter has to be associated with some kind of a slow change in the solar atmosphere well modelled by e.g. the He I index or the F10.7 as shown in Fig.5. The simplest explanation is a slow variation in the amount of the bright magnetic network outside active regions (Foukal and Lean 1988, Foukal et al. 1991).

Although the whole solar cycle can be modelled by these indices, there are subtle differences between the observations and the models which deem some consideration. In contrast to the calculation of *PSI* the scaling for F10.7 and He I is determined by linear regression of *S+PSI* during the decreasing part of cycle 21, thus only differences at different times are relevant. The most obvious difference is in 1980 where the irradiance is higher than the models would suggest. As NIMBUS/ERB shows the same behaviour during '80 this difference is most likely real and not an instrumental effect as suggested by Foukal and Lean (1988). Another difference is during the rising part of cycle 22 were in 1988 a slight phase shift between the model and the irradiance seems to exist with an earlier rise of the model.

Kuhn et al. (1988) and Kuhn and Libbrecht (1991) have reported observations of the limb brightness which can be directly compared with the total irradiance variation over the solar cycle. The observations are broadband, two-color photometric measurements of the brightness distribution in a narrow annulus 20 arcsec wide, just inside the solar limb. The solar limb flux interpreted from the two color measurements as ΔT and observed as a function of latitude can be divided into a 'facular' and 'temperature' part based on the assumption that the 'temperature' part is constant over the 4-month observing summer period and that the 'facular' part shows up as intermittent bright regions. Spots are not seen in this observation. The component of excess brightness moves toward the equator between 1983 and '85, and then re-appears at relatively high latitudes in 1987 and starts moving to the equator while it becomes brighter, following the butterfly diagram of solar activity. The excess brightness responsible for the 'temperature' effect ΔT is due to features which are not resolved by this observation, and it could be due to the bright network in and outside the active regions. The contribution of ΔT (facular and temperature) can account for the total irradiance decrease after 1983, and its increase since 1987 (Kuhn and Libbrecht 1991) as shown by the diamonds in Fig.6. This is, however, only the case if the 'temperature' part is not subjected to a facular type angular redistribution function; otherwise only about half of the irradiance variation can be explained by the ΔT observed at the limb.

Kuhn (1990) compared also the variation of ΔT with observed frequency shifts of the solar oscillations from the Big Bear Solar Observatory (Libbrecht and Woodard 1990) and can with a simple model explain the frequency dependence of the shifts by the corresponding ΔT perturbation, extending downward less than a few hundred kilometers below the photosphere. Physically, the frequency shifts are more likely explained by the strengthening of the network magnetic field (Goldreich et al. 1991), which in turn produces the ΔT at the limb. These independent results strongly support the idea that the observed ΔT is due to a global effect which varies with the solar cycle. Together with the explanation of the irradiance changes this suggests that it is the solar luminosity and not only the observed irradiance that evolves during the solar cycle.

The solar cycle variation and how the explanations are shared between *PSI* and the varying bright network is summarized in Table 1 for three different frequency ranges: the solar cycle in terms of peak-to-peak variation, 7-50 nHz (4.6 years to 230 days) and 50-250 nHz (230 to 46 days) in terms of variance. For F10.7, He I

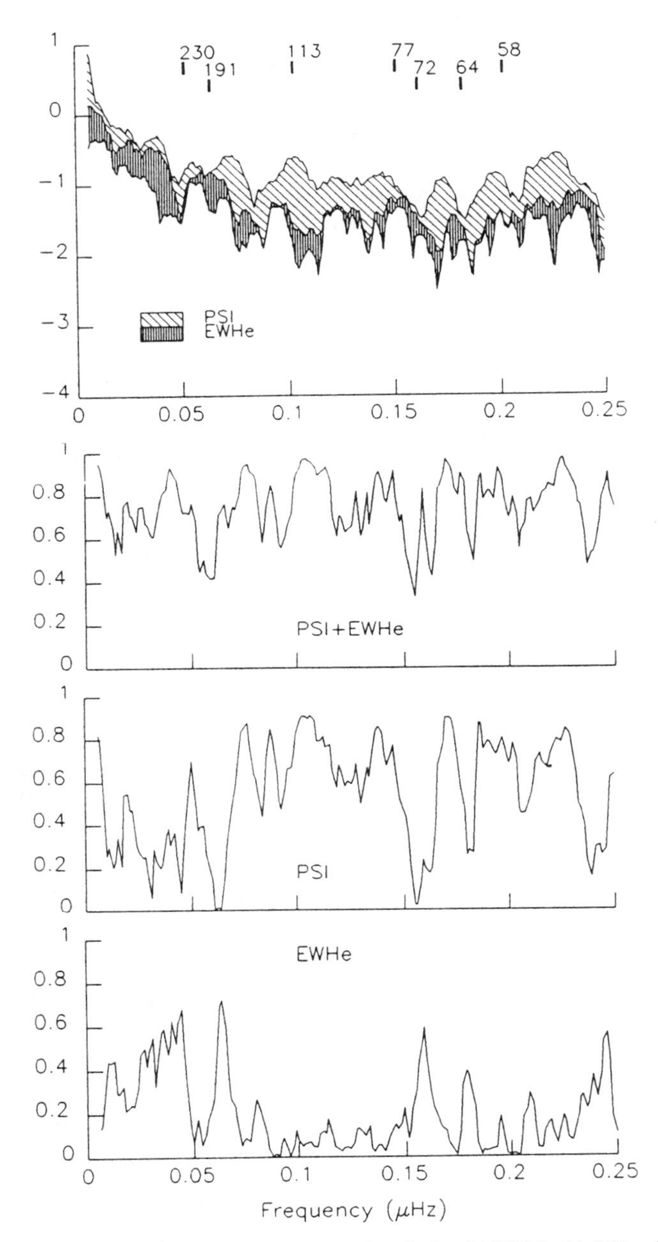

Fig.5 Results of the multivariate spectral analysis of ACRIM with *PSI* and He I index for the period from 1980 to 1989: power spectrum, total coherence squared, partial coherence squared for *PSI* and He I.

and *PSI* only that part of the variance is given which can explain total irradiance (as determined from multivariate spectral analysis). It is interesting to note that the share of the explained variance between F10.7 or He I and *PSI* expressed as a ratio is changing from 2 to $^1/_3$ for time scales from years to months. This supports in a more quantitative way the above findings that the variance at low frequencies is mainly due to the changes in magnetic network whereas at higher frequencies it is due to changes in spot occurence.

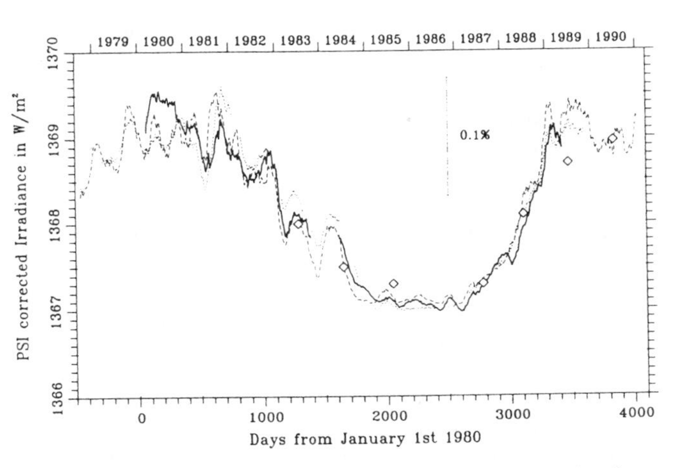

Fig.6 Comparison of the 81-day running mean of the *PSI* corrected irradiance (heavy solid line) with F10.7 (thin dashed), He I (thin dotted) and the limb observations (◊).

CONCLUSIONS

The variations of the total solar irradiance are interesting to solar physicists in several ways and have increased our knowledge about the Sun significantly during the last decade of precise measurements. For most of the variations at least qualitative explanations are available; the details of the underlying physics, however, are not yet fully understood and more simultaneous studies of the detailed features on the solar disk together with continued total solar irradiance observations are needed. A major result from the present data sets is, that the solar cycle modulation of the irradiance seems to be a true luminosity change of the Sun.

In addition, the presence of a distinct 11-year modulation of total irradiance suggests that longer-term variations may be significant for

TABLE 1 Solar Cycle variability for different frequency ranges (for solar cycle peak-to-peak, for other ranges integrated σ is given). The explained power (%) is calculated from multivariate analysis.

Range	Quantity	Variance ppm	Variance %	Ratio F10.7 or HeI to *PSI*
Solar Cycle	Irradiance	797		
	F10.7	1440		1.95
	He I	1447		1.96
	PSI	-738		
7-50 nHz (19-55 months)	Irradiance	1244		
	F10.7	559	44.9	1.38
	PSI	404	32.5	
	He I	483	38.8	1.09
	PSI	444	35.7	
50-250 nHz (3.8-19 months)	Irradiance	1894		
	F10.7	392	20.7	0.36
	PSI	1093	57.7	
	He I	303	16.0	0.27
	PSI	1112	58.7	

the Earth's climate. Eddy (1976) has argued that reduced solar irradiance during the Maunder minimum of solar activity (1640-1720) may have been responsible for the coldest period of the Little Ice Age when global temperature is estimated to have been as much as 1°C colder than today (e.g. Wigley and Kelly, 1990). Although the direction (low irradiance during low activity) is correct, the observed amplitude of the 11-year modulation of ~0.1% is about 5 times too small (see INTRODUCTION). Moreover, reconstructions of *S*, back to 1884 (Foukal and Lean, 1990), using *PSI* and the strong correlation between sunspot numbers and F10.7, show that the presently observed solar cycle amplitude was the largest in the whole series on the basis of sunspot and facular effects. During solar minimum, however, a surface network of bright magnetic elements remains on the Sun (Foukal et al. 1991). If this would have been removed during a Maunder Minimum period a lower irradiance would have been observed. From the model proposed by Foukal et al. (1991) the irradiance would be lowered by 1.8 Wm^{-2} if the quiet network is completely removed. The presence of bright magnetic elements is well indicated by e.g. the emission reversal in the core of the chromospheric Ca II H and K absorption lines. H and K fluxes are also observed from stars as measure of stellar activity. The HK index is defined as the flux in narrow passbands centered on the H and K emission cores compared in a ratio to the flux in the nearby continuum, and varied for the Sun between 0.174 during the solar maximum in 1980/82 and 0.166 during the minimum in 1985/87 (Livingston and White in Giampapa, 1990). More recent work (Lean et al., 1992) on the translation of solar Ca-II observations to the HK index indicate that the last solar cycle had more likely a variation between 0.175 and 0.195 (from Fig.1 of Lean et al., 1992; in the following the corresponding values are given in parentesis). Investigation of the activity of solar type stars (Baliunas and Jastrow, 1990) indicate that there is a group of stars which exhibit Maunder Minimum type 'non-activities' and have HK indices of ~0.146 or 2.5 (1.5) times the solar cycle 21 amplitude lower than its solar minimum value. From Fig.6 the 'magnetic' irradiance amplitude amounts to 2.1 Wm^{-2} or the Maunder Minimum value would have been 5.25 (3.05) Wm^{-2} below the minimum in 1985/87 if a linear relationship is assumed and if the 17th century Sun behaved like the low activity reference stars. This means a 0.38% (0.22%) lower irradiance (below the minimum of cycle 21) and is about double the value expected from only removing the network, which indicates that for a non-cycling Sun some non-network magnetic fields would have to be removed (Lean et al., 1992). As already stated by Balliunas and Jastrow (1990) this might 'explain' part of the lower temperatures during the Little Ice Age by activity related changes of the solar irradiance.

Acknowledgments: I thank Drs. R.C.Willson, JPL, Pasadena,and D.Hoyt, RDS, Greenbelt, for providing yet unpublished data. Thanks are extended to Drs. J.Pap, NOAA, Boulder, and J. Kuhn, Michigan State University, for many helpful discussions and for providing material before publication. Furthermore, the continuous support of this work at PMOD/WRC by the Swiss National Science Foundation is gratefully acknowledged.

REFERENCES

Andersen, B., Model Simulations of Solar Noise, in *Proc. Workshop on Diagnostics of Solar Osciallation Observations*, eds. P. Maltby and E. Leer (Inst. Theor. Astrophysics, Univ. Oslo), 15–23, 1991.

Baliunas, S., Jastrow, R., Evidence for long-term brightness changes of solar-type stars, *Nature, 348*, 520–523, 1990.

Chapman, G.A., Variations of Solar Irradiance due to Magnetic Activity, *Ann. Revs. Astron. Astrophys. 25*, 633–667, 1987.

Dodson, H.W., Hedeman, E.R., Time Variations in Solar Activity, in *Solar Terrestrial Physics, Part I*, eds. E.Dyer and C. deJager, (Reidel, Dortrecht) 151–172, 1970.

Eddy, J.A., The Maunder Minimum, *Science 192*, 1189–1202, 1976.

Foukal, P.V., Solar Luminosity over Timescales of Days to the Past few Solar Cycles, *Phil.Trans.R.Soc.Lond. A 330*, 591–599, 1990

Foukal, P.V., Vernazza, J., The Effect of Magnetic Fields on Solar Luminosity, *ApJ. 234*, 707–715, 1979.

Foukal, P.V., Lean, J., Magnetic Modulation of Solar Luminosity by Photospheric Activity, *ApJ. 328*, 347–357, 1988.

Foukal, P.V., Lean, J., An Empirical Model of Total Solar Irradiance Variation Between 1874 and 1988, *Science 247*, 556–558, 1990.

Foukal, P.V., Harvey, K., Hill, F., Do Changes in the Photosperic Network Cause the 11 Year Variation of Total Solar Irradiance?, *ApJ. 383*, L89–L92, 1991

Fröhlich, C., Foukal, P.V., Hickey, J.R., Hudson, H.S., Willson, R.C., Solar Irradiance Variability from Modern Measurements, in *Sun in Time*, eds. C.P.Sonnet, M.S.Giampapa, M.S.Matthews, Univ.of Arizona Press, Tucson, 11–29, 1991.

Fröhlich, C., Pap, J., Multi-Spectral Analysis of Total Solar Irradiance Variations, *Astron. Astrophys. 220*, 272–280 , 1989.

Giampapa, M.S., The solar – stellar connection, *Nature, 348*, 488–489 , 1990.

Goldreich, P., Murray, N., Willette, G., Kumar, P., Implications of Solar *p*-Mode Frequency Shifts, *ApJ. 370*, 752–762, 1991.

Hansen, J.E., Lacis, A.A., Sun and dust versus greenhouse gases: an assessment of their relative roles in global climate change, *Nature, 346*, 713–719, 1990.

Harvey, J., Helium 10830 Å Irradiance: 1975-1983, in *Solar Irradiance Variations on Active Region Time Scales*, eds. B.J.LaBonte, G.A.Chapman, H.S.Hudson, R.C.Willson, NASA Conf. Publ. CP-2310, 197–212, 1984.

Hickey, J.R., Alton, B.M., Kyle, H.L., Major, E.R., Observation of Total Solar Irradiance Variability from Nimbus Satellites, *Adv.Space Res. 8(7)*, 5–10, 1988.

Hoyt, D.V., Kyle, H.L., Hickey, J.R., Maschhoff, R.H., The Nimbus-7 solar total irradiance: A new algorithm for its derivation, *J.Geoph.Res. 97*, 51–63, 1992.

Hoyt, D.V., Eddy, J.A., An Atlas of Variations in the Solar Constant Caused by Sunspot Blocking and Facular Emissions from 1874 to 1981, *NCAR Techn.Note, NCAR/TN194+STR*, (NCAR, Boulder Co. U.S.A.), 1982.

Hudson, H.S., Willson, R.C., SMM Experiment: Initial Observations by the Active Cavity Radiometer, *Adv.Space Res. 1(13)*, 285–288, 1981.

Hudson, H.S., Willson, R.C., Sunspots and Solar Variability, in *Physics of Sunspots*, ed. L.Cram and J.Thomas, Sacramento Peak Obs.Publ., 434–445, 1982.

Kuhn, J.R., Libbrecht, K.G., Dicke, R., The Surface Temperature of the Sun and Changes in the Solar Constant, *Science 242*, 908–911, 1988.

Kuhn, J.R., Measuring Solar Structure Variations from Helioseismic and Photometric Observations, in *Progress in Seismology of the Sun and Stars*, eds. Y.Osaki and H.Shibahashi (Lecture Notes in Physics 367, Springer, Heidelberg), 157–162, 1990.

Kuhn, J.R., Libbrecht, K.G., Nonfacular Solar Luminosity Variations, *ApJ. 381*, L35–L37, 1991.

Libbrecht, K.G., Woodard, M.F., Observations of Solar Cycle Variations in Solar p-Mode Frequencies and Splittings in *Progress in Seismology of the Sun and Stars*, eds. Y.Osaki and H.Shibahashi (Lecture Notes in Physics 367, Springer, Heidelberg), 145–156, 1990.

Lean, J., Variations in the Sun's Radiatiove Output, *Rev.Geophys. 29*, 505–535, 1991.

Pap, J.M., Fröhlich, C., Multi-variate Spectral Analysis of Short Term Irradiance Variations,in *Proc.Solar Electromagnetic Radiation Study for Solar Cycle 22*, ed. R.F.Donnelly, SEL NOAA ERL, Boulder, 1992.

Wigley, T.M., Kelly, P.M., Holocene climatic change, ^{14}C wiggles and variations in solar irradiance, *Phil.Trans.R.Soc. A330*, 547–560, 1990.

Willson, R.C., Gulkis, S., Janssen, M., Hudson, H.S., Chapman, G.A., Observations of Solar Irradiance Variability, *Science 211*, 700–702, 1981.

Willson, R.C., Solar Irradiance Variations and Solar Activity, *J.Geophys.Res. 87*, 4319–4324, 1982.

Willson, R.C., Measurements of Solar Total Irradiance and its Variability, *Space Sci. Rev. 38*, 203–242, 1984.

Willson, R.C. Hudson, H.S., Solar Luminosity Variations in Solar Cycle 21, *Nature, 332*, 332–334, 1988.

Wolff, C., Solar Irradiance Changes Caused by g-Modes and Large Scale Convection, *Sol.Phys. 93*, 1–13, 1984.

Wolff, C., Hickey, J., Multiperiodic Irradiance Changes Caused by r-Modes and g-Modes, *Sol.Phys. 109*, 1–18, 1987.

C.Fröhlich, Physikalisch-Meteorologisches Observatorium Davos, World Radiation Center, P.O.Box 173, CH-7260 Davos-Dorf, Switzerland

Energetic Particle Influences on NO_y and Ozone in the Middle Atmosphere

CHARLES H. JACKMAN

Code 916, NASA/Goddard Space Flight Center
Greenbelt, MD 20771

Natural variations in the middle atmosphere can result from the penetration of energetic protons and electrons. These energetic particles produce NO_x (N, NO, NO_2) through interactions with the background atmosphere, primarily at polar latitudes. The NO_x species then produce other odd nitrogen compounds, NO_y (N, NO, NO_2, NO_3, N_2O_5, HNO_3, HNO_4, $ClONO_2$), fairly quickly in the stratosphere (on time scales of seconds to days). The long lifetime of the NO_y family (up to several months in the middle atmosphere) as well as the NO_y species' significant influence on stratospheric ozone abundance make the charged particle increases of NO_y important. Galactic cosmic rays produce NO_y in the lower stratosphere, solar protons produce NO_y in the middle and upper stratosphere as well as the mesosphere, and relativistic electrons produce NO_y in the upper stratosphere and mesosphere, each affecting the NO_y middle atmosphere budget directly. Production of NO_y constituents by solar protons has been associated with an observed polar ozone depletion during and after the August 1972 and the August, September, and October 1989 solar proton events and a polar NO increase after the July 1982 solar proton event. Auroral electron and photoelectron production of NO_x in the thermosphere and its subsequent transport downwards to the polar mesosphere and upper stratosphere is thought to be an important component of the NO_y budget in the middle atmosphere in the wintertime at high latitudes. It has been suggested that relativistic electrons can have a significant impact on the middle atmosphere NO_y budget over a solar cycle time period. The absolute flux of relativistic electrons to the middle atmosphere needs to be quantified more thoroughly to verify this suggestion.

INTRODUCTION

Recent analyses [Stolarski et al. 1991] indicate that total ozone has decreased in the past decade by several percent in the northern and southern hemisphere middle latitudes and by even more in the spring-time southern hemisphere polar latitudes. Such a disturbing result suggests that humankind (anthropogenic) activity is the cause of this decline. The natural changes in ozone, whether caused by solar ultraviolet variations, interannual dynamical changes, or energetic particle fluxes, need to be understood more completely in order to clearly separate anthropogenically-caused from naturally-caused ozone changes. To help understand one part of these natural atmospheric variations, we consider the influence of energetic protons and electrons on the middle atmosphere in this review paper.

Energetic protons and electrons influence the background middle atmosphere primarily at high geomagnetic latitudes by perturbing the chemistry and constituents. These charged particles produce ions, radioactive isotopes, and HO_x (H, OH, HO_2) and NO_x (N, NO, NO_2) through interactions with the atmosphere.

The charged particle influences on ions, radioactive isotopes and HO_x are important, however, long-term changes in ozone are primarily influenced by the enhanced NO_x levels from charged particle interaction with the middle atmosphere. For this reason, we concentrate on the NO_x increases from charged particle precipitation which result in an overall enhancement in odd nitrogen compounds, NO_y (N, NO, NO_2, NO_3, N_2O_5, HNO_3, HNO_4, $ClONO_2$) in the middle atmosphere. The lifetime of the NO_y family in the middle atmosphere is from hours in the mesosphere to one to two years in the stratosphere. Some of these effects on the middle atmosphere caused by NO_y species can be large and long-lived such as the August 1972 and the August, September, and October 1989 solar proton event disturbances [Heath et al. 1977; McPeters et al. 1981; Jackman and McPeters, 1987; Reid et al. 1991], but these large perturbations are infrequent. Other charged particle effects on the middle atmospheric NO_y abundance are continuous, but variable, such as the perpetual flow of galactic cosmic rays [Legrand et al. 1989]. Jackman [1992] discussed and reviewed the state of knowledge of the effects of charged particles on the middle atmosphere recently, thus this review

Interactions Between Global Climate Subsystems, The Legacy of Hann
Geophysical Monograph 75, IUGG Volume 15

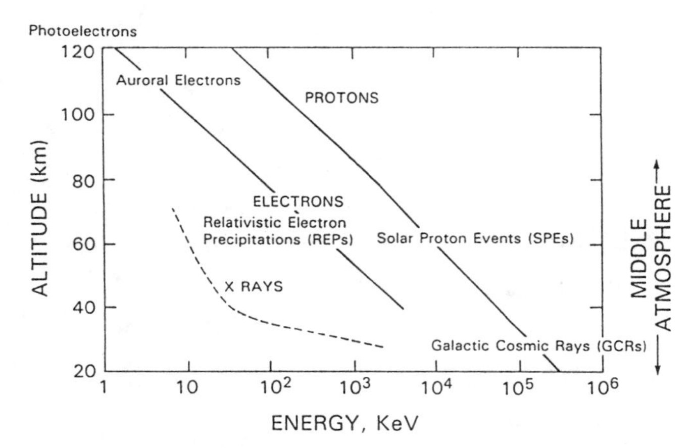

Fig. 1. Taken from Figure 1 of Jackman, Effects of energetic particles on minor constituents of the middle atmosphere, in press, *J. Geomag. and Geoelec.*, (1992), copyright by The Society of Geomagnetism and Earth, Planetary and Space Sciences. Altitude of penetration for protons, electrons, and X-rays vertically incident at the top of the atmosphere as a function of particle energy (adapted from Figure 2 of Thorne 1980).

will mainly consider developments in this field since that paper was written.

The middle atmosphere NO_y abundance is influenced directly by galactic cosmic rays which produce NO_y in the lower stratosphere, solar protons which produce NO_y in the stratosphere as well as the mesosphere, and relativistic electrons which produce NO_y in the upper stratosphere and mesosphere. Auroral electron and photoelectron production of NO_y in the thermosphere and its subsequent transport downwards to the mesosphere and upper stratosphere is thought to be important to the NO_y budget in the middle atmosphere for the wintertime. Both model results and measurements of charged particle influences on NO_y and ozone in the middle atmosphere will be discussed.

OVERVIEW OF CHARGED PARTICLE ENERGY DEPOSITION

A schematic diagram of the areas of influence by the various categories of charged particles and their associated products is shown in Figure 1. This graph was created by modifying Figure 2 from Thorne [1980]. For a given energy, X-rays penetrate further than electrons and electrons penetrate further than protons (see Figure 1). The X-rays (bremsstrahlung) result from the slowing down of the energetic electrons. The magnitude of energy deposition by the X-rays is usually at least three orders of magnitude smaller than the energy deposition of the associated parent electrons [Berger et al. 1974].

Photoelectrons are produced by extreme ultraviolet (EUV) with energies up to a few hundred eV throughout the thermosphere. Primary electrons with energies less than 500 eV do not, in general, penetrate below 120 km (see Figure 1). Photoelectrons are more similar to secondary electrons than primary electrons because these particles are produced by the in situ EUV ionization of background atmospheric constituents. The major region of atmospheric influence by the photoelectrons is the thermosphere (see upper left corner of Figure 1). Since photoelectrons are produced by EUV, the only latitudinal dependence of the photoelectron energy deposition arises from changes in the solar zenith angle.

Some auroral electrons have energies capable of penetration to the mesosphere (electron energies < 100 keV) with associated bremsstrahlung reaching the upper to middle stratosphere. The higher energy electron fluxes are indicated as relativistic electron precipitations (electron energies > 100 keV) and are capable of depositing energy in the mesosphere and even the upper stratosphere with associated bremsstrahlung reaching the middle to lower stratosphere. Both auroral and relativistic electrons mainly deposit their energy in the subauroral region (geomagnetic latitudes between 60° and 70°) and their altitudes of deposition are indicated in Figure 1.

Solar protons deposit their energy in the mesosphere and stratosphere and generally in the polar cap region (geomagnetic latitudes greater than 60°). Galactic cosmic rays deposit most of their energy in the lower stratosphere and upper troposphere at high latitudes; however, penetration of the higher energy galactic cosmic rays is possible all the way to tropical latitudes thus latitude dependent energy deposition distributions are required. The major altitudes of influence for solar proton events and galactic cosmic rays are indicated in Figure 1.

Protons, electrons, and associated bremsstrahlung produce odd nitrogen (NO_y) constituents through dissociation or dissociative ionization processes in which N_2 is converted to $N(^4S)$, $N(^2D)$, or N^+. Rapid chemistry is initiated after N_2 dissociation and most of the atomic nitrogen is rapidly converted to NO and NO_2. Charged particle produced NO_y constituents can then deplete ozone through the following catalytic reaction cycle:

$$NO + O_3 \rightarrow NO_2 + O_2$$
$$NO_2 + O \rightarrow NO + O_2$$

$$\text{Net:} \quad O_3 + O \rightarrow O_2 + O_2$$

GALACTIC COSMIC RAY INFLUENCE

The influence of galactic cosmic rays (GCRs) on the middle atmospheric NO_y abundance has been studied over the past two decades [Warneck 1972; Ruderman and Chamberlain, 1975; Nicolet 1975; Jackman et al. 1980; Thorne 1980; Garcia et al. 1984; Legrand et al. 1989]. The major production of NO_y results from nitrous oxide oxidation ($N_2O + O(^1D) \rightarrow NO + NO$), thus the GCR-related production of NO_y must be compared to that background source in order to put its NO_y budget contribution into perspective. Jackman et al. [1980] computed that GCRs were responsible for 2.7 to 3.7 x 10^{33} molecules of NO_y per year, whereas Crutzen and Schmailzl [1983] and Jackman et al. [1987] show that nitrous oxide oxidation produces about 2.3 to 2.6 x 10^{34} molecules of NO_y per year, respectively. Nitrous oxide oxidation produces odd

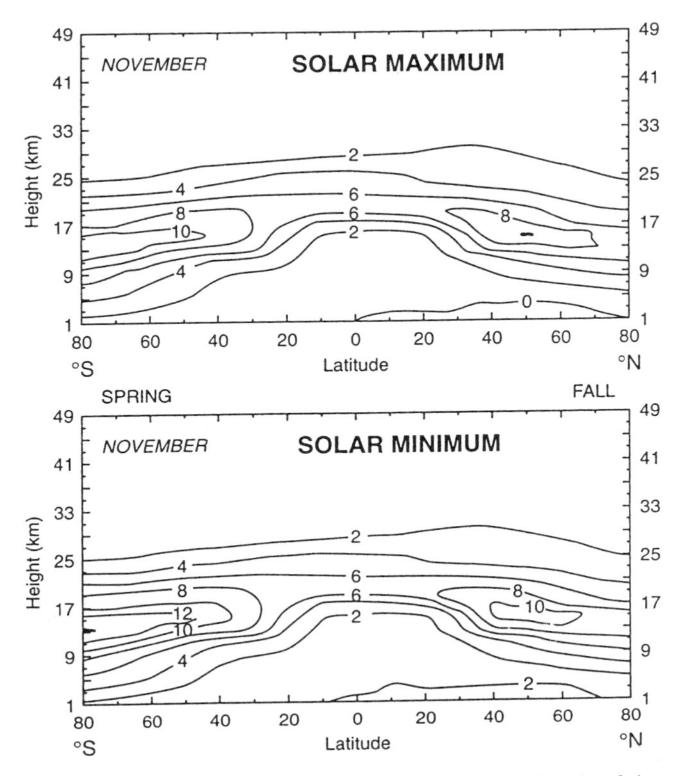

Fig. 2. Taken from Figure 4 of Legrand et al., A model study of the stratospheric budget of odd nitrogen, including effects of solar cycle variations, *Tellus, 41B*, 413-426, (1989), copyright by Munksgaard International Publishers Ltd. Computed percentage change in odd nitrogen from galactic cosmic rays at solar maximum and solar minimum.

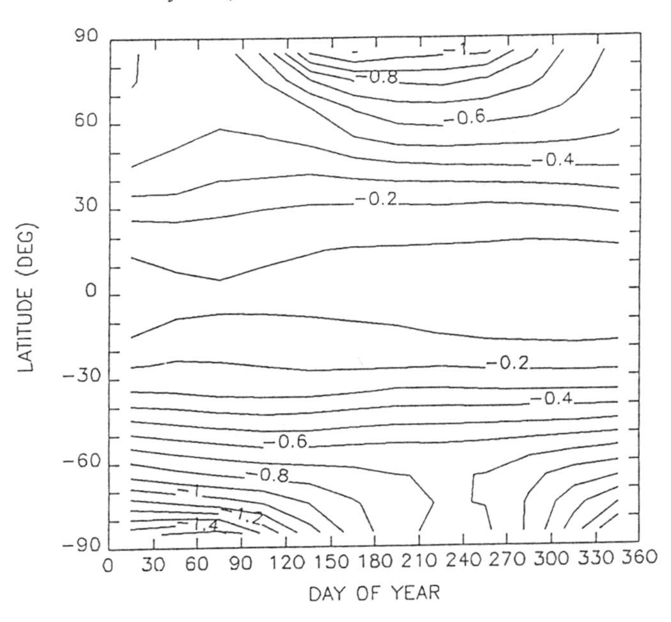

Fig. 3. Computed percentage total ozone change for a model simulation including galactic cosmic rays at solar minimum (maximum in GCRs) compared to a model simulation with no galactic cosmic rays. Contours range from -0.1% to -1.5% in 0.1% intervals.

nitrogen primarily in the middle stratosphere at low latitudes, while the GCR-induced odd nitrogen production peaks near the tropopause at high latitudes [see Figures 9a and 13 in Jackman et al. 1987].

Since the NO_y family has a lifetime of months in the middle and lower stratosphere, transport of NO_y created at higher altitudes and lower latitudes is significant. In Figure 2 [taken from Figure 4 of Legrand et al. 1989] we show the computed percentage change in NO_y which results from the GCR source of odd nitrogen. The lower stratospheric NO_y at high latitudes is calculated to be increased by about 10%. A solar cycle variation is apparent in the GCR flux with maximum flux during solar minimum and minimum flux during solar maximum.

The influence of GCRs on ozone over the 11-year solar cycle time period has been included in a two-dimensional (2D) model computation of Garcia et al. [1984]. Since the effects of ultraviolet (UV) and auroral flux variation were also included in that computation, no quantitative changes from only the GCRs were available. The Goddard Space Flight Center (GSFC) 2D model which extends from the ground to about 90 km [Jackman et al. 1990] was used to investigate the influence of GCRs on NO_y abundance and ozone amounts. The minimum flux in GCRs (solar maximum) from a GSFC model computation allows about 0.25% more total ozone near the poles than computed during the maximum flux in GCRs (solar minimum). Predictions from the GSFC model indicated about 1% less total ozone at polar latitudes for a model run including GCRs compared to a model run not including GCRs (see Figure 3). These model computations showed a seasonal as well as a strong latitudinal dependence with less than a 0.1% difference near the Equator between the two model runs just described.

SOLAR PROTON EVENT INFLUENCE

Direct constituent change in the middle atmosphere by particles has only been documented in the case of solar proton events (SPEs). SPEs are sporadic with durations up to several days and have a solar cycle dependence such that more SPEs occur closer to solar maximum.

The production of NO_y species by SPEs has been predicted since the mid-1970's [Crutzen et al. 1975]. The polar NO increase after the July 1982 SPE was inferred from the SBUV instrument to be about 6 X 10^{14} NO molecules cm^{-2} at polar latitudes [McPeters 1986], in good agreement with our calculated NO increase of 7 X 10^{14} NO molecules cm^{-2} in the polar cap [Jackman et al. 1990].

Polar ozone depletions associated with NO_y increases have been observed and modelled for the August 1972 SPE [Crutzen et al. 1975; Heath et al. 1977; Fabian et al. 1979; Maeda and Heath, 1980/81; McPeters et al. 1981; Solomon and Crutzen,

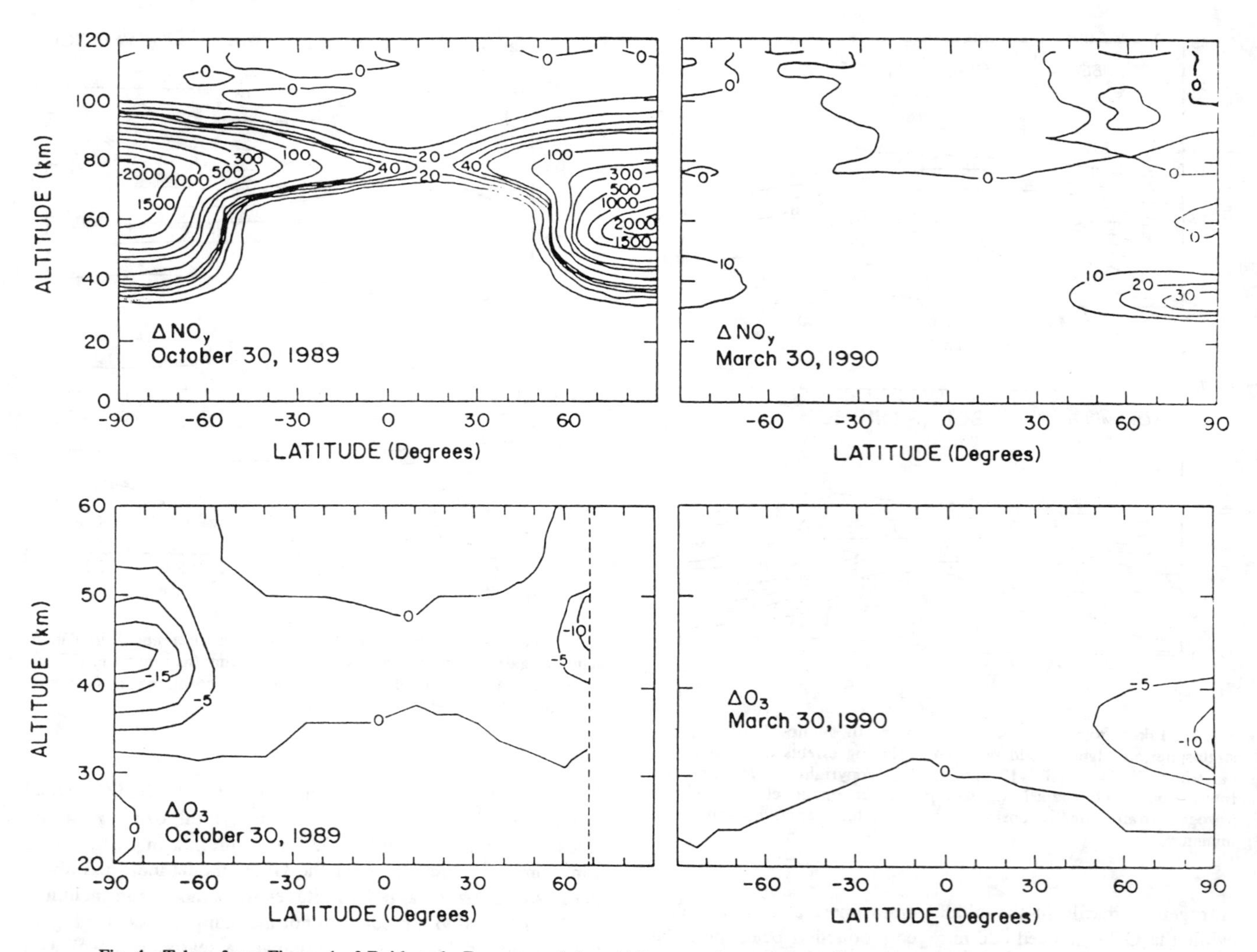

Fig. 4. Taken from Figure 4 of Reid et al., Response of the middle atmosphere to the solar proton events of August-December 1989, *Geophys. Res. Lett.*, *18*, 1019-1022, (1991), copyright by the American Geophysical Union. Contour plots of calculated percentage changes in NO_y and ozone at the end of October 1989 and March 1990 as compared to quiet geomagnetic conditions. Values of ozone percent change are shown only in the sunlit atmosphere.

1981; Reagan et al. 1981; Rusch et al. 1981; Jackman and McPeters, 1987; Jackman et al. 1990] and the August, September, and October 1989 SPEs [Reid et al. 1991]. These SPEs in 1972 and 1989 were very large events and substantial increases in NO_y have been computed to be associated with these events at polar latitudes in the middle to upper stratosphere [see, e.g., Jackman et al. 1990; Reid et al. 1991].

The model predicted percentage change for NO_y and the associated ozone depletion for October 30, 1989 and March 30, 1990 as a result of the August, September, and October 1989 SPEs are shown in Figure 4 [from Figure 4 of Reid et al. 1991]. NO_y on October 30, 1989 is predicted to increase by a maximum of over 20-fold in the polar mesosphere as a result of these SPEs. NO_y is also predicted to increase in the polar middle to upper stratosphere by 10% to 500% and the maximum associated ozone response is predicted to be a decrease of over 20%. By March 30, 1990, Reid et al. [1991] predicted increases in NO_y of over 30% with associated ozone decreases of over 10%.

Reid et al. [1991] used GOES-7 proton flux data in computing NO_y production rates. We have used IMP-8 proton flux measurements in our computations, which results in slightly higher NO_y production rates and, subsequently, a slightly larger ozone depletion. For example, on October 30, 1989, we predict a maximum ozone depletion of about 25% whereas Reid et al. [1991] predict a maximum ozone depletion of about 20% near 43 km and the Southern Hemisphere pole.

Our computations of NO_y and associated ozone percentage change at 75°N due to the 1989 SPEs are presented in Figure 5. The calculated NO_y changes are largest in the upper

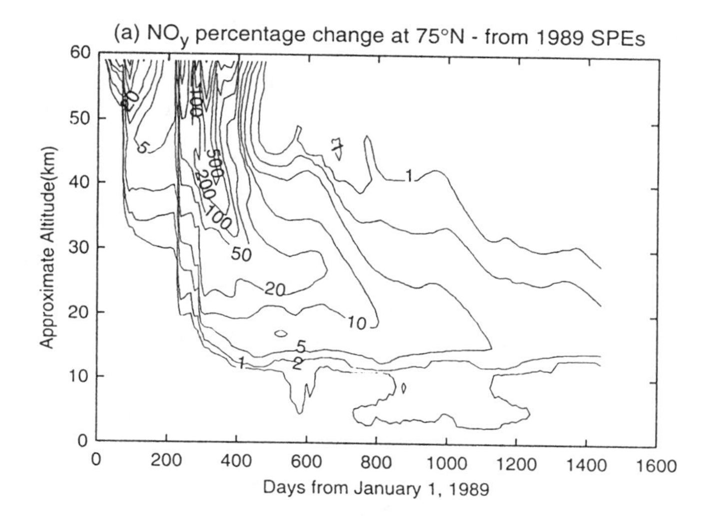

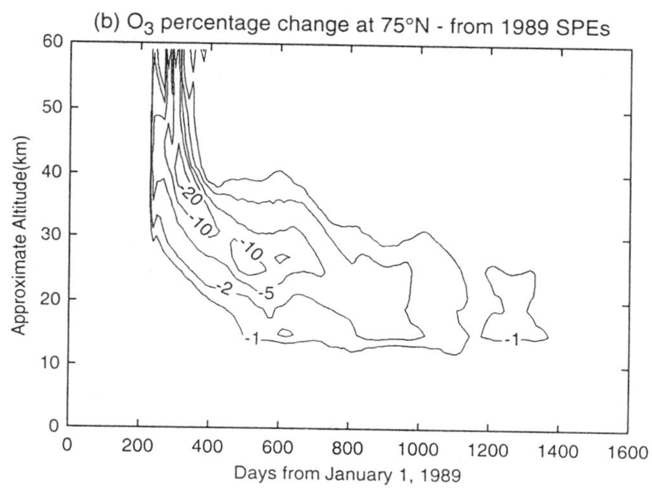

Fig. 5. Calculated percentage changes in (a) NO_y and (b) ozone at 75°N resulting from solar proton events in 1989 for four years starting from January 1, 1989. Contour intervals for NO_y are 1, 2, 5, 10, 20, 50, 100, 200, 500, 1000, 2000, and 5000%. Contour intervals for ozone are -1, -2, -5, -10, and -20%.

stratosphere and lower mesosphere with transport of the NO_y enhancements to lower altitudes occurring over the next couple of years. The simulated increases in NO_y of over 20% and associated decreases in ozone of over 5% persist for about a year and a half after the October 1989 SPEs.

Such predicted changes in profile ozone are reflected in predictions of total ozone amounts. We show our predicted percentage total ozone changes in 1990 as a result of the 1989 SPEs in Figure 6. Depletions of over 3% above 70° in the summer are predicted in both hemispheres. Due to the large variations in total ozone on days, weeks, seasons, and years time scales at the high latitudes, it would be unlikely that such SPE-related changes in total ozone could be measured unambiguously.

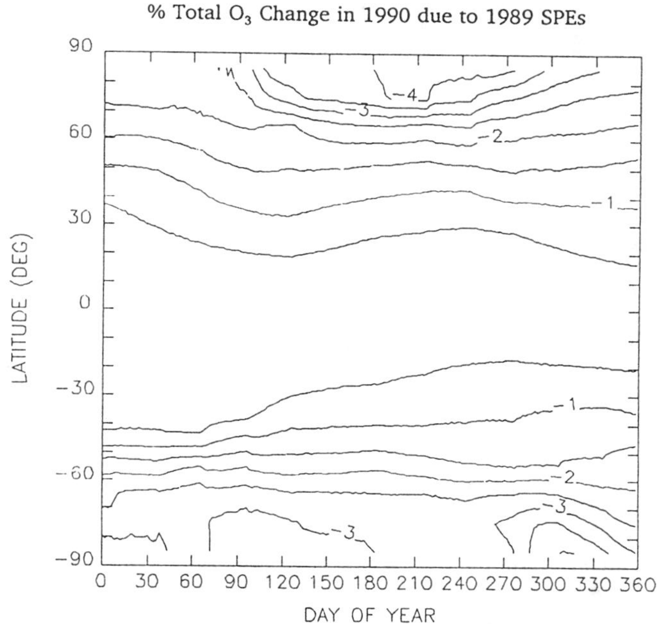

Fig. 6. Computed percentage total ozone change in 1990 which resulted from solar proton events in 1989. Contours range from -0.5% to -4% in 0.5% intervals

Measurements of ozone by SBUV-II (R. D. McPeters, private communication, 1991) and SAGE-II (J. M. Zawodny, private communication, 1991) have indicated a profile ozone decrease as a response to the 1989 SPEs, however, no published results are available. The SAGE-II ozone decreases appear to be correlated with substantial increases in NO_2, also measured by SAGE-II, after the 1989 SPEs (J. M. Zawodny, private communication, 1991). Zadorozhny et al. [1992], using rockets launched from a Soviet research vessel in the southern part of the Indian Ocean, measured large enhancements in NO as a result of the SPEs in October 1989. Since the 1989 SPEs were apparently large enough to cause an NO_y increase and an associated ozone decrease, it is important that the ozone decreases and NO_y increases be studied, as well as possible, through both measurements and model simulations to quantify the associated total ozone decrease.

RELATIVISTIC ELECTRON INFLUENCE

Relativistic electron precipitations (REPs) have been proposed in the past 15 years to be important in contributing to the polar NO_y budget of the mesosphere and upper stratosphere [Thorne 1977; Thorne 1980; Baker et al. 1987; Sheldon et al. 1988; Baker et al. 1988; Callis et al. 1991a]. The frequency and flux spectra of these REPs is still under discussion. Baker et al. [1987] show evidence of large fluxes of relativistic electrons at geostationary orbit measured by the Spectrometer for Energetic Electrons (SEE) instrument on board spacecraft 1979-053 and 1982-019. REPs, which are actually depositing energy into the middle atmosphere, have

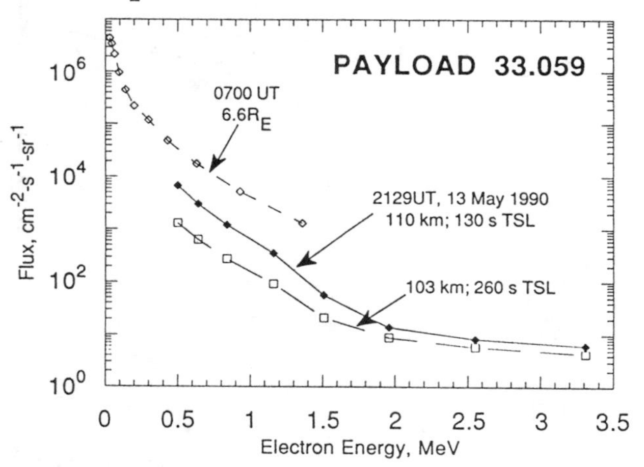

Fig. 7. Taken from Figure 6 of Herrero et al., Rocket measurements of relativistic electrons: New features in fluxes, spectra and pitch angle distributions, *Geophys. Res. Lett.*, *18*, 1481-1484, (1991), copyright by the American Geophysical Union. Comparison of spectra obtained with the high energy spectrometer to spectra obtained in-situ in geostationary orbit at $6.6R_E$. At energies higher than about 2 MeV, the flux decreases to the cosmic background level.

been measured by instruments aboard sounding rockets [Goldberg et al. 1984; Herrero et al. 1991]. These rocket measurements have typically indicated much smaller fluxes of relativistic electrons than measured by the SEE instrument.

The coincident measurement of relativistic electrons at $6.6R_E$ on the GOES satellite with a sounding rocket between altitudes of 70 and 130 km is shown in Figure 7 (taken from Figure 6 of Herrero et al. 1991). The relativistic electron flux was about a factor of 5 to 25 at $6.6R_E$ compared with the flux at 110 km on May 13, 1990. Other coincident measurements of relativistic electrons at $6.6R_E$ on the 1982-019 satellite and the low altitude (170-280 km) S81-1 payload have also been reported [Imhof et al. 1991]. These measurements indicate a relativistic electron flux of about a factor of 10 larger at $6.6R_E$ compared to the flux between 170 and 280 km over the period May 28, 1982 to December 5, 1982.

Callis et al. [1991a] use a relativistic electron flux of one-third of the flux measured at $6.6R_E$ in order to compute NO_y production rates over the 1979 to 1988 time period. Their computations show enhancements in total NO_y of over 40% in the southern hemisphere and about 40% in the northern hemisphere over the solar cycle as a result of relativistic electrons [see Figure 20 of Callis et al. 1991a]. The associated calculated global total ozone decrease from these relativistic electrons is illustrated in Figure 8 [from Figure 16c of Callis et al. 1991b]. Figure 8 has four lines plotted: 1) The large dashed line shows a 2D model simulation which includes increasing CH_4, N_2O, and chlorine from chlorofluorocarbon and hydrochlorocarbon trace gases; 2) The dash-dot line illustrates a 2D simulation including the increasing trace gases as well as the solar ultraviolet (UV) variation; 3) The dash-with-three-dots line shows a 2D simulation with increasing trace gases, solar UV variation, and relativistic electron precipitations; and 4) The solid line illustrates a 2D simulation which includes the increasing trace gases, solar UV variation, REPs, and dilution of the Antarctic ozone hole. The REP's simulated global total ozone effects are clearly the largest of these simulated natural and anthropogenic components shown in Figure 8.

Since relativistic electrons are computed to cause such large variations in total ozone, it is reasonable to assume that a large REP event should cause a measurable decrease in ozone in the upper stratosphere. Callis et al. [1991b] argue that such ozone depletion has already been observed near 1 mbar for longitudes between 60°W and 60°E [see Plate 1 of Callis et al. 1991b]. Another investigation (Aikin, 1992) has failed to find any REP-caused ozone depletion. More work is necessary in order to definitively establish a strong correlation between REPs and middle atmosphere ozone depletion.

Relativistic electrons could be important in determining global ozone abundances *if* the electron flux entering the middle atmosphere is 10% or more of the flux measured at $6.6R_E$. It is still not clear which REP events are more typical of REPs which deposit their energy in the earth's atmosphere, the large fluxes measured at geostationary orbit or the somewhat smaller fluxes measured by the sounding rockets. More coincident relativistic electron flux measurements between rocket and satellite instruments are necessary to determine the magnitude and frequency of REP events.

AURORAL ELECTRON AND PHOTOELECTRON INFLUENCE

The influence of auroral electrons and photoelectrons on the NO_y budget of the middle atmosphere through transport of NO_x from the thermosphere has been studied for the past two

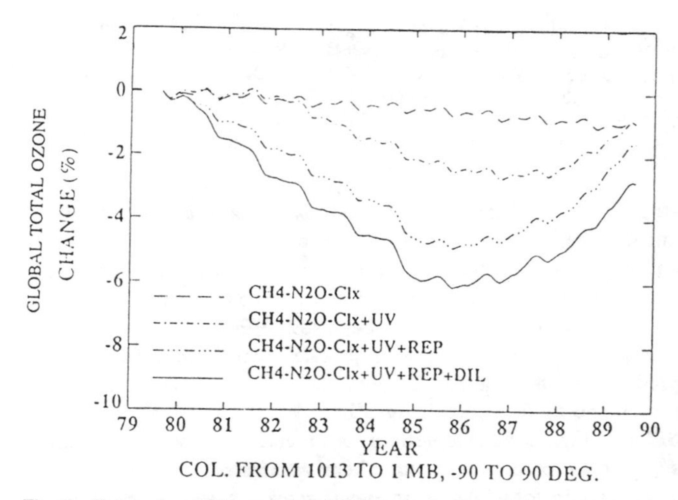

Fig. 8. Taken from Figure 16c of Callis et al., Ozone depletion in the high latitude lower stratosphere: 1979-1990, *J. Geophys. Res.*, *96*, 2921-2937, (1991b), copyright by the American Geophysical Union. Predicted percentage changes in global (90°S to 90°N) total column ozone as calculated for four scenarios explained in the text.

decades [Strobel 1971; McConnell and McElroy, 1973; Brasseur and Nicolet, 1973; Jackman et al. 1980; Solomon 1981; Solomon et al. 1982; Frederick and Orsini, 1982; Garcia et al. 1984; Solomon and Garcia, 1984; Russell et al. 1984; Brasseur 1984; Legrand et al. 1989]. Both auroral electrons and photoelectrons are capable of dissociating N_2 to form huge amounts of atomic nitrogen in the thermosphere. Transport of this NO_x to the mesosphere and upper stratosphere is possible, but certain conditions must be present.

The lifetime of NO_x in the thermosphere and mesosphere is short (less than a day) in the daytime and it is only during the long period of polar night at high latitudes, when several weeks of darkness is typical, that significant downward transport of NO_x is possible. Solomon et al. [1982] undertook a detailed 2D model study of the thermosphere - middle atmosphere coupling. Two of these model computations are shown in Figure 9 [taken from Figures 8 and 17 of Solomon et al., 1982]. They calculated enhancements of over an order of magnitude in the NO_x mixing ratio distribution in the upper mesosphere when auroral electron and photoelectron production of NO_x was included compared to a computation when both auroral electron and photoelectron production of NO_x were not included. These large enhancements of NO_x in the mesosphere caused by auroral electrons and photoelectrons are especially significant in the northern hemisphere which was most recently shrouded in polar night (see Figure 9).

Measurements by the limb infrared monitor of the stratosphere (LIMS) of one significant species of NO_x, NO_2, have also indicated that large enhancements of NO_x in the mesosphere (above 1 mbar) are possible during polar night [see Figure 5 in Russell et al. 1984]. Larger NO_2 values are indicated at higher latitudes, in qualitative agreement with model predictions.

Solomon and Garcia [1984] have computed the response of ozone in the middle atmosphere to changes in production of NO in the thermosphere over the course of an 11-year solar cycle. Their model simulations compare well with BUV and SBUV instrument measurements of ozone change at the 0.75 and 1.0 mbar levels. More study is required to determine how much of the thermospheric enhancement of NO_x is transported to lower altitudes and, also, how much of an effect is expected and measured in stratospheric ozone.

Bremsstrahlung from auroral electrons penetrate to the stratosphere and lower mesosphere and are also capable of producing NO_x directly in the middle atmosphere. Although the flux of bremsstrahlung is about three orders of magnitude smaller than the flux of auroral electrons, the in-situ production of middle atmospheric NO_x could be important for large auroral storms. Previous work [see, e.g., Goldberg et al. 1984] suggests that bremsstrahlung from auroral and relativistic electrons can be the dominate ion source in the upper stratosphere for the subauroral region during certain periods. The frequency and magnitude of this bremsstrahlung flux needs to be quantified to establish its regional and global contribution to the middle atmospheric NO_y budget.

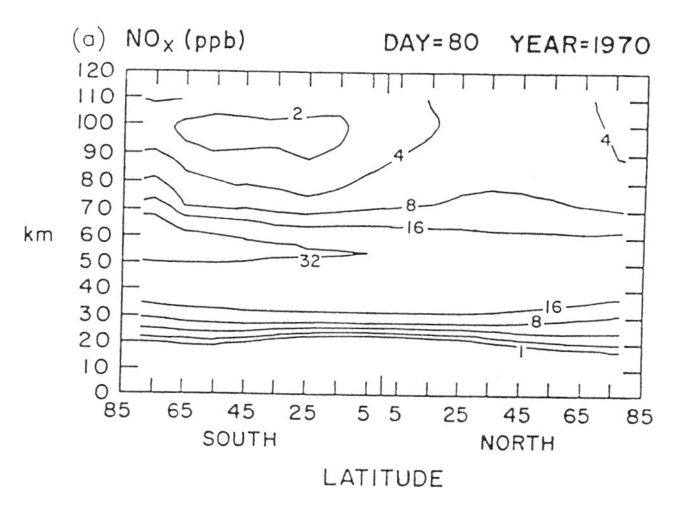

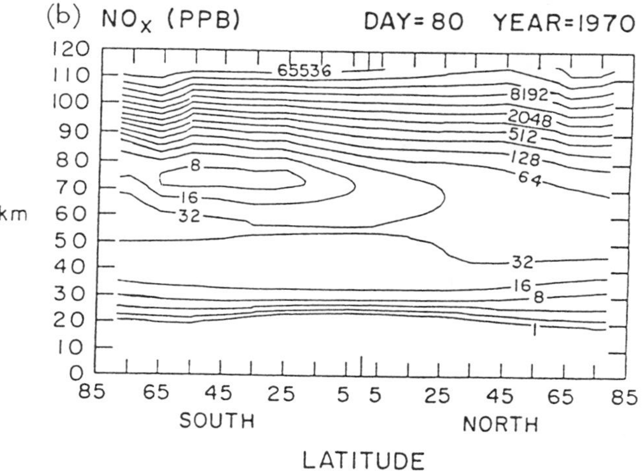

Fig. 9. Taken from Figures 8 and 17 of Solomon et al., Photochemical coupling between the thermosphere and the lower atmosphere, 1, Odd nitrogen from 50 to 120 km, *J. Geophys. Res.*, *87*, 7206-7220, (1982), copyright by the American Geophysical Union. (a) Computed NO_x mixing ratio distribution obtained after 80 days of integration assuming a zero flux at the upper boundary and neglecting production of odd nitrogen by thermospheric ion chemistry. (b) Calculated NO_x mixing ratio distribution obtained after 80 days of integration including photoelectrons and a parameterized model aurora.

CONCLUSIONS

Galactic cosmic rays are computed to cause changes in the NO_y abundance in the lower stratosphere by as much as 10% over a solar cycle. Infrequent large solar proton events, such as those observed in August 1972 and August, September and October 1989 are computed to cause substantial changes in the NO_y abundance in the middle and upper stratosphere. A long-lived ozone decrease during the August 1972 SPE has been calculated and verified by measurements to be important at polar latitudes in the middle to upper stratosphere. Relativistic electron precipitations could be important in modulating the NO_y abundance of the middle atmosphere but require more study and measurements of relativistic electrons in the earth's atmosphere. Auroral electrons and photoelectrons cause enhancements in NO_x amounts in the thermosphere which then can be transported to the mesosphere during polar night. More study is required to quantify the stratospheric NO_y and ozone change from auroral electron and photoelectron precipitation. The importance of bremsstrahlung from auroral and relativistic electrons in producing NO_x in the middle atmosphere has not been clearly established and needs further investigation.

Acknowledgements. The author thanks Richard D. McPeters (NASA Goddard Space Flight Center) for information regarding SBUV-II measurements of ozone response to the 1989 SPEs, Joseph M. Zawodny (NASA Langley Research Center) for information regarding SAGE-II measurements of ozone and NO_2 responses as a result of the 1989 SPEs, and Arthur C. Aikin (NASA Goddard Space Flight Center) for information regarding SBUV measurements of ozone as a result of REPs. The author also thanks Frederico A. Herrero (NASA Goddard Space Flight Center) and an anonymous reviewer for constructive comments on this manuscript.

REFERENCES

Aikin, A. C., Stratospheric evidence of relativistic electron precipitation, *Planet. Space Sci.*, *40*, 413, 431, 1992.

Baker, D. N., J. B. Blake, D. J. Gorney, P. R. Higbie, R. W. Klebesadel, and J. H. King, Highly relativistic magnetospheric electrons: A role in coupling to the middle atmosphere?, *Geophys. Res. Lett.*, *14*, 1027-1030, 1987.

Baker, D. N., J. B. Blake, D. J. Gorney, P. R. Higbie, R. W. Klebesadel, and J. H. King, Reply, *Geophys. Res. Lett.*, *15*, 1451-1452, 1988.

Berger, M. J., S. M. Seltzer, and K. Maeda, Some new results on electron transport in the atmosphere, *J. Atmos. Terr. Phys.*, *36*, 591-617, 1974.

Brasseur, G., and M. Nicolet, Chemospheric processes of nitric oxide in the mesosphere and stratosphere, *Planet. Space Sci.*, *21*, 939-961, 1973.

Brasseur, G., Coupling between the thermosphere and the stratosphere: The role of nitric oxide, MAP Handbook, Volume 10, 1984.

Callis, L. B., D. N. Baker, J. B. Blake, J. D. Lambeth, R. E. Boughner, M. Natarajan, R. W. Klebesadel, and D. J. Gorney, Precipitating relativistic electrons: Their long-term effect on stratospheric odd nitrogen levels, *J. Geophys. Res.*, *96*, 2939-2976, 1991a.

Callis, L. B., R. E. Boughner, M. Natarajan, J. D. Lambeth, D. N. Baker, and J. B. Blake, Ozone depletion in the high latitude lower stratosphere: 1979-1990, *J. Geophys. Res.*, *96*, 2921-2937, 1991b.

Crutzen, P. J., and U. Schmailzl, Chemical budgets of the stratosphere, *Planet. Space Sci.*, *31*, 1009-1032, 1983.

Crutzen, P. J., I. S. A. Isaksen, and G. C. Reid, Solar proton events: Stratospheric sources of nitric oxide, *Science*, *189*, 457-458, 1975.

Fabian, P., J. A. Pyle, and R. J. Wells, The August 1972 solar proton event and the atmospheric ozone layer, *Nature*, *277*, 458-460, 1979.

Frederick, J. E., and N. Orsini, The distribution and variability of mesospheric odd nitrogen: A theoretical investigation, *J. Atmos. Terr. Phys.*, *44*, 479-488, 1982.

Garcia, R. R., S. Solomon, R. G. Roble, and D. W. Rusch, A numerical response of the middle atmosphere to the 11-year solar cycle, *Planet Space Sci.*, *32*, 411-423, 1984.

Goldberg, R. A., C. H. Jackman, J. R. Barcus, and F. Soraas, Nighttime auroral energy deposition in the middle atmosphere, *J. Geophys. Res.*, *89*, 5581-5596, 1984.

Heath, D. F., A. J. Krueger, and P. J. Crutzen, Solar proton event: Influence on stratospheric ozone, *Science*, *197*, 886-889, 1977.

Herrero, F. A., D. N. Baker, and R. A. Goldberg, Rocket measurements of relativistic electrons: New features in fluxes, spectra and pitch angle distributions, *Geophys. Res. Lett.*, *18*, 1481-1484, 1991.

Jackman, C. H., Effects of energetic particles on minor constituents of the middle atmosphere, in press, *J. Geomag. and Geoelec.*, 1992.

Jackman, C. H., and R. D. McPeters, Solar proton events as tests for the fidelity of middle atmosphere models, *Phys. Scr.*, *T18*, 309-316, 1987.

Jackman, C. H., J. E. Frederick, and R. S. Stolarski, Production of odd nitrogen in the stratosphere and mesosphere: An intercomparison of source strengths, *J. Geophys. Res.*, *85*, 7495-7505, 1980.

Jackman, C. H., P. D. Guthrie, and J. A. Kaye, An intercomparison of nitrogen-containing species in Nimbus 7 LIMS and SAMS data, *J. Geophys. Res.*, *92*, 995-1008, 1987.

Jackman, C. H., A. R. Douglass, R. B. Rood, R. D. McPeters, and P. E. Meade, Effect of solar proton events on the middle atmosphere during the past two solar cycles as computed using a two-dimensional model, *J. Geophys. Res.*, *95*, 7417-7428, 1990.

Legrand, M. R., F. Stordal, I. S. A. Isaksen, and B. Rognerud, A model study of the stratospheric budget of odd nitrogen, including effects of solar cycle variations, *Tellus*, *41B*, 413-426, 1989.

Maeda, K., and D. F. Heath, Stratospheric ozone response to a solar proton event: Hemispheric asymmetries, *Pure Appl. Geophys.*, *119*, 1-8, 1980/1981.

McConnell, J. C., and M. B. McElroy, Odd nitrogen in the atmosphere, *J. Atmos. Sci.*, *30*, 1465-1480, 1973.

McPeters, R. D., C. H. Jackman, and E. G. Stassinopoulos, Observations of ozone depletion associated with solar proton events, *J. Geophys. Res.*, *86*, 12071-12081, 1981.

McPeters, R. D., A nitric oxide increase observed following the July 1982 solar proton event, *Geophys. Res. Lett.*, *13*, 667-670, 1986.

Nicolet, M., On the production of nitric oxide by cosmic rays in the mesosphere and stratosphere, *Planet. Space Sci.*, *23*, 637-649, 1975.

Reagan, J. B., R. E. Meyerott, R. W. Nightingale, R. C. Gunton, R. G. Johnson, J. E. Evans, W. L. Imhof, D. F. Heath, and A. J. Krueger, Effects of the August 1972 solar particle events on stratospheric ozone, *J. Geophys. Res.*, *86*, 1473-1494, 1981.

Reid, G. C., S. Solomon, and R. R. Garcia, Response of the middle atmosphere to the solar proton events of August-December 1989, *Geophys. Res. Lett.*, *18*, 1019-1022, 1991.

Ruderman, M. A., and J. W. Chamberlain, Origin of the sunspot modulation of ozone: Its implications for stratospheric NO injection, *Planet. Space Sci.*, *23*, 247-268, 1975.

Rusch, D. W., J.-C. Gerard, S. Solomon, P. J. Crutzen, and G. C. Reid, The effect of particle precipitation events on the neutral and ion chemistry of the middle atmosphere, 1, Odd nitrogen, *Planet. Space Sci.*, *29*, 767-774, 1981.

Russell, J. M., III, S. Solomon, L. L. Gordley, E. E. Remsberg, and L. B. Callis, The variability of stratospheric and mesospheric NO_2 in the polar winter night observed by LIMS, *J. Geophys. Res.*, *89*, 7267-7275, 1984.

Sheldon, W. R., J. R. Benbrook, and E. A. Bering, III, Comment on "Highly relativistic magnetospheric electrons: A role in coupling to the middle atmosphere?," *Geophys. Res. Lett.*, *15*, 1449-1450, 1988.

Solomon, S., One- and two-dimensional photochemical modeling of the chemical interactions in the middle atmosphere (0-120 km), Cooperative Thesis No. 62, University of California and National Center for Atmospheric Research, 1981.

Solomon, S., and P. J. Crutzen, Analysis of the August 1972 solar proton event including chlorine chemistry, *J. Geophys. Res.*, *86*, 1140-1146, 1981.

Solomon, S., P. J. Crutzen, and R. G. Roble, Photochemical coupling between the thermosphere and the lower atmosphere, 1, Odd nitrogen from 50 to 120 km, *J. Geophys. Res.*, *87*, 7206-7220, 1982.

Solomon, S., and R. R. Garcia, Transport of thermospheric NO to the upper stratosphere?, *Planet. Space Sci.*, *32*, 399-409, 1984.

Stolarski, R. S., P. Bloomfield, R. D. McPeters, and J. R. Herman, Total ozone trends deduced from Nimbus 7 TOMS data, *Geophys. Res. Lett.*, *18*, 1015-1018, 1991.

Strobel, D. F., Odd nitrogen in the mesosphere, *J. Geophys. Res., 76*, 8384-8393, 1971.

Thorne, R. M., Energetic radiation belt electron precipitation: A natural depletion mechanism for stratospheric ozone, *Science, 195*, 287-289, 1977.

Thorne, R. M., The importance of energetic particle precipitation on the chemical composition of the middle atmosphere, *Pure Appl. Geophys., 118*, 128-151, 1980.

Warneck, P., Cosmic radiation as a source of odd nitrogen in the stratosphere, *J. Geophys. Res., 77*, 6589-6591, 1972.

Zadorozhny, A. M., G. A. Tuchkov, V. N. Kikhtenko, J. Lastovicka, J. Boska, and A. Novak, Nitric oxide and lower ionosphere quantities during solar particle events of October 1989 after rocket and ground based measurements, *J. Atmos. Terr. Phys., 54*, 183-192, 1992.

C. Jackman, Code 916, NASA/Goddard Space Flight Center, Greenbelt, MD 20771.

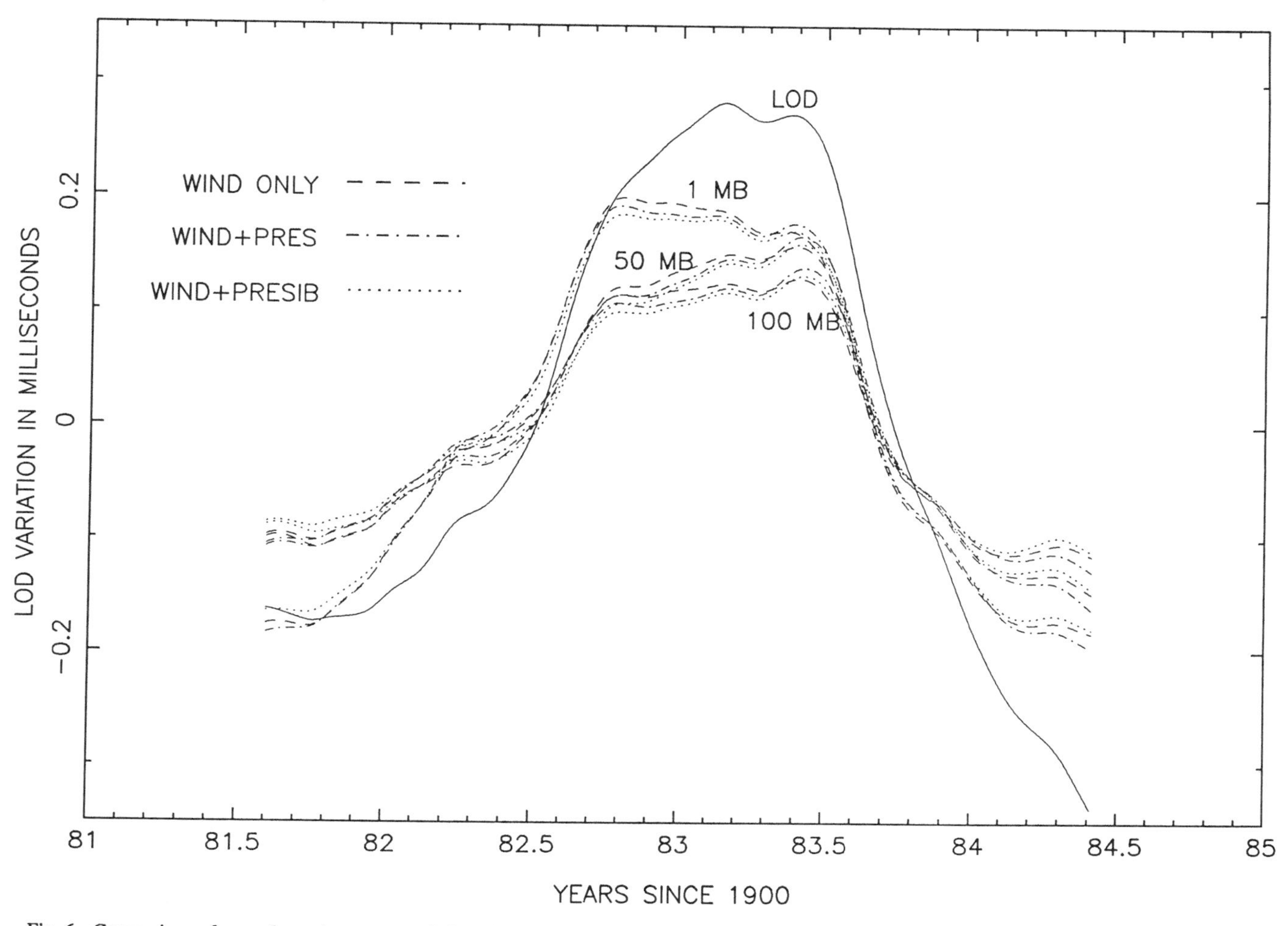

Fig. 6. Comparison of annual running mean variations in LOD (Λ_β–solid line) and AAM contributions [wind term only (dashed), wind plus pressure (dot-dashed) and wind plus pressure with the inverted barometer (dotted)] analyzed up to 100, 50 and 1 mb; a linear term has been removed from all series to account for the decadal variation.

noted a similar lead with respect to LOD for AAM variations in the stratosphere, computed from monthly station data for the period 1964-84 using an idealized model of the QBO.

Fig. 8 displays the variance explained by the AAM data sets, as a function of lag with respect to the interannual LOD series. For AAM integrated to 100 and 50 mb the explained variance is roughly symmetric about the zero axis, with no discernable lag or lead between the two data types. When the stratospheric data is included, however, the LOD appears to lag by about 20 days, with the AAM explaining a maximum of 90% of the interannual LOD variation. A similar analysis using data from the European Centre for Medium-Range Forecasts (not shown) indicates a smaller lag for the LOD, with AAM explaining a maximum of 92% of the variance. If the difference of 2% is considered to be a lower bound on the uncertainty in the explained variance, a null hypothesis of zero lag between the two data types cannot be rejected on the basis of the data considered here (see Fig. 8c). Clearly, further analysis over an extended period of data is required.

The use of an admittance (or gain) factor to arbitrarily scale the AAM series with respect to the LOD results in a substantial increase in the variance explained, particularly for the 100 mb and 50 mb data sets (see Table 1). A systematic underestimation of the interannual AAM variation by a factor of more than 1.3 is highly unlikely, however, and the larger admittances found for these data would probably not be obtained in a study incorporating longer time series. In particular, the high admittance found for the 100 mb data appears to result from the superposition of AAM variations in the troposphere and stratosphere during this episode (Fig. 7), which may be related to the near coincidence between maxima in the stratospheric QBO and the ENSO cycle at this time [Chao, 1989; Fig. 4]. Higher admittances are also found for the

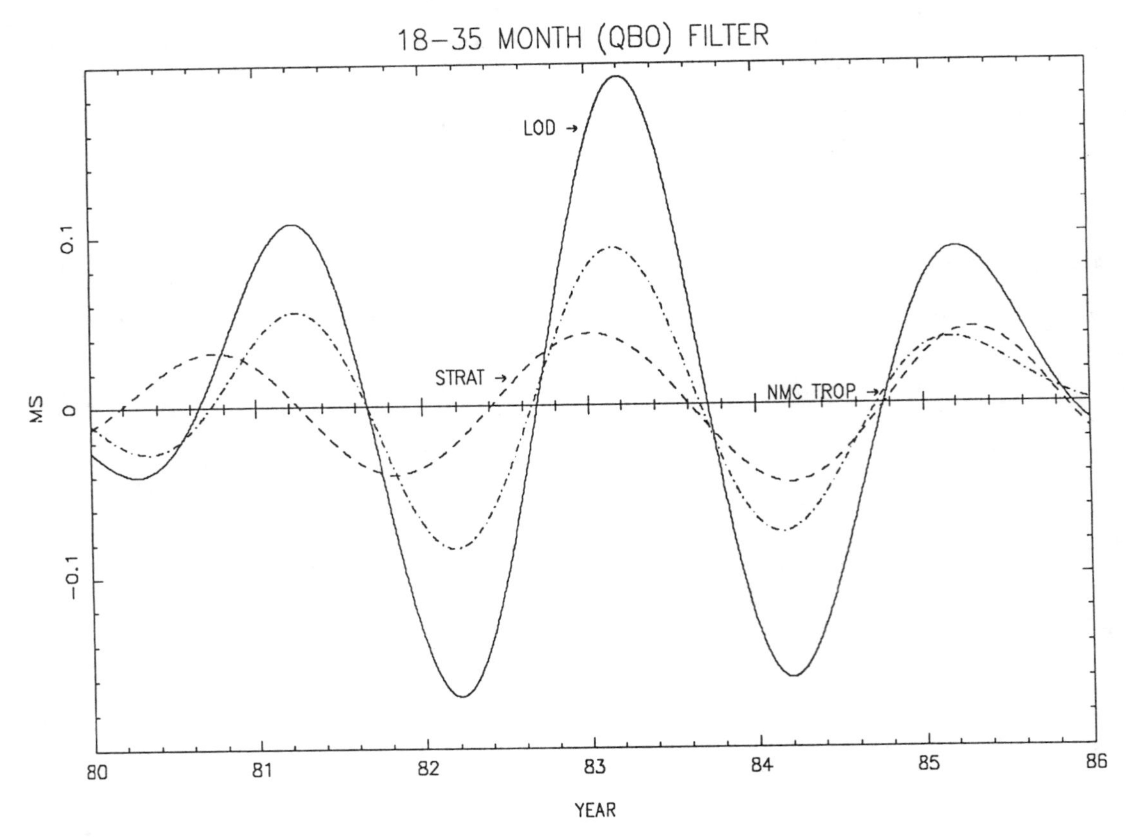

Fig. 7. Comparison of LOD and AAM contributions Fourier-filtered in the 18-35 month band.

series at all three levels incorporating the inverted barometer (IB) assumption, possibly indicating some difficulty with the application of this model to interannual variations.

Atmospheric forcing was thus found to be the dominant cause of interannual LOD variation during the largest ENSO event for which detailed records exist, with AAM integrated to the 1 mb level accounting for 89% of the variance when no scaling is applied. While independent determinations of stratospheric AAM were not available for the period of this study, the rms difference between interannual variations in the 50 mb NMC and EC wind plus pressure excitation was 18.2 μs, indicating that a substantial part of the unexplained LOD residual (64 μs) could be due to "noise" in the AAM data. Due to the scarcity of observational data over large regions of the tropics involved in the ENSO cycle, "systematic" AAM errors (i.e. missing or inaccurate data common to all centers), which cannot be detected by inter-center comparisons, could also contribute to the discrepancy [cf. Dickey et al., 1992c]. Finally, simplified estimates of angular momentum changes associated with the oceanic circulation and mass distribution anomalies documented by Wyrtki [1985] during the 1982-83 El Niño are of a sufficient magnitude to generate LOD variations comparable in size to the observed residual [Dickey et al., 1992a]. Thus ocean effects, though small, cannot be ignored in detailed determinations of the Earth's angular momentum budget on interannual time scales.

5. Historical Study of Interannual LOD Variations

To investigate the behavior of interannual LOD fluctuations before 1962, we considered the historical LOD series centered at the mid-point of each calendar year as described in Section 2. Fig. 1(b) shows the span of data beginning in 1860, to which we applied five-year and ten-year moving averages in order to reveal the long-period structure of the series associated with core-mantle interaction [Dickey et al., 1992a]. To study the shorter period interannual variations, we follow the same procedure implemented earlier, defining $\Lambda_\beta(t)$ as the difference between a one-year and a five-year moving average and comparing it with the MSOI. Comparison of the two time series [Fig. 9(a)] shows only little resemblance at first, with better correlations found later as the quality of the geodetic measurements improves. Comparison of the Λ_β variations with their error estimates (Fig. 9(b)) shows that the Λ_β signals are typically smaller than the estimated errors prior to 1930; a favorable signal-to-noise ratio allows for comparison for the period 1930 forward. Table 2 summarizes the correlation between Λ_β and MSOI for different time spans. If the entire period is considered (1885-1988) a correlation of 0.28 is obtained

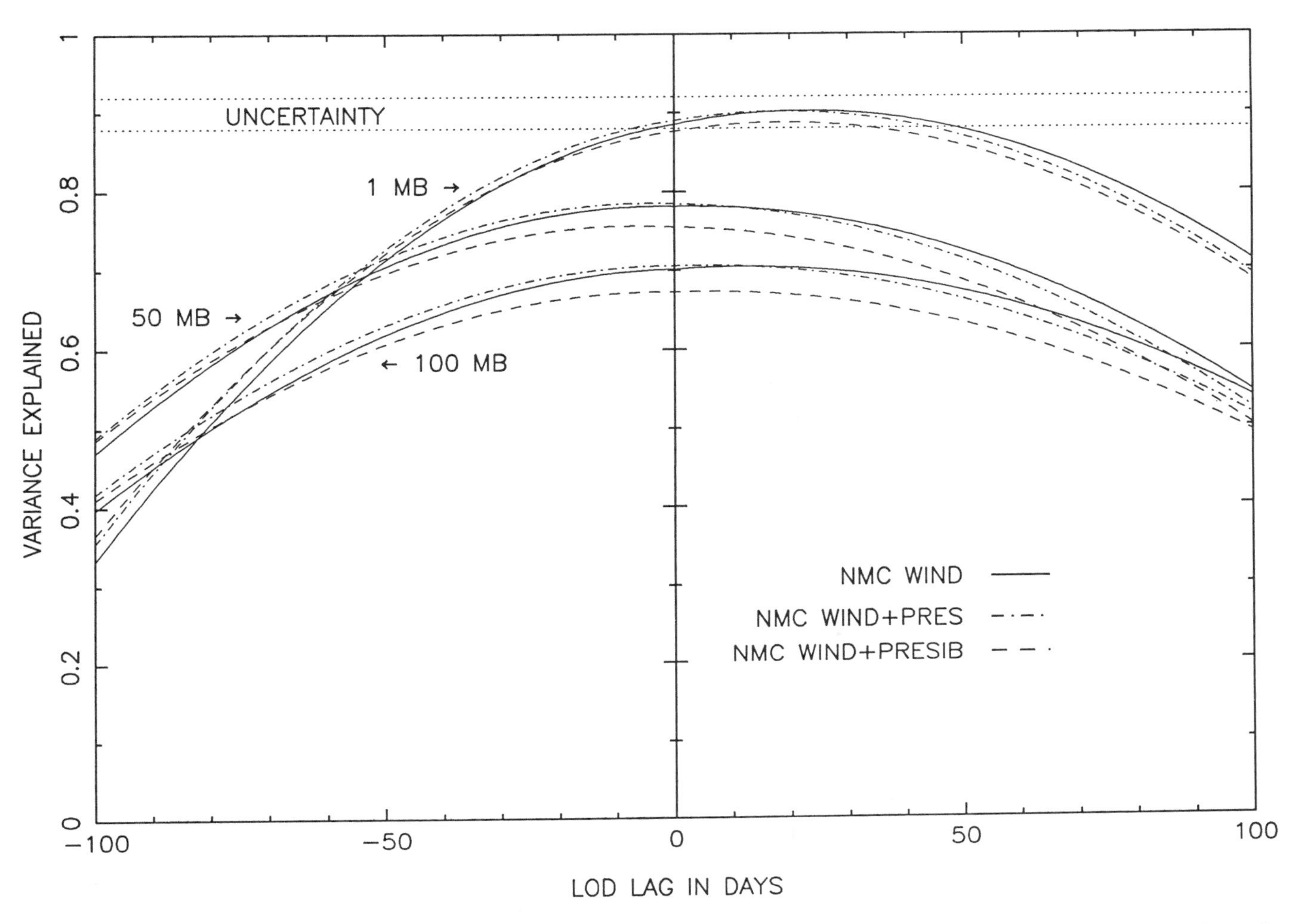

Fig. 8. Variance of LOD explained by wind only (solid line), wind plus pressure (dashed-dotted line), and wind plus pressure with the inverted barometer assumption (dashed line), integrating AAM up to 100 mb, 50 mb and 1 mb. The uncertainty in the explained variance is indicated based on a comparison with EC data (not shown).

which is statistically significant at the two-sigma level (0.26). If one confines the study to the pre-1930 data (1885-1929) where the uncertainty in LOD is comparable to or greater than the interannual LOD signature, one finds no significant correlation (0.13 with $\sigma = 0.20$). For the period 1930-1988 for which there is a favorable signal-to-noise ratio, the correlation increases to 0.49. Further, if one restricts the study to the era of atomic time, the correlation increases to 0.58.

This study indicates that interannual LOD is an excellent proxy-index for atmospheric and climatic changes where the data has a sufficient signal-to-noise ratio. Re-analysis of the lunar occulatation data should be considered in light of this investigation. Improved determinations are expected through a re-analysis of the existing data set with an improved optical reference frame (FK5) and lunar and planetary ephemeris (DE200/LE200) [Morrison, priv. comm., 1990]. A previous study by Salstein and Rosen [1986] also investigated the interannual variability of the Earth rotation using the data set of McCarthy and Babcock [1986]. They inferred that the day is typically longer during the year following an ENSO warm event than otherwise for the period 1860 through 1980. While this result was shown to be statistically significant for the entire period using a Student's t test, the large errors associated with Λ_β determinations during the pre-1930 span (Fig. 9(b)) cast doubt on the utility of Λ_β variations as a climate proxy during this earlier period.

6. Summary

The role of the atmosphere in producing interannual variations in $\Lambda(t)$ was investigated using both AAM data and indices representing the strength of the Southern Oscillation. Recent studies [Rasmusson et al., 1990; Barnett, 1991, Keppenne and Ghil, 1991] have demonstrated that in addition to strong

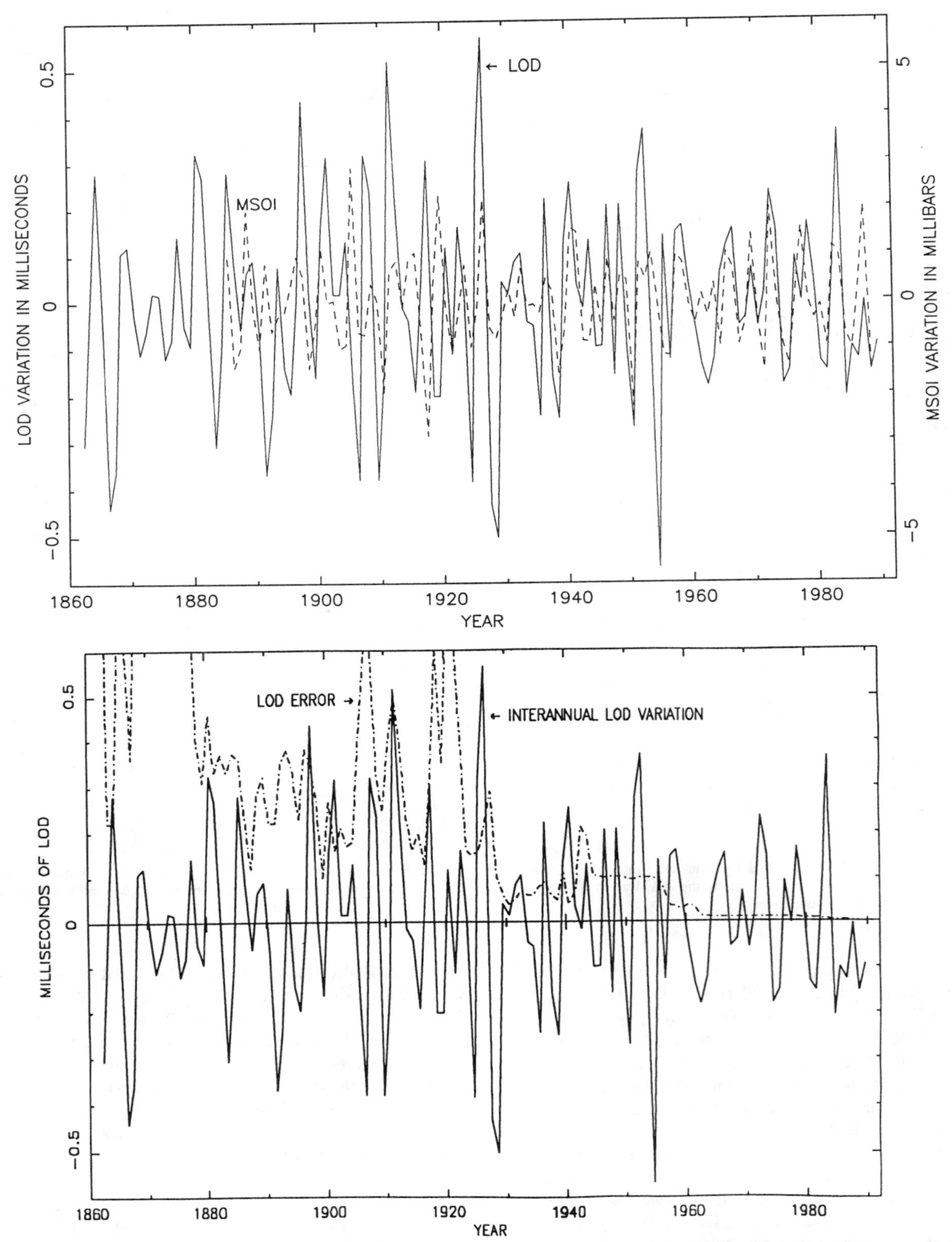

Fig. 9. (a) The interannual LOD variation (solid line) for the time series shown in Fig. 1(b) obtained by differencing the one and five-year averaged series shown with a similarly smoothed MSOI time series (dashed line). The correlation between the $\Lambda_\beta$series and MSOI series is listed in Table 2 for various subperiods. *(b)* The interannual length-of-day series (Λ_β–solid line) of (a) (top frame) compared with its associated uncertainty (dot-dashed line). Note that uncertainties are typically as large or larger than the Λ_β variations before 1930.

TABLE 1. Study of the 1982-83 ENSO Event
LOD Variance Explained by Atmospheric Angular Momentum

	Components Used	Admittance/ Gain	Residual (ms) LOD-AAM	Variance Explained
Total	W	1.0	0.0655	88.5%
(1000 to 1 mb)	W+P	1.0	0.0641	89.0%
—see text	W+P (IB)	1.0	0.0680	87.6%
for data	W	1.317 ± 0.010	0.0476	93.9%
description	W+P	1.318 ± 0.010	0.0453	94.5%
	W+P (IB)	1.385 ± 0.010	0.0435	94.9%
Contribution	W	1.0	0.0903	78.1%
to 50 mb	W+P	1.0	0.0896	78.4%
(1000-50 mb)	W+P (IB)	1.0	0.0955	75.5%
	W	1.833 ± 0.007	0.0241	98.4%
NMC Analysis	W+P	1.823 ± 0.007	0.0235	98.5%
	W+P (IB)	1.933 ± 0.008	0.0243	98.4%
Contribution	W	1.0	0.1055	70.1%
to 100 mb	W+P	1.0	0.1047	70.6%
(1000-100 mb)	W+P (IB)	1.0	0.1105	67.2%
	W	2.095 ± 0.012	0.0361	96.5%
NMC Analysis	W+P	2.087 ± 0.012	0.0343	96.8%
	W+P (IB)	2.240 ± 0.012	0.0336	97.0%

Yearly moving average detrended for both the LOD and AAM series
Total LOD variance = 0.1930 msec
Time period considered: 1981.5–1984.5
W = Wind terms
P = Pressure terms
IB = Inverted barometer effect assumed

variability in the 3 to 7 year period range, the ENSO cycle is also characterized by a distinct quasi-biennial component. We note here that this bimodality is a common characteristic of both the LOD and Modified Southern Oscillation Index (MSOI) time series during the period studied. The AAM data for the same period (1976-1991) exhibit low frequency variability but not, however, a statistically significant QB oscillation. This series contains data integrated to 50 mb, encompassing the tropospheric but not the full stratospheric contribution to the global AAM. While the troposphere is characterized by strong variability on the QB time scale, the QB signal in the zonal winds appears to be dominated by baroclinic modes at wavenumbers 1 and 2 [Yasunari, 1989], making difficult the detection of a QBO in zonally and vertically-averaged data from the troposphere.

TABLE 2. Maximum Correlation of LOD with Modified MSOI

Period Examined	Correlation	Standard Error*
1885-1988	0.28	0.13
1885-1929	0.13	0.20
1930-1988	0.49	0.16
1957-1988	0.58	0.21

* The standard errors of the correlation coefficient were computed following the procedure outlined in Sciremammano [1979].

Interannual variations in Λ were compared with the MSOI for the period 1962-1991, showing the largest correlation (0.67) when Λ lags the MSOI by one month. Restricting the analysis to the period 1972-86 gives a maximum correlation of 0.79 with Λ lagging the MSOI by two months, consistent with growth and development of an El Niño. The higher correlation and longer lag obtained for the shorter period may reflect the dominant influence of the intense 1982-83 El Niño.

The availability of an AAM data set integrated up to 1 mb for the period 1980-86 permits an investigation of the stratosphere's role on the interannual time scale and allows for a case-study of the unusually strong and well-observed 1982-83 El Niño. Atmospheric forcing was observed to be the dominant cause of the interannual LOD variation, with the AAM integrated up to 1 mb accounting for 89% of the variance during this event (see Table 1). The stratosphere was found to play an important role in the Earth's angular momentum budget on interannual time scales, accounting for up to an additional 20% of the LOD variance during the case study. The difference between the interannual AAM variations from the NMC and EC analyses integrated up to 50 mb (18.2 µsec rms) is a considerable fraction of the remaining LOD–AAM residual (64 µsec), implying that "noise" in the AAM data could account for a considerable portion of the residual. Possible sources of the remaining discrepancy include systematic AAM error and a contribution from the oceans.

To investigate the behavior of interannual LOD fluctuations before 1962, we utilized an LOD series beginning in 1860 (Fig. 1(b)). We find that the LOD data prior to 1930 (as currently analyzed) may not be accurate enough to be a useful proxy for interannual climate variability, as the uncertainties are generally as large or larger than the magnitude of the observed interannual variations. After 1930, however, the LOD estimates are accurate enough to be used as a proxy index of global wind fluctuations and should be of interest to meteorologists, oceanographers and climatologists. A re-analysis of the lunar occultation data using both an improved optical reference frame and lunar ephemeris is desirable.

Acknowledgments. We acknowledged interesting discussions with A. P. Freedman, R. S. Gross, D. D. McCarthy, L. Morrison, R. D. Rosen, D. A. Salstein, and J. A. Steppe. The authors thank A. Babcock, and D. McCarthy for providing us with length-of-day series beginning in 1656, D. A. Salstein and R. D. Rosen for supplying the stratospheric AAM data used in our analysis, and R. Gross for comments on the original manuscript. This paper presents the results of one phase of research carried out at the Jet Propulsion Laboratory, California Institute of Technology, sponsored by the National Aeronautics and Space Administration.

REFERENCES

Arpe, K., Effective Atmospheric Angular Momentum Functions Computed at the European Centre for Medium-Range Weather Forecasts, *IERS Technical Note 2, Earth Orientation, Reference Frame*

Determinations, and Atmospheric Excitation Functions; Observatoire de Paris, 81, 1989.

Barnes, R. T. H., R. Hide, A. A. White, and C. A. Wilson, Atmospheric Angular Momentum Fluctuations, Length of Day Changes and Polar Motion, *Proc. R. Soc. Lon., 387,* 31-73, 1983.

Barnett, T. P., The Interaction of Multiple Time Scales in the Tropical Climate System, *J. Climate, 4,* 269, 1991.

Bloomfield, P., *Fourier Analysis of Time Series:An Introduction*, John Wiley & Sons, New York, 1976.

Boer, G. J., Earth-Atmosphere Exchange of Angular Momentum Stimulated in a General Circulation Model and Implications for the Length-of-Day, *J. Geophys. Rev., 95,* 5511-5531, 1990.

Carter, W. E., and D. S. Robertson, Polar Motion and UT1 Time Series Derived from VLBI Observations. *IERS Technical Note 5, Earth Orientation, Reference Frame Determinations, and Atmospheric Excitation Functions,* Observatoire de Paris, 25, 1990.

Cazenave, A. (ed.), *Earth Rotation: Solved and Unsolved Problems, NATO Advanced Institute Series C: Mathematical and Physical Sciences Vol. 187,* D. Reidel, Boston, 137-162, 1986.

Chao, B. F., Interannual Length-of-Day Variations with Relation to the Southern Oscillation/El Niño, *Geophys. Res. Lett., 11,* 541-544, 1984.

Chao, B. F., Correlation of Interannual Lengths-of-Day Variation with El Niño/Southern Oscillation, 1972-1986, *J. Geophys. Res., 93,* B7, 7709-7715, 1988.

Chao, B. F., Length-of-Day Variations Caused by El Niño/Southern Oscillation and the Quasi-Biennial Oscillation, *Science, 243,* 923-925, 1989.

Chen, W. Y., Assessment of Southern Oscillation Sea-Level Pressure Indices., *Mon. Weather Rev., 110,* 800-807, 1982.

Dickey, J. O., X X Newhall, and J. G. Williams, Earth Orientation from Lunar Laser Ranging and an Error Analysis of Polar Motion Services, *J. Geophys. Res., 90,* 9353-9363, 1985.

Dickey, J. O., and T. M. Eubanks, The Application of Space Geodesy to Earth Orientation Studies, Space Geodesy and Geodynamics (eds.) by A. J. Anderson and A. Cazenave, Academic Press, New York, 221-269, 1986.

Dickey, J. O., T. M. Eubanks, and J. A. Steppe, Earth Rotation Data and Atmospheric Angular Momentum. *Earth Rotation; Solved and Unsolved Problems, NATO Advanced Institute Series C: Mathematical and Physical Sciences, 187,* edited by A. Cazenave, D. Reidel, Boston, 137-162, 1986.

Dickey J. O., T. M. Eubanks, and R. Hide, Interannual and Decade Fluctuations in the Earth's Rotation, *Variations in the Earth's Rotation, Geophysical Monograph Series of the American Geophysical Union,* Washington, D. C., D. McCarthy (ed.), 157, 1990.

Dickey, J. O., S. L. Marcus, R. Hide, and T. M. Eubanks, Analysis of Interannual and Decadal Fluctuations in the Length of Day and their Geophysical Interpretation, in preparation, 1992a.

Dickey, J. O., S. L. Marcus and R. Hide, Global Propagation of Interannual Fluctuations in Atmospheric Angular Momentum, *Nature,* 357, 484-488, 1992b.

Dickey, J. O., S. L. Marcus, J. A. Steppe and R. Hide, The Earth's Angular Momentum Budget on Subseasonal Time Scales, *Science,255,* 321-324, 1992c.

Eubanks, T. M., J. O. Dickey, and J. A. Steppe, The 1982-83 El Niño, the Southern Oscillation, and Changes in the Length of Day, *Trop. Ocean and Atmos. Newsletter, 29,* 21-23, 1985.

Eubanks, T. M., J. A. Steppe, and J. O. Dickey, The El Niño, the Southern Oscillation and the Earth's Rotation, in *Earth Rotation; Solved and Unsolved Problems, NATO Advanced Institute Series C: Mathematical and Physical Sciences, 187,* ed. A. Cazenave, 163-186, D. Reidel, Hingham, Mass., 1986.

Gross, R. S., and J. A. Steppe, A combination of Earth orientation data: SPACE90, in *IERS Technical Note 8: Earth Orientation and Reference Frame Determinations, Atmospheric Excitation Functions, up to 1990* (Annex to the IERS Annual Report for 1990), Observatoire de Paris, Paris, France, 1991.

Hide, R., Towards a Theory of Irregular Variations in the Length of the Day and Core-Mantle Coupling, *Phil. Trans. Roy. Soc., A284,* 547-554, 1977.

Hide, R., Rotation of the Atmospheres of the Earth and Planets, *Phil. Trans. R. Soc. Lond. A313,* 107-121, 1984.

Hide, R., Presidential Address: The Earth's Differential Rotation, *Quart. J. Roy. Astron. Soc., 278,* 3-14, 1986.

Hide, R., Effective Atmospheric Angular Momentum Functions Computed by the U. K. Meteorological Office Using Data Processed at the European Centre for Medium-Range Weather Forecasting. *IERS Technical Note 2, Earth Orientation, Reference Frame Determinations, and Atmospheric Excitation Functions,* Observatoire de Paris, 85, 1989.

Hide, R., and J. O. Dickey, Earth's Variable Rotation, *Science, 253,* 629, 1991.

Horel, J. D., and J. M. Wallace, Planetary-Scale Atmospheric Phenomena Associated with the Southern Oscillation, *Mon. Weather Rev., 109,* 813-828, 1981.

Keppenne, C. L., and M. Ghil, Adaptive Spectral Analysis of the Southern Oscillation Index, in *Proceedings of the XVth Annual Climate Diagnostics Workshop,* 30-35, U. S. Department of Commerce, NOAA, 1991.

Lambeck, K., *The Earth's Variable Rotation,* Cambridge Univ. Press, London and New York, 1980.

Lambeck, K., *Geophysical Geodesy, The Slow Deformation of the Earth,* Clarendon Press, Oxford, 1988.

Lambeck, K., and A. Cazenave, The Earth's Rotation and Atmospheric Circulation, I, Seasonal Variations, *Geophys. J. R. Astron. Soc., 32,* 79-93, 1973.

McCarthy, D. D., and A. K. Babcock, The Length of Day Since 1656, *Phys. Earth and Planet. Inter., 44,* 281-292, 1986.

Morabito, D. D., T. M. Eubanks, and J. A. Steppe, Kalman Filtering of Earth Orientation Changes. *The Earth's Rotation and Reference Frames for Geodesy and Geodynamics,* A. K. Babcock and G. A. Wilkins, Kluwer Academic Publishers, Dordrecht, 257-268, 1988.

Moritz, H., and I. I. Mueller, *Earth Rotation: Theory and Observation,* The Ungar Publishing Co., New York, 1987.

Morrison, L. V., An Analysis of Lunar Occultations in the Years 1943 to 1974 for Corrections to the Constants in Brown's Theory, the Right Ascension System of the FK4, and Watt's Lunar-Profile Datum, *Mon. Not. R. Astron. Soc., 187,* 41-82, 1979a.

Morrison, L. V., Re-Determination of the Decade Fluctuations in the Rotation of the Earth in the Period 1861-1978, *Geophys. J. R. Astron. Soc., 38,* 349-360, 1979b.

Munk, W. H., and G. J. F. Macdonald, *The Rotation of the Earth,* Cambridge University Press, 1960.

Naito, I., Y. Goto, and N. Kikuchi, Effective Atmospheric Angular Momentum Functions Computed from the JMA Data. *IERS Technical Note 2, Earth Orientation Reference Frame Determinations and Atmospheric Excitation Functions,* Observatoire de Paris, 83, 1989.

Newhall, X X, J. O. Dickey, and J. G. Williams, Earth Rotation (UT0-UTC) from Lunar Laser Ranging, *IERS Technical Note 5, Earth Orientation, Reference Frame Determinations and Atmospheric Excitation Functions,* Observatoire de Paris, 41, 1990.

Philander, S. G. H., El Niño Southern Oscillation Phenomena, *Nature, 302,* 295-301, 1983.

Philander, S. G. H., *El Niño, La Niña, and the Southern Oscillation,* Academic Press, New York, 1990.

Rasmusson, E. U., and J. M. Wallace, Meteorological Aspects of the El Niño/Southern Oscillation, *Science, 222,* 1195-1202, 1983.

Rasmusson, E. M., X. Wang, and C. F. Ropelewski, The Biennial Component of ENSO Variability, *J. Mar. Systems, 1,* 71-96, 1990.

Robertson, D. S., J. Campbell, W. E. Carter, and H. Schuh, Daily Earth Rotation Determinations from IRIS Very Long Baseline Interferometry, *Nature, 316,* 424, 1985.

Rosen, R. D., D. A. Salstein, T. M. Eubanks, J. O. Dickey, and J. A. Steppe, An El Niño Signal in Atmospheric Angular Momentum and Earth Rotation, *Science, 225,* 411-414, 1984.

Rosen, R. D., and D. A. Salstein, Contribution of Stratospheric Winds to Annual and Semi-Annual Fluctuations in Atmospheric Angular Momentum and the Length of Day, *J. Geophys. Res., 90,* 8033-8041, 1985.

Rosen, R. D., K. Arpe, A. J. Miller, and D. A. Salstein, Accuracy of Atmospheric Angular Momentum Estimates from Operational Analyses, *Monthly Weather Rev., 115,* 1627-1639, 1987.

Rosen, R. D., D. A. Salstein, T. M. Wood, Discrepancies in the Earth-Atmosphere Angular Momentum Budget, *Journal of Geophysical Research, 95,* 265-279, 1990.

Salstein, D. A., Effective Angular Momentum Function for Earth Rotation and Polar Motion from the United States NMC analysis. *International Earth Rotation Service Technical Note 2,* Observatoire de Paris, 77-79, 1989.

Salstein, D. A., and R. D. Rosen, Earth Rotation as a Proxy for Interannual Variability in Atmospheric Circulation, 1860-Present, *J. Clim. and Appl. Meteorol., 25,* 1870-1877, 1986.

Sciremammano, Jr., F., A Suggestion for the Presentation of Correlations and Their Significant Levels, *J. Phys. Ocean., 9,* 1273-1276, 1979.

Stephanic, M., Interannual Atmospheric Angular Momentum Variability 1963-1973 and the Southern Oscillation, *J. Geophys. Res., 87,* 428-432, 1982.

Steppe, J. A., S. H. Oliveau, and O. J. Sovers, Earth Rotation Parameters from DSN VLBI, *IERS Technical Note 5, Earth Orientation, Reference Frame Determinations, and Atmospheric Excitation Functions,* Observatoire de Paris, 13, 1990.

Swinbank, R., The Global Atmospheric Angular Momentum Balance Inferred from Analyses made during FGGE, *Q. J. R. Meteorol. Soc., 111,* 977, 1985.

Wahr, J. M., Friction and Mountain Torque Estimates from Global Atmospheric Data, *J. Atmos. Sci., 41,* 190, 1984.

Wahr, J. M., The Earth's Rotation, *Ann. Rev. Earth Planet Sci., 16,* 231-249, 1988.

Wolf, W. L., and R. B. Smith, Length-of-Day Changes and Mountain Torque During El Niño, *J. Atmos. Sci., 44,* 3656-3660, 1987.

Yasunari, T., Possible Link of the QBOs Between the Stratosphere, the Troposphere and the Sea Surface Temperature in the Tropics, *J. Meteorol. Soc. Japan, 67,* 483, 1989.

J. O. Dickey and S. L. Marcus, Mail Stop 238-332, Jet Propulsion Laboratory, Mail Stop 238-332, California Institute of Technology, Pasadena, California 91109-8099 USA.

T. M. Eubanks, U. S. Naval Observatory, Mail Code TSEO, Bldg. 52, Room 222, 34th & Massachusetts Ave. NW, Washington, D. C. 20392.

R. Hide, University of Oxford, Dept. of Physics, c/o NERC Unit, Oceanography Group, Room 1, The Observatory, Clarendon Laboratory, Parks Road, Oxford OX1 3PU, England, UK.